Advances in Global Change Research

Volume 64

More information about this series at http://www.springer.com/series/5588

Mart A. Stewart • Peter A. Coclanis
Editors

Water and Power

Environmental Governance and Strategies for Sustainability in the Lower Mekong Basin

Springer

Editors
Mart A. Stewart
Western Washington University
Bellingham, WA, USA

Peter A. Coclanis
Global Research Institute
University of North Carolina-Chapel Hill
Chapel Hill, NC, USA

ISSN 1574-0919 ISSN 2215-1621 (electronic)
Advances in Global Change Research
ISBN 978-3-319-90399-6 ISBN 978-3-319-90400-9 (eBook)
https://doi.org/10.1007/978-3-319-90400-9

Library of Congress Control Number: 2018952630

Cover illustration: Phat Chandara

This Springer imprint is published by the registered company Springer Nature Switzerland AG.
The registered company address is: Gewerbestrasse 11, 6330 Cham, Switzerland

Preface

Among the many serious environmental problems the world faces today, those related to freshwater resources are clearly among the most difficult to address successfully. This is particularly true to transboundary waters; for in such cases, more than one polity is involved, which, generally speaking, renders consensus more difficult to reach. Further complicating matters, transboundary water disputes often create or exacerbate divergent interests within each polity. Nowhere today are transboundary water disputes more fraught—or more contested—than in South and Southeast Asia where developments on mighty rivers such as the Brahmaputra and the Mekong are posing severe environmental threats, development dilemmas, and governance challenges to massive groups of inhabitants and their leaders.

It is thus not surprising that such issues and concerns have not only captured widespread attention but also galvanized important and influential constituencies both within South and Southeast Asia and around the world. Despite other pressing environmental concerns in the latter region, the seasonal haze problem in Indonesia-Singapore-Malaysia and threats posed by potential dam building on the Nu-Salween (Thanlwin) River system in China and Myanmar, for example, the varied and complex threats to the Mekong, are increasingly seen as the most important and certainly the most high-profile environmental concerns in the region. Given the profound significance of the Mekong in the region's life, it is easy to understand why.

The Mekong River is, of course, the longest river in Southeast Asia (roughly 4500 kilometers in length) and the twelfth longest in the world. It rises on the Himalayan Plateau and flows through China (where it is known as the Lancang), then between Laos and Myanmar, and, briefly, between Laos and Thailand, before entering Cambodia and connecting with the Tonle Sap (the largest natural freshwater lake in Southeast Asia, whose enormous ecological vitality depends on the Mekong water flow). It then separates into several distributaries to form the Mekong Delta. The two principal distributaries, known as the Bassac and the Mekong, then enter Vietnam, forming various other distributaries, all of which ultimately empty into the South China Sea.

Although the entire Lancang-Mekong system is important, the system as a whole constitutes the largest inland fishery in the world; for people living along the lower reaches of the Mekong, the river has historically been particularly vital. Not only is the Lower Mekong the most important flow of water in the region, but the river is also absolutely central to the livelihoods of the vast majority of people in Cambodia and Vietnam who reside near its basin. For starters, out of roughly 60 million people who live in or near the Lower Mekong Basin, about three-quarters rely directly on agriculture and the natural resources of the Mekong system for food and livelihoods. The Tonle Sap is the largest freshwater fishery in the region and produces as much as two-thirds of the protein in Cambodia's food supply; the Mekong Delta of Vietnam is one of the most productive agricultural areas and premier rice granaries in the world.

Abundant resources and enormous productivity notwithstanding, nothing has ever come easily in the Lower Mekong. The extremely dynamic environments characteristic of the region—environments marked by drastic differences in seasonal water flows and alluvial soil deposits—have required intense attention to the nuances of environmental change in order for the region's inhabitants to successfully wrest livelihoods from the Mekong's ever varying riparian environments and microenvironments.

A number of changes in recent decades have made the Lower Mekong an even more dynamic and challenging place in which to make a living. For example, alterations in the ways in which the region has been linked to larger flows of commodities and capital have had a huge impact on the region, most notably in the case of Vietnam, whose reemergence as a major rice-exporting country has linked it inextricably to global markets and their vicissitudes. Moreover, dam building along the upper reaches of the Mekong—a key concern in this volume—threatens both rice production in the Lower Mekong and the seasonal surges of water that make the Tonle Sap such a productive fishery. Such decreases in water flows are occurring at the same time as sea level increases because of global warming, adding new threats in the Mekong Delta: droughts, on the one hand, and saltwater intrusion, on the other hand.

While on the subject of global warming, it should be noted, alas, that Cambodians and Vietnamese living in the Lower Mekong region—especially farming and fishing populations with little margin for error in their livelihood strategies—are among the peoples in the world who will be most adversely affected by climate-caused environmental change. Since only a small fraction of global greenhouse gases come from this region, larger environmental justice issues are part of the discussion of policies and adaptations as well.

Even from the brief discussion above, the Mekong's many threats and challenges come through loud and clear. Because the problems posed to, on, and along the river involve numerous nation-states, including the great power, China, the stakes are especially high. In *Water and Power: Environmental Governance and Strategies for Sustainability in the Lower Mekong Basin*, a talented international group of scholars, scientists, policy practitioners, and NGO professionals explores a range of issues relating to the most salient environmental, developmental, and governing challenges on the Mekong. As the volume's title suggests, questions regarding governance loom especially large, for if the Mekong is to survive and the populations living in the river's basin to thrive, new forms of governance and strategies for sustainability must be developed.

Water and Power begins with a frame-setting introduction by retired diplomat David Brown, who has been writing perceptively about the Mekong for years. The main body of the volume is divided into three substantive sections devoted in turn to historical perspectives on the Lower Mekong; issues relating to livelihood strategies, environmental threats, and adaptation strategies; and various aspects of river governance, with individual authors treating questions of governance at different levels of refraction and in different registers. The 18 individual chapters in these 3 sections treat various parts of the river basin—from Yunnan to the lower delta—and, in so doing, provide readers with empirical depth and theoretical breadth on a variety of issues related to the future economic and environmental sustainability of the entire Lancang-Mekong system. The result, we believe, is a fresh and innovative collection of essays, which, taken together, offer rich detail and much-needed new perspectives on some of the most important and seemingly intractable environmental and development issues in contemporary Asia. As such, it constitutes a worthy successor and complement to the previous collection in Springer's Advances in Global Change Research series, *Environmental Change and Agricultural Sustainability in the Mekong Delta*, edited by Mart A. Stewart and Peter A. Coclanis (New York and Heidelberg: Springer, 2011), which volume has been widely read and has made a number of helpful interventions in ongoing discussions and debates regarding the past, present state, and future prospects of one of the world's greatest river systems.

Like the 2011 volume, *Water and Power* developed out of a stimulating conference convened in the Lower Mekong region. Whereas the first volume grew out of a conference hosted by the Can Tho University in Can Tho, Vietnam, in March 2010, *Water and Power* got its start at a meeting hosted by the Royal University of Phnom Penh in Cambodia in March 2017. In each case, the host institution partnered on the meeting with the editors' home institutions in the United States: Western Washington University and the University of North Carolina at Chapel Hill.

Anyone who has ever organized or co-organized a scholarly conference knows how important the people on the ground are in determining the degree of success achieved. In both cases, Mart Stewart first developed a network of colleagues and institutional connections by way of a Fulbright Senior Scholar and two Fulbright Senior Specialist appointments in Vietnam and Cambodia, as well as many years of curriculum development workshops, teaching, and field research. A talented group of colleagues at the Can Tho University facilitated the first conference, and then an equally talented and hardworking group at the Royal University of Phnom Penh maintained the high standards set 7 years earlier in Can Tho. At RUPP the Department of Natural Resource Management and Development was our principal partner, and we would like to thank Dr. Seak Sophat, who heads this department, not only for playing the lead role in coordinating the proceedings in Phnom Penh but also for his many kindnesses to conference participants (not to mention for his fine, coauthored contributions to *Water and Power*). We would like to acknowledge the important roles played by many others at RUPP, including Phat Chandara, Samet

Sok, Sok Serey, and especially Rathkunthea Rim, along with other colleagues there, and an energetic group of RUPP graduate students, in organizing and conducting the conference. Although the conference was held in Phnom Penh, people and resources at both Western Washington University and UNC-Chapel Hill proved to be instrumental in getting the conference off the ground and, later, in making this book possible. Western Washington University gave Mart Stewart some leave, as a result of which he had the time to organize the conference, solicit research papers, develop the program, connect with a publisher, and coax the conference papers into almost publishable form. The key partner at UNC-Chapel Hill was the Global Research Institute (GRI), part of UNC Global, and the funding they provided made the conference possible. The editors would like to acknowledge the roles played by several people at UNC in particular in moving this project along. Chief International Officer Ronald P. Strauss was supportive throughout. Terry Tamari, coordinator of the Global Research Institute, provided many kinds of support. Narvis Green, director of Finance and Human Resources at UNC Global, was both creative and diligent (even relentless) in finding ways to make the budget work. Speaking of budgets, both the conference and this volume benefited greatly from the generosity of one of the GRI's great friends, Mr. Wang Guangfa, chair of the Fazheng Group (Beijing). We would also like to thank Patricia Watson for her superb copyediting, the expert managers of this project at Springer Scientific, Truptirekha Das Mahapatra and Selvaraj Ramabrabha, and our outstanding editor at Springer, Margaret Deignan.

Chapel Hill, NC, USA Peter A. Coclanis
Bellingham, WA, USA Mart A. Stewart

Framing the Future of the Mekong When Nothing Is Certain But Change – An Introduction

The Mekong Delta floodplain comprises all the low-lying wetlands from the Cambodian town of Kratie south to the river's nine mouths in southernmost Vietnam, including the Tonle Sap Basin. Over 145,000 km^2 altogether, the area had a population in 2010 of more than 30 million persons, with roughly two-thirds in Vietnam and one-third in Cambodia (Mekong River Commission 2011). The Mekong Delta is extraordinarily productive and hydrologically complex. In many areas a highly engineered system of dikes, canals, and sluices manage the flow and distribution of water. However, the delta faces severe and mutually exacerbating challenges that, taken together, amount to a slowly unfolding environmental disaster that will have enormous impacts on the way delta dwellers have lived for centuries. These changes cannot be avoided; however, bold policies can enable sensible adaptation.

When I set out to write about the delta's future in 2016 (see Brown 2016a, b, c, d), I imagined a story about adaptation to climate change. I knew that I should also mention the impact of the dam construction upriver. I learned that there is a third threat to the well-being of the people who live in the Mekong floodplain: unsustainable and inappropriate agricultural practices.

Climate change impacts are becoming evident in the delta (and so becoming less controversial). In 2007, the World Bank judged Vietnam's portion of the Mekong Delta, as well as the deltas of the Ganges (Bangladesh) and the Nile (Egypt), to be an area "most threatened" by climate change. The World Bank's analysts primarily considered sea level rise and population density (Dasgupta et al. 2007). And indeed, the sea level is rising and may rise a full meter in this century. That's significant in Vietnam's delta, an area where the average elevation is just 2 m. Climate change is also bringing significant change in weather patterns. The monsoon rains seem already to be less regular and to arrive later in the year. Temperature is expected to rise to levels that stress current crop varieties.

In addition, and perhaps of greater significance, are the impacts of dam construction upstream. There is, first of all, the Lancang cascade, high dams in Yunnan province that in the rainy season impound large amounts of water and in the dry season release it slowly to produce electricity. Second, there are roughly 200 dams built, under construction, or planned on the tributaries of the Mekong. Many of the largest projects are on the so-called 3S system—the Sekong, Sesan, and Sepong Rivers, which arise in Vietnam's central highlands and join the Mekong mainstream in Cambodia near Stung Treng. Like the Chinese cascade, these tributary dams impound great quantities of water during the rainy season and release it slowly during the dry season.[1] Third, there is the Mekong mainstream dam cascade, the human interventions that have captured most of the attention, particularly because these dams will impede the migration of many species of fish from spawning grounds in the Tonle Sap and lower down the Mekong.

Under even the most optimistic scenarios—assuming that many migratory fish will be able to climb fish ladders and their offspring will survive, and that nonmigratory "blackfish" will fill some ecological niches left by less adaptable migratory species—there will still be a huge impact on the diet of delta dwellers. Further, all of the dams trap river-borne sediment, which has for eons enriched farmers' fields downstream. In a word, the more dams, the less available calories, either from fish or from vegetables. The good news about the mainstream dams is that they will not impound much water. They are "run of the river" dams that rely on a fairly steady (rather than seasonal) flow to drive their turbines. Even so, say Vietnamese scientists, the combined impacts of climate change and upstream dam construction already amount to a more than 50% reduction in the amount of nutrient-laden silt that is reaching Vietnam's Mekong Delta (Thanh Nien News 2016). A weaker flood pulse and rising sea level combine to increase salt intrusion there in the dry season. Early in 2016, during the worst drought in 90 years, salt concentrations lethal to rice were measured as far upstream as Cần Thơ, Vietnam's delta metropolis, 90 km from the river mouth.

And yet, the Mekong Delta's problems are not just a matter of climate change and dams. Particularly in the delta area in Vietnam, unsustainable farming practices have been the norm. Ever since they migrated into the Mekong Delta in the eighteenth century, the Vietnamese have been transforming its landscape, digging canals and draining swampy areas. The French colonial authorities brought in machines, and the work went faster. The greatest changes took place after 1975, however, when engineers from Hanoi built dikes to transform two very large wetland areas, the Plain of Reeds and the Long Xuyen Quadrangle. By perfectly controlling the water level, they made it possible to grow three or even three and a half crops of rice there each year. Lower down the Mekong's many branches, the Vietnamese elaborated a system of dikes and sluices to prevent saline intrusion. This huge project of hydrological engineering aimed at increasing rice production to ensure Vietnam's food security. By the mid-1990s, Vietnam had an annual rice surplus of three million tons. By 2012, Vietnam was briefly the world's number one rice exporter; that year it supplied over 8 million tons of rice to the world market.

For years, however, some experts had questioned the sustainability and real costs of the absolute priority Hanoi gave to rice production. Many farmers were unhappy because they were required to grow rice year round, even though they could earn

[1] When all present and proposed dams on the Mekong's main tributaries are considered, dry-season flows will increase by 63% over baseline, and wet season flows will decline by 22% (Piman et al. 2013).

more growing other crops. The warehouses of the state-owned marketing company filled up with low-quality rice that was difficult to sell at a profit. Each crop required increasing inputs of fertilizer and pesticides. Further, near the seacoast, where brackish water is a growing seasonal problem, the land was sinking as great quantities of freshwater were withdrawn from aquifers deep underground. In some places the land has been subsiding faster than the sea level has been rising. Thus, it is increasingly difficult to keep saltwater from invading coastal areas during the dry season and to flush out salt and alkaline minerals when the monsoon rains come.

Beginning in 2011, senior officials from concerned Vietnamese ministries and Vietnamese scientists met with experts from the Netherlands to discuss foreseeable impacts on the Mekong Delta area out to the year 2100. From these talks emerged a strategy of strategic retrenchment on a large scale: the Mekong Delta Plan (Socialist Republic of Vietnam and the Kingdom of the Netherlands 2013). Vietnam's government is moving to implement elements of the plan. In coastal areas, conversion to brackish-water aquaculture is well advanced. In this area it will rely mainly on "soft" barriers, chiefly mangrove forests, to limit coastal erosion. The dike and sluice system will be refocused on defending higher elevations in the central and upstream areas of Vietnam's delta. Even on prime rice land, farmers will be allowed to grow other crops in the dry season. And finally, efforts will be made to restore functionality of the Plain of Reeds and Long Xuyen Quadrangle as aquifers that can store freshwater during the flood season and release it during dry months.

Upstream in Cambodia, there's a very big but so far less studied worry: the Tonle Sap Basin. The lower Mekong floodplain is one continuous environmental region. What happens upstream in Cambodia impacts Vietnam as well as Cambodia, so farmers and fishermen in both nations ought to be very worried about the Tonle Sap's future ability to store freshwater and then release it. The mechanics of the great lake's seasonal filling and flushing are well understood. When the Mekong floods, water flows into the Tonle Sap via the Sap River. Swollen also by local monsoon rains, the volume of the Tonle Sap grows by 30 times. It is a remarkably fecund ecosystem. And then, as the Mekong water level falls, the flow reverses and the Tonle Sap empties, releasing freshwater to nourish crops in the areas downstream.

So far, there appears to have been little system wide modeling focused on the possibility that the multiplication of dams upstream could compromise the great lake's filling and flushing mechanism. According to a 2013 review, "No studies have predicted how hydrological alterations may permanently compromise the role that existing floodplain wetlands play in reducing flood peaks and providing storage that can naturally supply streams with water during periods of low or no rainfall" (Hecht and Lacombe 2014). Dr. Alan Potkin (2017) reached a similar conclusion: "The question … is whether [dry-season] flows will remain sufficiently low that the Mekong stage at the Tonle Sab river confluence continues to enable the extraordinary flow reversal upon which the hydroecology of the [floodplain] certainly depends, and perhaps the fisheries far upstream as well." Studies by Cochrane, Arias et al. (2014) provide evidence that current and proposed construction of dams upstream, particularly on the Mekong's major tributaries in Laos, Thailand, and Vietnam, foreshadows "drastic alterations to the hydrological pulse and subsequent ecological features in the Tonle Sap." According to Dr. Sok Saing Im, he and colleagues are working on "a study to identify effective counter measures to reduce impacts of flow regime changes by considering diversion of early Mekong flood … to the Tonle Sap" (personal communication, March 13, 2017). And since 2012, a team of scientists has been at work on what may, in the sphere of public policy, prove to be the definitive assessment of development impacts in the Mekong River Basin, the "Council Study."

Until now, almost all attention has focused on prevention: persuading Laos and its enablers to give up building dams on the Mekong mainstream. It now seems likely that, for economic reasons, after Pak Beng and Don Sahong, only one or two more mainstream dams will be built (Cronin and Weatherby 2015). Still, negative impacts are already huge, particularly on fish migration and the volume of silt transported downstream. Ironically, these lower Mekong mainstream dams will not much alter the hydrology of the Mekong Delta. As discussed above, it is dam construction on tributaries and in the Yunnan gorges, which impound far greater volumes of water in the wet season and release it in the dry season, that is altering the flood pulse and thus impacting the annual rhythm of life.

Now in Cambodia and in Vietnam, the focus must shift to adaptation to new realities. There is a vital need for closer cooperation between Vietnamese and Cambodian experts, between Vietnamese and Cambodian government officials, and between Vietnamese and Cambodian journalists. The problems impacting the lower Mekong are not "solvable" in the conventional sense. However, bold policies can enable sensible adaptation. With a long-range vision, for example, an expanded version of Vietnam's Mekong Delta Plan, determined national and local leadership, and the backing of aid donors like the World Bank, the delta's future may at least be manageable.

Freelance Analyst/Writer
Fresno, CA, USA

David Brown

References

Brown, D. (2016a, October 3). *Will climate change sink the Mekong Delta?* (Mongabay Series: Mekong Dams). http://news.mongabay.com/2016/10/will-climate-change-sink-the-mekong-delta/

Brown, D. (2016b, October 6). *Vietnam sweats bullets as China and Laos dam the Mekong* (Mongabay Series: Mekong Dams). http://news.mongabay.com/2016/10/vietnam-sweats-bullets-as-china-laos-dam-the-mekong/

Brown, D. (2016c, October 13). *Mother nature and a hydropower onslaught aren't the Mekong Delta's only problems* (Mongabay Series: Mekong Dams). http://news.mongabay.com/2016/10/mother-nature-and-a-hydropower-onslaught-arent-the-mekong-deltas-only-problems

Brown, D. (2016d, October 18). *A plan to save the Mekong Delta* (Mongabay Series: Mekong Dams). http://news.mongabay.com/2016/10/a-plan-to-save-the-mekong-delta/

Cochrane, T. A., Arias, M. E., & Piman, T. (2014). Historical impact of water infrastructure on water levels of the Mekong River and the Tonle Sap system. *Hydrology and Earth Systems Sciences, 18*, 4529–4541.

Cronin, R., & Weatherby, C. (2015). *Letters from the Mekong: Time for a new narrative on Mekong Hydropower* (Stimson Center issue brief). https://www.stimson.org/content/letters-mekong-time-new-narrative-mekong-hydropower

Dasgupta, S., Laplante, B., Meisner, C., Wheeler, D., & Yan, J. (2007, February). *The impact of sea level rise on developing countries: A comparative analysis* (World Bank Policy Research Working Paper 4136). https://doi.org/10.1596/1813-9450-4136

Hecht, J., & Lacombe, G. (2014). *SOK 5: The effects of hydropower dams on the hydrology of the Mekong Basin* (State of knowledge series 5). Vientiane: CGIAR Research Program on Water, Land and Ecosystems.

Mekong River Commission. (2011). Basin development plan programme: Planning atlas of the lower Mekong River Basin: Cambodia, Lao PDR, Thailand, Vietnam. http://www.mrcmekong.org/assets/Publications/basin-reports/BDP-Atlas-Final-2011.pdf

Piman, T., Lennaerts, T., & Southalack, P. (2013). Assessment of hydrological changes in the lower Mekong Basin from basin-wide development scenarios. *Hydrological Processes, 27*(15), 2115–2125.

Potkin, A. (2017). *Are fisheries impacts of Riparian deforestation in Siphandone already eclipsing hydropower effects?* Paper presented to International Conference on Environmental Change, Agricultural Sustainability and Economic Development in the Lower Mekong Basin, Royal University of Phnom Penh, Cambodia.

Socialist Republic of Vietnam and the Kingdom of the Netherlands. (2013, December). *Mekong Delta plan: Long-term vision and strategy for a safe, prosperous and sustainable delta.* https://data.opendevelopmentmekong.net/dataset/e988f21c-9129-45f6-8be5-148677ba0480/resource/a520f1b1-7e84-4f65-9da6-dbb6a90d254e/download/mekongdeltaplan.pdf

Thanh Nien News. (2016, May 2). *Mekong Delta loses half of silt to upstream dams: Scientists.* http://www.thanhniennews.com/society/mekong-delta-loses-half-of-silt-to-upstream-dams-scientists-61739.html

Contents

Contributors

Michitake Aso Department of History, State University of New York, Albany, NY, USA

Olivier Aznar Clermont Auvergne University, AgroParisTech, INRA, Irstea, VetAgro Sup, Territoires, Clermont-Ferrand, France

David J. H. Blake Independent Scholar, Taunton, Somerset, UK
David Brown Freelance Analyst/Writer, Fresno, CA, USA

Andrea H. Claassen Department of Fisheries, Wildlife, and Conservation Biology, University of Minnesota, Minneapolis, MN, USA

Peter A. Coclanis Global Research Institute, University of North Carolina-Chapel Hill, Chapel Hill, NC, USA

Stéphanie Defossez French Research Institute for Sustainable Development, UMR GRED, Montpellier, France

Christophe Déprés Clermont Auvergne University, AgroParisTech, INRA, Irstea, VetAgro Sup, Territoires, Clermont-Ferrand, France

Dinh Diep Anh Tuan Research Institute for Climate Change, Can Tho University, Can Tho, Vietnam

Goto Akira Department of Hydrological and Environmental Engineering, Utsunomiya University, Utsunomiya, Japan

W. Nathan Green Department of Geography, University of Wisconsin-Madison, Madison, WI, USA

Huynh Thi Phuong Linh Ecole Française d'Extrême-Orient, Hồ-Chí-Minh City, Vietnam

Kjeld Ingvorsen Department of Bioscience, Aarhus University, Aarhus, Denmark

Olivier M. Joffre WorldFish, Mekong Regional Office, Phnom Penh, Cambodia

Knowledge, Technology and Innovation Group, Wageningen University, Wageningen, The Netherlands

Kim Lavane College of the Environment and Natural Resources, Can Tho University, Can Tho, Vietnam

Kimchhin Sok Ecosystem Services and Land Use Research Centre, Royal University of Agriculture, Phnom Penh, Cambodia

Lam Van Thinh College of the Environment and Natural Resources, Can Tho University, Can Tho, Vietnam

Le Hoang Viet College of Environment & Natural Resources, Can Tho University, Can Tho, Vietnam

Malyne Neang ECOLAND Research center, Royal University of Agriculture, Phnom Penh, Cambodia

Danny Marks Department of Asian and International Studies, City University of Hong Kong, Kowloon Tong, Hong Kong

Philippe Méral IRD, UMR GRED, Montpellier, France

Nguyen Dinh Giang Nam United Graduate School of Agriculture Science, Tokyo University of Agriculture and Technology, Tokyo, Japan

College of Environment and Natural Resources, Can Tho University, Can Tho, Vietnam

Nguyen Hieu Trung College of Environment and Natural Resources, CanTho University, Can Tho, Vietnam

Nguyen Hong Duc College of the Environment and Natural Resources, Can Tho University, Cần Thơ, Vietnam

Research Institute for Climate Change, Can Tho University, Can Tho, Vietnam

Nguyen Huu Chiem College of Environment & Natural Resources, Can Tho University, Can Tho, Vietnam

Nguyen Vo Chau Ngan College of Environment & Natural Resources, Can Tho University, Can Tho, Vietnam

Osawa Kazutoshi Department of Hydrological and Environmental Engineering, Utsunomiya University, Utsunomiya, Japan

Phat Chandara Department of Natural Resource Management and Development, Faculty of Development Studies, Royal University of Phnom Penh, Phnom Penh, Kingdom of Cambodia

Didier Pillot SupAgro, Montpellier, France

Klaus Schmitt Deutsche Gesellschaft für Internationale Zusammenarbeit (GIZ) GmbH, Manilla, Philippines

Seak Sophat Department of Natural Resource Management and Development, Faculty of Development Studies, Royal University of Phnom Penh, Phnom Penh, Kingdom of Cambodia

Seng Vang Massachusetts Institute of Technology, Cambridge, MA, USA

ASEAN-US S&T Advisor, Phnom Penh, Cambodia

Jenny Leigh Smith Hong Kong University of Science and Technology, Hong Kong, Hong Kong

Soriya Yin Massachusetts Institute of Technology, Cambridge, MA, USA

Mart A. Stewart Western Washington University, Bellingham, WA, USA

Olivier Tessier Ecole Française d'Extrême-Orient, Hồ-Chí-Minh City, Vietnam

Nguyen Thanh Loc College of the Environment and Natural Resources, Can Tho University, Can Tho, Vietnam

Tran Sy Nam College of Environment & Natural Resources, Can Tho University, Can Tho, Vietnam

Li Zhang Center for China's Neighbor Diplomacy Studies and Institute of Southeast Asian Studies, Yunnan University, Kunming, China

Part I
History and Environment in Southeast Asia

Chapter 1
The Mekong Imaginary: From Apocalypse Then to Anthropocene Now

Mart A. Stewart

Abstract The Mekong River is the 12th longest river in the world, around 4500 km, and until recently has been remarkable among its class of rivers for not being "developed." Though intensely used by people who live near, on, and in the Mekong, and though several substantial irrigation projects historically have structured portions of the Mekong Basin and the Mekong Delta, until recently the Mekong has been free of major hydroelectric dams and regional irrigation projects. Just how this relatively undeveloped but highly used river has been imagined by observers can be summarized by what cultural studies specialists call an "imaginary" – a complex of perceptions and expressions that collectively represent, in this case, a geographical phenomenon. The 1978 American film *Apocalypse Now*, about the Vietnam War, provided an enormously influential depiction of a long stretch of the Mekong – by way of its fictional stand-in, the Nung River, on which most of the action in the film took place. This film was deeply influenced in turn by the classic novella *Heart of Darkness* by the Polish-English writer Joseph Conrad – what Edward Said has decisively explained as an important anti-imperialist imperialist work of literature. In the 1990s and the first decade of this century, several dam-building projects, planned and now being constructed, have begun to domesticate the river for high-level exploitation. These projects and the responses to them have also encouraged a regional understanding of the Mekong, on a larger scale and as a basin and region rather than as a single thread of access into a place of unmitigated darkness. Paradoxically, hydroelectric dam projects and visions for regional exploitation of the Mekong – and struggles over who gets a say and how to manage that exploitation – have also encouraged regional approaches to managing fisheries, riverine environmental, and conservation efforts. This chapter discusses these collective imaginings with the purpose of encouraging a more reflective understanding of a geography that is often taken for granted in efforts to manage the Mekong.

M. A. Stewart (✉)
Western Washington University, Bellingham, WA, USA
e-mail: Mart.Stewart@wwu.edu

M. A. Stewart, P. A. Coclanis (eds.), *Water and Power*, Advances in Global Change Research 64, https://doi.org/10.1007/978-3-319-90400-9_1

Keywords Imaginary, Mekong River · Mekong Region · Hydroelectric dams · Mekong Exploration Commission · Apocalypse Now · Heart of Darkness · Edward Said · Post-colonial · Anthropocene

1.1 The Mekong Imaginary

The Mekong River is the 12fth longest river in the world, around 4500 km, and until recently has been remarkable among its class of rivers for not being developed. Though intensely used by people who live near, on, and in the Mekong, and though several substantial irrigation projects historically have structured portions of the Mekong Basin and the Mekong Delta, until recently the Mekong has been comparatively free of major hydroelectric dams and regional irrigation projects. At the same time that the Mekong River has resisted totalizing schemes of development, it has also resisted easy summaries of just what it is; "many Mekongs" have historically made the efforts of both scientists and planners to generalize about the river or even significant portions of it a formidable task. But visions of the Mekong have nonetheless emerged, not as much in efforts to analyze and explain it as indirectly in depictions and representations of the river, or in assumptions about what can be done with it.

Just how this relatively undeveloped but highly used river has been imagined can be summarized by what cultural studies specialists call an "imaginary" – a complex of mediated perceptions and expressions that collectively represent, in this case, a geographical phenomenon. An imaginary develops and functions differently than a calculated scientific concept, on the one hand, or a selected representative image, on the other, and is not expressed explicitly in the same way as either of these. An imaginary is also an ideal way to capture social and cultural components of a regional space as they interact with ecological ones. It is an especially resilient conceptualization when these components, as they are metabolized by expressions of the regional space, are always in flux (Anderson 1991; Wilson 2000). The significant large-scale imaginaries that have emerged from and sometimes guided plans for using the river have usually originated with outsiders to the region, but always in an interaction with Mekong environments – in tangible on-the-ground as well as imagined experiences. How these add up in particular expressions, especially the ones that have had staying power, can be identified as a Mekong imaginary. This chapter discusses several of these collective imaginings with the purpose of encouraging a more reflective understanding of what is often taken for granted in efforts to comprehend and manage the Mekong.

Accounts of the Mekong by explorers and travel writers in the nineteenth and for most of the twentieth century were either too infrequent or not widely enough read to suggest an imaginary. But they point in the direction of seeing the Mekong River watershed as a corridor of navigation that resisted in the most implacable of terms the efforts of outsiders to navigate it. The accounts of the first scientific expedition of the Mekong, the Mekong Exploration Commission of 1866–1868, conducted by a small group of French navy officers and supporting staff, set the terms for later depictions of the river – if not a place of dark challenges, at least of exotic inscrutability and of unnavigable physical challenges. Their account of their exploration not only had the official sanction of the French government but also had an afterlife in late nineteenth- and early twentieth-century depictions of South East Asia and the Mekong. During its 2 years the commission was able to map about 2000 km of the river and, in explorations of its tributaries and hinterland, was able to map another 5000 km of the region. But they were often thwarted by the physical challenges of simply moving up the river, by its cataracts, whirlpools, and monsoon-swollen currents, as well as by diseases (malaria, dysentery, leech infections), accidents, cultural conflicts (they changed one of their routes to circumvent a local war), and ultimately, utter destitution. They suffered the death of their commanding officer and the permanent illness and debility of several others (Garnier 1885, trans. 1996; Osborne 1975).

The larger imperial struggle between England and France for control in Southeast Asia provided a context for the expedition's resolve, and also created political challenges that both helped and hindered them, but the real challenges to this imperial enterprise were on the ground. Though these explorers were able more or less to track the Mekong into Yunnan in China, they ultimately failed in their goal of mapping the upper reaches of the Mekong. The Mekong also forcefully eluded their intention of developing a river route for trade with China and subverted the myth of navigability that had originally supported this goal. The Mekong was not simply unnavigable but too unnavigable for explorers with considerable surveying and engineering expertise to imagine a route, though the endurance of the expedition in attempting to do so went quite beyond a commitment to this goal. The Mekong Exploration Commission was a costly but magnificent failure.

One of the chief officers of this exploration, Lieutenant Francis Garnier, later recalled the irrational seduction worked on him by the blank spaces on the map of the Mekong and then by his experience attempting to navigate up the river:

> For my part I attached a special importance to continuing to track its winding and bizarre course. Ever since we had entered regions untouched by European enquiry, each meander of the Mekong as added to my map seemed an important geographical discovery. Nothing could distract me from this abiding concern. … It came to possess me like a monomania. *I was mad about the Mekong.* (quoted in Keay 2005, 293; italics mine)

The Mekong Exploration Commission did not, like its fictional counterparts (see below), travel into the darker regions of their souls at the same time that they attempted to navigate the Mekong – or at least they weren't as explicitly aware that they did. But they encountered challenges that European rationality could not penetrate. And they lost their moorings to this rationality –were mad about the Mekong – as they fought their way upriver.

1.2 Apocalypse Then

In more recent times, the 1978 American film *Apocalypse Now*, about the Vietnam War, provided an enormously influential depiction of a long stretch of the Mekong – by way of its fictional stand-in, the Nung River, on which most of the action in the film took place. The screenplay for this film was deeply influenced in turn by the classic novella *Heart of Darkness* by the Polish-English writer Joseph Conrad – what Edward Said has decisively suggested has been our greatest work of anti-imperialist imperialist literature (Said 1993, 19–31). Conrad's (1899) novella was about a journey up the Congo River, but in *Apocalypse Now* the Nung and the Congo are conflated in environmental terms, and also in the way that people native to the environs of the river are depicted.

Other, and better, films about the Vietnam/American War were made in the next 20 years, and at the same time the Mekong imaginary underwent a transformation – from a place of conflict, darkness, and savagery to one of commercial possibilities, as well as a last stand against environmental degradation. Since the late 1980s, when then Thai Prime Minister Chatichai Choonhavan famously urged transformation of the region from a "battlefield into a marketplace," several dam-building projects – planned and now being constructed – have begun to domesticate the river for high-level exploitation (Hirsch 2001, 22). These projects and the responses to them have also encouraged a regional understanding of the Mekong, on a larger scale and as a basin and region rather than as a single thread of access into a place of unmitigated darkness. Paradoxically, hydroelectric dam projects and visions for regional exploitation of the Mekong – and struggles over who gets a say and how to manage that exploitation – have also encouraged regional approaches to managing fisheries, riverine development, and conservation efforts.

Americans of the Vietnam War generation first came into contact with the Mekong River as soldiers, as a place fraught with environmental challenges as well as military ones. The American public saw dramatic snippets of this struggle in their living rooms, in television news reports. But the first fully imagined vision of the Mekong by an American during this period came by way of the award-winning and influential 1978 Francis Ford Coppola film, *Apocalypse Now*. This film was nominated for eight Academy Awards and won two, as well as a slate of other awards, and often makes the "Top 100" lists of films in the twentieth century. For many years, in spite of its several nearly absurd renderings of combat conditions in the Vietnam War, this was a film taught Americans – apart from veterans – how to understand the war. The "apocalypse" in the film was the journey into the heart of darkness, into the jungle, and into regions yet undeveloped by the presence of Westerners and modern technology, as well as into the regions, Coppola postulated, of this war's dark soul. A film that was deliberately more metaphor than history also took the physical and environmental conditions of conducting a ground war in Southeast Asia seriously. The film was made mostly in the Philippines, but before difficult environments for filmmaking could be replicated easily in the computer lab, and by way of a commitment by the director to creating a film in which tropical heat and lushness were a constant presence. (The film took a year to make and had several meltdowns, including a heart attack by one of the stars, Martin Sheen, and altogether created a drama substantial enough to yield a successful documentary about the making of this film, *Hearts of Darkness: A Filmmaker's Apocalypse*, which Coppola produced in 1991.)

The screenplay for the film was written by Coppola and the war journalist Michael Herr and had two main sources. One was Herr's short, brilliant, and in some places near-hallucinogenic essays from the war zone, collected into an award-winning book, *Dispatches* (1977). The second was Joseph Conrad's late-nineteenth century novella *Heart of Darkness*, about a similar journey up the Congo River into the interior of the African continent – to the "heart of darkness." By the time Coppola began imagining *Apocalypse Now*, Conrad's novella had a well-established reputation as a classic anti-imperialist statement, a fictional condemnation of the greed and exploitation of colonialism, and one of those books that, after World War II, when many nations in Africa acquired independence from colonial rule, every educated observer should read.

In the novella, a fictional but somewhat autobiographical narrator, Marlow, tells a story of his journey up the Congo River to a group of sailors on a boat anchored securely in the Thames. He begins by describing the Congo as a fluid passageway into the very heart of Africa, "resembling an immense snake uncoiled," and his interest in this destination, "as a snake would a bird" (Conrad 1899, 10). His perspective moves from this birds-eye summary to a more tangible on-the-ground description, as he explains how he came to be the captain on a ivory-trading boat, far up this river on a mission to find and assist Kurtz, a talented and eccentric ivory trader. Though the Congo is in principle an easy passage, compared to the dense and tangled jungle stretching out with near infinitude from its banks, it becomes darker and an increasingly formidable challenge – from human as well as from natural forces – as Marlow's journey proceeds: he marches up the river to the station where he is to take over his captaincy, only to find the boat at the bottom of the river because of an accident; putting it back into action requires several months of repairs and waiting for parts and materials after a mysterious fire destroys some of these; as the expedition approaches Kurtz's location, the boat is enveloped by a low fog, a hail of small arrows from invisible attackers falls upon the boat, and the helmsman is felled by a spear; palpable fear and uncertainty, intensified by the discovery of a row of posts with the severed heads of natives atop them near Kurtz's station house, make both Marlow and especially his men increasingly skittish; when they arrive, Kurtz is ill and near death, and they make their escape with him – but then he dies, whispering as his last words, "the horror, the horror." It is a challenging journey that becomes horrific as it proceeds.

Conrad juxtaposes the "civilizing mission" of European trade and imperialism with the dark, impenetrable, and sinister savagery of Africa and African natives in several ways throughout the novella – Marlow uses the steam whistle of his boat to frighten off natives who are attacking or about to attack them, for example. But he also makes clear that the "whited sepulchre" of this mission was by no means immaculate. The corruption, venality, and savagery of the colonizers and ivory traders in the novella are continuous with the African hinterland and its inhabitants. The journey up the Congo into the heart of darkness is not simply an adventure story and a travelogue, in other words, but in this resonant and repeatedly analyzed novella it is a journey into the rotten core of European imperialism, into the heart of the exploitation and the cultural system that promoted it.

Apocalypse Now departed from this narrative in several ways but preserved the core story and many of the defining details – even Kurtz's debilitated exhortation, "the horror, the horror," near the end, this time by Marlon Brando's Colonel Walter Kurtz. Most important was the depiction of the Nung – again, as a stand-in for the Mekong – as a place of increasing darkness and danger as this film's version of Marlow, Martin Sheen's Captain Benjamin Willard, moves up the river with his beleaguered and dwindling crew. This imaginary of the Mekong as a place of sinister resonance (spiked by a soundtrack that included Jimi Hendrix) and of dark and challenging environments has other sources and depictions, whether Coppola knew of them or not: the exploration journals of the nineteenth century and of colonial accounts of Southeast Asia by Westerners, such as the Mekong Exploration Commission mentioned above; and Graham Greene's classic *The Quiet American* (1955), which depicted the Vietnamese countryside and the Mekong Delta as a place of conflict and danger – and in fact with very few places of safety in steamy Saigon as well. The moral landscapes of Marguerite Duras's autobiographical novel *The Lover*, published in 1984, shortly after Coppola's film but before the documentary about making it, also reinforced this imaginary. This provocative and engaging novel, by someone who had lived on the periphery of polite colonial society in Vietnam in the 1920s and that portrayed the south of Vietnam as a place of heated transgressions (and, in places, revulsions), also promoted a larger imaginary of the Mekong Delta and the south Vietnam as a site of emotionally violent encounters and as an exotic (and erotic), mysterious, and ultimately unknowable place. Some of this imaginary was classic European Orientalism, both a product and a driver of European imperializing encounters with Asia and all places where they "othered" natives: Asia and inscrutable (and, often, wily) Asians always shrouded by a veil of mystery, and ultimately quite dangerous as well. An awareness of "darkness" beyond the civilizing bounds of European reason and science, and a geography in which it is embedded in tangible terms, was, as Conrad most profoundly understood, fundamental to the European encounter with the regions they sought to colonize in Africa and Asia. Generalizations about this geography were also fundamental to the imperial mission (Said 1993, 47–48).

What all of these depictions have in common is a Western vision of the watery landscapes of Southeast Asia as places of natural beauty and sometimes innocence but also of corruption, danger, and, often, death for white people. Even before Coppola made his film, the Nigerian writer Chinua Achebe had attacked this vision and the anti-imperialist credentials of *Heart of Darkness*, and pointed out the subterranean racism in the novella – the dark and sinister environments were bright and sustaining ones, he explained, for the Fang people living along the Congo – and West African forests are depicted in the same terms in Achebe's masterpiece, *Things Fall Apart*. What went for Conrad's depiction of the environments along the Congo also went for his depictions of the peoples along the river – they are simplified and condescended to in Conrad's novella, in a way that was common to imperialist representations everywhere (Achebe 1978; Watts 1983).

Coppola lifts up and interprets Conrad's vision superbly into cinematic form, but also does not escape it. To Captain Willard and his crew, the inhabitants of this boundaryless region beyond the American presence in Saigon and up the Nung are always suspect, never portrayed in terms that would give them character, and are often treated with savage violence by the American characters in the film. The recent comments of Vietnamese American Pulitzer-Prize-winning novelist Viet Thanh Nguyen about *Apocalypse Now* – a film whose vision, he says, has been an inspiration but also a source of pain – confirms its skewed view of the inhabitants of the Nung and its environments. For Vietnamese Americans, he says, the pivotal moment in the film is a scene where Vietnamese farmers are massacred by Willard's skittish crew because of a movement within their sampan – later determined to be caused by a puppy. "People just like me were being slaughtered," he said; "I felt violated" (Streitfeld 2016). Coppola's geography, like Conrad's, obliterates the humanity of those who live on and around the river, or at least submerged them beneath the dominant narrative. This scene, absolutely consistent with the imaginary of a Southeast Asian river that Coppola created, is horrific to a Vietnamese viewer and revealed a view of the Mekong and those who lived on it that was conditioned by a world very much outside of Vietnam.

The influential postcolonialist theorist Edward Said takes a more thoughtful approach to *Heart of Darkness* and its descendants in his 1990s study of the relationship between literature and colonialism, *Culture and Imperialism*. Said acknowledges the greatness of Conrad's novella and acknowledges its legitimate status as a condemnation of imperialism. But he also accepts Achebe's critique of the novella, as well, and defines *Heart of Darkness* as both anti-imperialist and imperialist: a work of art that we can study to understand how imperialism might have looked to someone from an imperialist culture, but that requires correction from someone with his or her feet on the ground in Africa. Conrad's narrative of a journey up a river into the heart of darkness is much more than this, Said adds – it journeys also into the instability and constant remaking of things that is life itself (Said, 19). At the same time, this novella – and the film that borrowed extensively from it for its screenplay – creates a geography of "horror" that for outsiders can be entered and fled only through the dark river upon which it also depends.

1.3 Anthropocene Now

Along with the end of war and the return of relatively stable political structures in the countries through which the Mekong passes, and the emergence of development visions that consider the region as a whole, observers of the region who study natural environments have also accomplished a major shift in how "nature" is imagined and conceptualized. Notions of the Mekong as a transboundary "region" may well have required political stability and confirmation of national boundaries for its viability as a workable geographical concept in the first place (Sneddon and Fox 2012b, 153). The Mekong is now portrayed as a general region, and the lower reach of it as a basin rather than merely linked threads of water. The "Mekong Region" in principle includes all of the Mekong, but in practice scientists and policy mavens generally apply it to focus on the Mekong from Yunnan Province in China to the Mekong Delta – with a focus on the lower Mekong and especially the ecologically rich and complex subregions of the Tonle Sap tributary and the Mekong Delta. Scientists now look at discrete ecosystems within the watershed but also assume that the watershed is a unified region. Development and policy experts talk about management of the Mekong in regional terms that include river, watershed, and sometimes neighboring regions and that cross national boundaries as a space for transboundary management strategies.

Mekong water continues to be materially essential to definitions of the river – either up it or down it, and now throughout its watershed as well. But economic development schemes, efforts to manage its complex ecosystems, and regional geopolitical strategies have meant that the meaning of Mekong water has been redefined. At the same time, new definitions of darkness in this geography – the threat of ecological destruction and social displacement – are portrayed as originating more from outside the region rather than as something endemic to it. The geographers who have been most prominent in noticing this redefinition of the Mekong, Chris Sneddon and Coleen Fox, summarize: "In the Mekong region, water, beyond its fundamental necessity, constitutes both a critical component of people's livelihood strategies as well as [now] the focal point of a variety of interrelated developmental, geopolitical, and biophysical processes governed primarily through state actions" (Sneddon and Fox 2012b, 156; also see Sneddon and Fox 2006, 2008a, b, 2012a; Sneddon 2013). Water, shaped by political ecology, has created a region of both possibility and destruction, where once "the horror" was at home.

Not that the narrators of regional geographical spaces did not earlier dream of modernizing this region in ways that would transform darkness into economic possibility and replace a tangle of forests and waterways with landscapes of moderated control. In the 1960s and 1970s a regional project plan that would replicate in Southeast Asia one of the most comprehensive (and emblematic) water management projects in the United States, the 1930s Tennessee Valley Authority, was developed by the U.S. government; President Lyndon B. Johnson made available $2 billion in aid to facilitate the project in 1965 (Ekbladh 2002, 26). But this scheme could not gain a foothold in the contentious Mekong watershed of the time, and it was relegated to a mere historical note – but a note that emphasizes once again, as Karl Wittfogel recognized a half-century ago in his study of hydraulic civilizations in China, the relationship between the control of water and the flow of political power (Wittfogel 1957, 1981; Bennedikter 2014).

More recent schemes for studying or developing the Mekong or for conserving its ecological and social resources have also recognized this relationship, and the deep interconnection between what might be done with water and structures and expressions of political power – of environmental governance. One example will suffice: the organized response of several kinds of nongovernmental organizations to plans for dams on the mainstream of the Mekong that coalesced around the formation of a coalition called Save the Mekong about a decade ago. This coalition was formed from a number of international, national, and local organizations with a long interest in environmental struggles in Southeast Asia, to campaign against the building of the dams. Their protest against plans for damning the Mekong required a full discussion of just what "development" might mean in the Mekong region and how to balance economic and commercial development with the conservation of ecological integrity and local livelihoods. Since its inception in 2009, the coalition has built an argument around the disruption and loss of fish migrations and millions of fisheries livelihoods that "depend upon the Mekong for their food security and income" (Save the Mekong Coalition 2010).

The coalition has been especially critical of a major actor in the environmental governance of the Mekong and another promoter of a regional, transboundary view of the Mekong, the Mekong River Commission (MRC). The MRC was first organized as a transboundary river basin organization that would attempt to moderate cooperation among Vietnam, Cambodia, Laos, and Thailand on the sustainable management of the water resources of the Mekong River. It has claimed neutrality as an institutional arbitrator of water resources management in the region, a position that the Save the Mekong Coalition and others have increasingly attacked for ignoring many of the complex development debates of the region and focusing only on what different states in the region should or should not do in terms of hydropower development. MRC programs have been funded by donors from Japan, Australia, and the Scandinavian countries (Sweden, Norway, Denmark, and Finland), and critics claim that it has too often therefore been pulled away from the various scales and kinds of economic decision making, development policy formulation, and environmental institutions and processes of the states it purports to work with. Critics have also claimed that concentrating on the management of hydropower development has distracted the MRC from proposing policy on the regional and basin level scale it claims to support, and to concentrate only on the river as a stream of water energy whose capture needs to be cooperatively managed. What is significant here is that though the MRC and the Save the Mekong Coalition differ sharply in how the environments of the Mekong region should be governed and managed, they more or less agree on a regional approach to the river. Both directly and indirectly, they all agree on identifying the Mekong as a watershed, basin, and region rather than merely course and confluence. They understand the geography of the region in the same way – no longer as a tunnel into primitive darkness but as a complex basin of opportunities, both for economic development and for conservation (and, often, both at the same time) (Wong 2010, 7, 24–44).

The promotion of this particular Mekong imaginary among the many possible ones has acquired another dimension in recent times – one that is often defined from the air and that posits the Mekong as a point of action on a global rather than a local or regional scale. This satellite imaginary, an acknowledgment of the global dimension of environmental problems, and the internationalization of the political ecology of the region – now as often articulated by international nongovernmental organizations and partnerships between international agencies and local bodies as by local political leaders – have made the Mekong a global river. In general, an awareness of the effects of climate change on the residents of the Mekong region, where more than 70% of the population depend on agriculture or fisheries to make a living, has contributed to the emergence of a Mekong that is not only a region but also a resonant field of action in a global environment. This Mekong imaginary has been pushed along by an activist discourse that is increasingly – and for good reason – alarmist about the prospects of the Mekong watershed for surviving the barrage of changes imposed upon it.

One of the ways that environmentalists have conceptualized environmental change of the kind and on the scale that is now occurring in the Mekong – what Bill McKibben in a 1990 book about global warming called the "end of nature" – is by appropriating the concept of the Anthropocene from its yet uncertain scientific context and using this to explain our modern relationship with nature. The Anthropocene is a proposed epoch that begins when human activities started to have a significant global impact on Earth's geology and ecosystems and that accommodates that historian John McNeill calls the defining characteristic of those changes in the twentieth century, a "great acceleration" (McNeil and Engelke 2016)." Neither the International Commission on Stratigraphy nor the International Union of Geological Sciences has yet officially approved the term as a recognized subdivision of geological time (it would be demarcated from the previous epoch, the Holocene). But this has not deterred scholars and activists who want to emphasize that we've reached the point where we're living on the earth in a new way, and the value of the Anthropocene as a framing concept that not only marks decisively the end of nature but, for some, also encourages an understanding of the emergence of a new "nature" and strategies for adaptation and hope for the future.

Some among "Generation Anthropocene" celebrate a full embrace of the Anthropocene in a kind of neoliberal ecstasy that throws out even a qualified idea of wilderness and that turns its back on conservation efforts – and that also assumes that we can engineer and manage environments as adeptly as we have exploited them. Others among the more conservation-minded resign themselves to an acceptance of the Anthropocene and hope for whatever can be salvaged from the destruction that it identifies. They embrace the Anthropocene in a full acknowledgment of the embeddedness of history in nature, but with a vision of history and the future that is shadowed by the apocalypse. Others reject the new framing entirely and remain committed to what they believe to be more genuine or authentic ways for humans to live in what might remain of nature, and to hope for and appreciate emergent ecosystems that transcend human conceits about ending them. And others see the Anthropocene as an unnecessary and clunky, and perhaps distracting, concept that nonetheless in shorthand reminds us that we now live on a different planet, and it doesn't look good (Macfarlane 2016; Kress and Stine 2017; McNeill 2015; Stewart 2018).

When we follow through on a full commitment to "the end of nature" (and the possible beginning of another "nature") or to thinking in general about nature in the Anthropocene, the "globe" becomes not just a spatial scale in which environmental action happens but also an important narrative frame, part of the imaginary of a place and how it is explained. Though the nonscientific discourse of Anthropocene citizens (as one writer has called them) has not yet made itself fully explicit in discussions of the environmental problems of the Mekong, these discussions take for granted several points of contact with this discourse: that development will create opportunities for some but poses grave challenges to the rich biodiversity of the region, and some will pay a very heavy livelihood price for development. This discourse also acknowledges, at least indirectly, that there is no turning back the clock, that the Mekong has already been changed dramatically, and that these changes are driven by changes on a global level. It acknowledges that there is no return (to the heart of darkness or a pristine past) and that perhaps the hydroelectric dams that have been built, are being built, or that are planned have already structured how we think about protecting nature, strategies for agricultural sustainability, the use of shrinking supplies of fresh water, and the protection of wildlife, forests, and fish resources of the Mekong region. In other words, the Mekong is now Anthropocenic, nature modified profoundly, and the river we now have is the one we're thinking about – rather than attempting to restore it to what it was – either through wistful memories and protections of patches of the pristine or through antidam politics. The essays that follow all brush up against this acknowledgment in one way or another, and several of them identify a perspective and even make suggestions about kinds of management and governance that can be applied on different scales to best negotiate an Anthropocenic Mekong Region – and its many human-scale subregions. These essays demonstrate one way to refine as well as identify points of stability in this fluid exchange between cultural construction and physical reality that has been the history of encounters with the Mekong: discrete, nuanced, abundantly documented, and with boots on the ground, they provide new information about the Mekong – and, as least indirectly, suggest what we can do about it. They destabilize anything we might identify as a common imaginary at the same time as they seek to work toward common ground in how we understand and imagine the Mekong.

Apocalypse Now remains a great accomplishment of movie making that held powerful sway in the ongoing national conversation in the United States in the late twentieth century about the Vietnam/American War. As we now talk about the Mekong in much different ways, and as another Mekong imaginary has begun representing the river both as a site of biodiversity and as a contentious marketplace rather than a site of conflict or imperialist soul searching, and as a region as much as a river, the film seems almost absurdly self-absorbed. Simply making this comparison, by taking seriously the Mekong as a mid-twentieth century heart of darkness, illuminates the extent to which imagination shapes what we understand as geography. But we may not have entirely extracted ourselves from this imaginary, either, as scientists and scholars have reimagined the Mekong in an age of environmental apocalypse, with conflicts over the fate of this relatively undeveloped river as a set piece in how we imagine our future place on earth. Here, a final observation by Edward Said in *Culture and Imperialism* about *Heart of Darkness* is relevant:

> By accentuating the discrepancy between the official "idea" of empire and the remarkably disorienting actuality of Africa, Marlowe unsettles the reader's sense not only of the very idea of empire, but of something more basic, reality itself. For if Conrad can show that all human activity depends on controlling a radically unstable reality in which words approximate only by will or convention, the same is true of empire, of venerating the idea, and so forth. With Conrad, then, we are in a world being made and unmade more or less all of the time. What appears stable and secure – the policeman at the corner, for instance – is only

> slightly more secure than the white men in the jungle, and requires the same continuous (but precarious) triumph over an all-pervading darkness, which by the end of the tale is shown to be the same in London and in Africa. (Said 1993, 29)

As a relatively undeveloped river, the Mekong has been a rich tableau for the geographic imagination, one that has yielded itself up to fantasies about pristine beauty and primitive savagery both, as well as palm-rubbing visions of economic development and exploitation or optimistic visions of recovery and sustainability. These have attempted to depict or explain the Mekong with comprehensive and totalizing geographies, either as a tortuous challenge to imperialism or a basin of both opportunity and ecological challenges. Other visions of the Mekong, more local, more in touch with the quotidian rhythms of life on this watershed, more immersed in experience in the river, and not so motivated by the neo-liberal schemes in which development experts would like them to participate have not often received the attention or acquired the geographical spaces that they've earned. These visions have at times been rendered revulsive, irrelevant, unstable, or invisible by regional geographies, and at other times have been subsumed by them. This is not simply a question of scale, though the "environmentalism of the poor" is more easily rendered on smaller scales; regional management geographies need to be constructed upon the local and the everyday environmental behaviors and hopes of Mekong region inhabitants, rather than attempting merely to absorb them into larger schemes (Nixon 2011). The Mekong River that we are studying and for which we're suggesting policy solutions and about which we take for granted when we talk about strategies of sustainability at a time of environmental change, continues to be as much an unstable cultural and historical construction as a physical reality. And these Mekongs all have in common what Said explained was most fundamental about Conrad's vision of the relationship between outsiders and insiders in an earlier imperial world – fluidity, change, "a world being made and unmade more or less all of the time."

References

Achebe, C. (1978). An image of Africa. *Research in African Literatures, 9/1*, 1–15.

Anderson, Benedict, 1991. *Imagined communities: Reflections on the origin and spread of nationalism* (Rev., ed.). New York: Verso.

Bennedikter, S. (2014). *The Vietnamese hydrocracy and the Mekong delta: Water resources development from state socialism to Bureaucratic Capitalism*. Berlin/London: LIT Verlag.

Conrad, J. (1899). *The heart of darkness*. London: Blackwood's Magazine.

Ekbladh, D. (2002). 'Mr TVA': Grass-roots development, David Lilienthal, and the rise and fall of the Tennessee Valley authority as a symbol for U.S. Overseas Development, 1933–1973. *Diplomatic History, 3*, 335–374.

Garnier, F. (1885). Voyage d'Exploration en Indo-Chine (Paris); transl. in 2 vols. (Bangkok 1996).

Hirsch, P. (2001). Globalization, regionalization, and local voices: The Asian development bank and rescaled politics of environment in the Mekong Region. *Singapore Journal of Tropical Geography, 3*, 237–251.

Keay, J. (2005). The Mekong exploration commission, 1866–68: Anglo-French rivalry in South East Asia. *Asian Affairs, 36*(3), 289–312.

Kress, W. J., & Stine, J. K. (2017). *Living in the Anthropocene: Earth in the age of humans*. Washington, DC: Smithsonian Books.
Macfarlane, R. (2016). *Generation Anthropocene: How Humans Have Altered the Planet Forever*. London: The Guardian.
McNeil, J. R., & Engelke, P. (2016). *The great acceleration: An environmental history of the anthropocene since 1945*. Cambridge, MA: Harvard University Press.
McNeill, J. (2015, October 27). Interview with John McNeill, *E-International Relations*. http://www.eir. Accessed 16 Apr 2017.
Nixon, R. (2011). *Slow violence and the environmentalism of the poor*. Cambridge, MA/London: Harvard University Press.
Osborne, M. (1975). *River road to China: The Mekong river expedition, 1866–1873*. New York: Liveright Publishing.
Said, E. (1993). *Cultural and imperialism*. New York: Vintage Books.
Save the Mekong Coalition. (2010). *Save the Mekong* 2010. Available from. http://www.savethemekong.org. Accessed 12 Apr 2016.
Sneddon, C. (2013). Water, governance and hegemony. In L. Harris, J. Goldin, & C. Sneddon (Eds.), *Contemporary water governance in the global south: Scarcity, marketization and participation* (pp. 13–24). New York: Routledge.
Sneddon, C., & Fox, C. (2006). Rethinking transboundary waters: A critical hydropolitics of the Mekong basin. *Political Geography, 25*, 181–202.
Sneddon, C., & Fox, C. (2008a). River-basin politics and the rise of ecological and transnational democracy in Southeast Asia and Southern Africa. *Water Alternatives, 1*(1), 66–88.
Sneddon, C., & Fox, C. (2008b). Power, development and institutional change: Participatory governance in the Lower Mekong basin. *World Development, 35*(12), 2161–2181.
Sneddon, C., & Fox, C. (2012a). Inland capture fisheries and large river systems: A political economy of Mekong fisheries. *Journal of Agrarian Change, 12*(2/3), 279–299.
Sneddon, C., & Fox, C. (2012b). Water, geopolitics, and economic development in the conceptualization of a region. *Eurasian Geography and Economics, 53*(1), 143–160.
Stewart, M. (2018). Islands, edges, and globe: The environmental history of the Georgia Coast. In P. Sutter & P. Pressley (Eds.), *Coastal nature, coastal culture: Environmental histories of the Georgia Coast* (p. 2018). Athens: University of Georgia Press.
Streitfeld, D. (2016, June 21). For Viet Thanh Nguyen, Author of 'The Sympathizer,' a Pulitzer but No Peace. *New York Times*.
Watts, C. (1983). 'A Bloody Racist': About Achebe's view of Conrad. *The Yearbook of English Studies, 13*, 196.
Wilson, R. (2000). *Reimagining the American Pacific: From South Pacific to Bamboo Ridge and beyond*. Durham: Duke University Press.
Wittfogel, K. (1957, 1981). Oriental despotism: A comparative study of total power. New York: Vintage Books.
Wong, S. M. T. (2010). *Making the Mekong: Nature, region, postcoloniality*. Unpublished PhD dissertation, Ohio State University.

Chapter 2
Hunger and Governance: The Food Supply in Cambodia, 1979–1980 and Beyond

Jenny Leigh Smith

Abstract In the 4 years the Khmer Rouge was in power, after seizing control of Phnom Penh, an estimated 1.7 million people were killed by state terror, starvation or disease. In 1979, Democratic Kampuchea was overthrown by its neighboring communist rival, the Socialist Republic of Vietnam, and Heng Samrin was appointed the leader of the new People's Republic of Kampuchea (PRK). Access to food was the first and most pressing concern for the regime's first 13 months in power. While the Killing Fields and other sites of mass murder have become notorious symbols of the Khmer Rouge, nearly half of all deaths between 1975 and 1979 were from starvation or opportunistic diseases related to acute malnourishment. In Democratic Kampuchea hunger and malnutrition were intentional, manufactured crises created by the government in order to assure compliance from civilians. Choices made by the government that succeeded these 4 years of horror also deeply affected efforts to recovery from famine. They affected agricultural production, and frustrated or delayed a return to normalcy for an estimated one million Cambodians who were internally displaced or who became international refugees between 1979 and 1985. The history of Cambodia's experience with food scarcity, famine prevention, and distributing and maintaining humanitarian supplies to civilians help indicate the future directions the country would follow in its style of governance and management toward a broad range of natural and human resources.

Keywords Hunger · Malnutrition · Humanitarian aid · Oxfam refugees · Khmer Rouge · Capability · History of development

J. L. Smith (✉)
Hong Kong University of Science and Technology, Hong Kong, Hong Kong
e-mail: jenny.smith@ust.hk

M. A. Stewart, P. A. Coclanis (eds.), *Water and Power*, Advances in Global Change Research 64, https://doi.org/10.1007/978-3-319-90400-9_2

2.1 Introduction

Contemporary Cambodia has had an eventful and traumatic recent history. This history casts a long shadow over the challenges the country faces today, especially in the realm of government and civil society. This chapter focuses on the history of Cambodia during some of its most difficult and formative years, 1975–1985. During this time two very different regimes ruled the country; first, between 1975 and 1979 was Democratic Kampuchea, a deadly and isolationist communist state led by the Khmer Rouge, and second, between 1979 and 1985, the Vietnamese and Soviet supported People's Republic of Kampuchea, led by Heng Samrin, in close consultation with the Vietnamese government and a cabinet of powerful ministers. Both governments experienced—and at times manufactured—large-scale social and political crises. The most serious of these crises in 1979, in the 1st months of Heng Samrin's leadership, were those facing the nation's food supply.

Most of this chapter focuses on the events of 1979 and 1980, when Samrin's government was fighting for international legitimacy as well as domestic stability. Preventing famine and alleviating the devastating chronic malnutrition that had become widespread during the DK years was the first and most crucial challenge Samrin's government faced. In order to understand how most Cambodians came to be underweight, malnourished and suffering from potentially deadly infectious diseases in 1979, I first provide an overview of the state of food, agriculture, and governmental health policies during the Khmer Rouge years. I then discuss the evolving policies of the early PRK government, specifically their policies related to the largest Western charity they dealt with in this year; Oxfam. By 1981, the acute crisis of hunger and malnutrition had passed; and government officials as well as outside observers assumed that, in spite of the continued Vietnamese occupation, the PRK was on a path to independence, development, and stability. Yet there remained significant and telling indications that all was not well in the PRK, and that food insecurity might continue to plague the country. Indeed, just 1 year before Vietnam ended its decade-long occupation of the country, the PRK experienced another near-famine when harvests failed during the winter of 1988, and a massive influx of outside aid was again necessary to prevent deaths from starvation. These vulnerabilities in the food supply that were exposed so sharply in the 1980s remain to the present day.

Indeed, it is not just the resource of food that is vulnerable because of the present administration's insistence on centralized control; almost every aspect of governance that might fall under the heading of resource management, from supevision of fisheries to land redistribution to the creation of hydroelectric dams along the Mekong River falls under this threat. In their 1992 book *Hunger and Public Action*, the economists Amrtya Sen and Jean Drèze claim that ensuring food security for a population and preventing famine are a crucial test case of effective governance. Drèze and Sen focus on the ability of a government to expand the *capability* of its population. Here the term capability is used to mean the ability of the population to remain alive, healthy and productive. Although Sen and Drèze are both experts in famine prevention and anti-hunger policy, *Hunger and Public Action* makes a broader argument about human well-being, focused not just on access to food and clean water, but also medical care, sanitation, and a safe, healthy environment (Sen and Drèze 1992). They argue that governments have the responsibility to foster both public and private initiatives that expand citizen *capability*. Their arguments about the responsibilities of the state versus private organizations are nuanced, but they can be summarized as following: if the state cannot provide for basic entitlements citizens need in order to shore up and expand their own capabilities, the state then has the obligation to allow charities, international organizations and for-profit institutions the opportunity to provide for these basic rights.

Since colonial independence, Cambodia's access to resources that will expand human capability have been in doubt. This is true in spite of the fact Cambodia has survived several brutal forms of governance during these years. It is also true despite the fact modern-day Cambodia has had a relatively stable communist authoritarian leader in place since 1989. The Khmer Rouge regime, in power from 1975 to 1979, was certainly the most deadly time, but this does not exonerate the present regime for failing to provide for the populations' capabilities. Many of the challenges and uncertainties present-day Cambodia faces are part of a much longer legacy of rule that began in 1979, with the founding of the People's Republic of Kampuchea. The ministers and other officials (almost all of them former Khmer Rouge fighters) who found themselves in power in 1979 did not initially set out to create a governance model that diminished and threatened the capabilities of their population, yet over a fairly brief amount of time this is exactly what happened. This paper explores the 1st years of governance in the PRK, and shows that many of the challenges the country deals with today had their roots in the rapid-fire and reactive pattern of decision making that was common during this period.

2.2 The Khmer Rouge Years

The lethal violence of the Khmer Rouge has been accounted for elsewhere; here I provide a very brief overview of the disastrous results their 4 years in power yielded, as they relate to food access and the nutrition levels of the population (Chandler 1998; Kiernan 1996, 2003; Short 2007; Tyner and Rice 2015; Pran 1999; Murtha 2014; Widyono 2008). After a bloody 5-year civil war in postcolonial Cambodia, one faction of soldiers, the Khmer Rouge, won the support of Cambodia's crown prince Sihanouk, and after months of fighting managed to capture the tactically crucial capital city of Phnom Penh in April of 1975. With the capture of Phnom Penh, the Khmer Republic ended, and the country was renamed Democratic Kampuchea, or DK. The Khmer Rouge proclaimed a collective, anonymous group of military leaders, known simply as 'the brothers' as the DK's new government. By 1976 Pol Pot would emerge as their infamous and brutal leader, but for most of the time the DK existed, the outside world understood the country to be ruled by an anonymous brotherhood. As a ragged group of poorly trained soldiers, many of them still teenagers, the Khmer Rouge seemed like an unlikely group of victors and leaders. Yet within days it became clear appearances were deceiving. The Khmer Rouge's first orders were to evacuate all urban areas, Phnom Penh being the most significant. During the Civil War the capital city had become a collecting place for displaced persons. Built to support half a million residents, by 1975 the population had swelled to between two and three million residents. The majority of these people lived with relatives or in makeshift tent cities, and by 1975 food and medical supplies in the city had been critically low for months.

Phnom Penh at capture provides an opportunity to examine how the Khmer Rouge leadership addressed a major food crisis. In what would become a characteristic response, the Khmer Rouge used the humanitarian crisis to shore up its own power and weaken Cambodian civil society rather than as an opportunity to ease suffering and improve living conditions. If employing Sen and Dreze's metric of ensuring and expanding human capability, the Khmer Rouge failed at its first test. The two to three million residents of Phnom Penh represented over one fourth of the total population of the country in 1975; resettling so many impoverished people so rapidly had a significant negative impact on almost every community to which urban residents were sent, reducing access to food, disrupting work opportunities and creating housing shortages. The Khmer Rouge attempted to resolve some of these crises administratively, by abolishing private property and setting up communal dwellings based on labor brigades, but this did not alter the popular perception that quality of life and opportunity had suffered terribly as a result of the Khmer Rouge coming to power. While food supplies in the cities had been critically low during the Civil War period, evacuating the cities did nothing to alleviate food shortages, they simply created less visible and more complicated food shortages that the DK's primitive road and transport system were incapable of addressing.

Urban residents were resettled to different parts of the countryside, and upon resettlement were almost always classified as 'new people' a vague category the Khmer Rouge invented to denote urban, minority and/or middle class citizens whose political loyalty to the new regime was suspect. While small corners of the countryside maintained a tentative normalcy, especially in the 1st years of the DKs existence, the majority of Cambodian citizens became more rather than less food insecure after Phnom Penh was vacated. Relocation was just the first act of community destabilization: in their 1st year in power the Khmer Rouge banned money, private property, commercial fishing, and gathering wild or roadside foods for personal consumption. All of these measures were intended to decrease independence and self-determination of citizens and increase their dependence on state institutions like feeding kitchens for survival. In 1976 the regime further limited cooking by individual households by confiscating cooking pots, pans and utensils, and by opening feeding kitchens that were attached to collective work sites. These policies contributed to a dramatic increase in food insecurity, and created a permanent sense of emergency and crisis in the country, eliminating even the potential for Democratic Kampuche to return to a sense of normalcy. There are no accurate statistics on the rate of increase of malnutrition in these years, but memoirs and informal accounts estimate that malnutrition increased dramatically in 1975–1976, and a bad harvest in late 1976 further intensified shortages, with tens of thousands of citizens dying from outright starvation in both of these years in multiple locations around the country. The majority of the population that survived the Khmer Rouge's lethal regime spent at least one, and more commonly 2–3 years malnourished and suffering from concomitant diseases.

The Khmer Rouge spent nearly 4 years in power. Their most famous historical legacy remains the mass terror and violent deaths they inflicted on hundreds of thousands of Cambodian citizens, many of whom died violently in the infamous Tuol Sleng prison and the mass-grave killing fields at Choeung Ek and other sites. Yet up to half of Cambodians who lost their lives during these years were not killed by bullets or bayonets, but by chronic malnourishment and opportunistic diseases that accompanied malnutrition. The food shortages that affected Democratic Kampuchea during the rule of the Khmer Rouge are not associated with a particular site or date. In DK, the starvation deaths of the late 1970s are more accurately interpreted as a slowly unfolding act of mass slaughter and civil war.

2.3 1979

The bloody rule of the Khmer Rouge ended with the invasion of Cambodia by Vietnamese troops in January, 1979. Occupying Vietnamese forces encountered a country in crisis; one they were ill equipped to aid. Within the 1st month of occupation the Vietnamese army documented through reports and photographs Cambodians who exhibited classic signs of acute malnourishment: marasmus and kwashiorkor (marasmus is extreme underweight, kwashiorkor is a protein deficiency associated with edema). Vietnamese forces also documented a growing and impoverished refugee population along Cambodia's border with Thailand. Although Vietnam had couched its invasion of Kampuchea in terms of a humanitarian intervention, it was not immediately prepared to supply the Cambodian population with the large-scale supplies of food and medical supplies the country so desperately needed.

Vietnam responded to critical food shortages it discovered in Cambodia (now renamed the People's Republic of Kampuchea, or PRK) in two ways, both of which were effective first steps, especially considering Vietnam's limited financial and material resources. The first was by rationing its own army's food and medical supplies so that some foods and supplies earmarked for the Vietnamese military, particularly rice, could be diverted to the Cambodian civilian population. Coming as it did during a time of relative domestic food scarcity, the Vietnamese military rationing program represented a significant sacrifice for the country, and indicates Vietnam's commitment to defeating the Khmer Rouge and safeguarding the Cambodian civilian population. The second form of aid Vietnam sought in the name of Cambodia was donations from its most powerful ally, the Soviet Union. Although it took a few months to arrive, by the summer of 1979 food aid in the form of shipments of grain (corn grits, corn meal and rice, purchased by the Soviet Union from India and African suppliers) was forthcoming. Donations from the USSR continued through mid-1980.

Late in the spring of 1979, Vietnam and the new government of the People's Republic of Kampuchea also made a wider international appeal. Due to a complicated web of Cold War alliances, the PRK was not recognized as a country by most nations, and many countries, including the United States and much of the Western Bloc, suspected Vietnam of exaggerating the crisis in the country in order to secure unnecessary humanitarian donations that would actually support Vietnam's occupation of the country. This widespread mistrust politicized and delayed donations from governments and large international bodies like the International Committee of the Red Cross. The United Nations was the most vital of these reluctant international bodies. Its UNHCR branch pledged to help Cambodian refugees who crossed the border into Thailand. For much of the spring and summer of 1979, the small food allowance this group was given from UNHCR stores was the only food aid arriving in northern Cambodia, which was cut off from the shipping port in the South the Soviet Union used. The UNHCR's policy directly contributed to a dramatic increase in the refugee population just over the Cambodian border in Thailand, which remained a humanitarian hotspot for years after the Khmer Rouge left power and served as a lifeline for some Khmer Rouge holdout groups.

One international charity, Oxfam, ultimately decided to ignore Cold War politics and commit to a Cambodian humanitarian campaign. This decision to engage with the PRK ultimately inspired the Cambodian government to evolve its own policies governing outside assistance. Oxfam's commitment to the People's Republic of Kampuchea was largely due to the work of Oxfam's Technical Officer for Asia, Jim Howard, who made a ten day visit to the country in late August of 1979. His trip to Phnom Penh and the countryside surrounding the capital convinced him a nationwide famine was imminent and even likely. In his trip report he stated the situation in hospitals and orphanages in Phnom Penh was "as bad as anything I ever saw (in India)…Hospitals with little food, no medicines, no linen or dressings or soap, children with kwashiorkor so severe…for them and indeed for everyone, the first need is for food "(Howard 1979a). Howard's solicitous hosts were newly appointed ministry officials, in particular the Minister of Health, one of the few medical doctors to have survived the Khmer Rouge, and the Vice Minister of Foreign Affairs, Keo Preseth. Howard liked both of these officials and felt confident Oxfam and other charities would be welcome to collaborate with the government to meet the pressing needs of the people. In his report to Oxfam, Howard wrote "(I) felt we had become very close friends over the ten days in PP… they need our friendship badly and we must give it, sensitively, generously and humbly and it will be a most valuable aspect of our aid." (Howard 1979a). Howard left convinced that in spite of ideological differences, the PRK was eager to work with Oxfam to bring food, medicine and other humanitarian supplies into the country.

While he was in Phnom Penh, Jim Howard stayed at a hotel called the Samrika, one of the only establishments in the country to offer both air conditioning and regular meal service (albeit only "rice, rice, rice and more rice") (Howard 1979a). At the Samrika, Howard became friendly with the journalist John Pilger, who was in the PRK to film a television documentary, *Year Zero*, about the experience of Cambodia under the Khmer Rouge. Based on his experience in the PRK, Pilger decided to shift his documentary to focus on the current plight of Cambodians, focusing on the food shortages and malnutrition he had witnessed throughout the country. Like Howard, Pilger was convinced that food was the country's most pressing need, and that without outside aid, one to two million Cambodians would be facing starvation by early 1980, a statistic with little foundation but one that was repeated by Howard, Pilger, and many others throughout the fall of 1979 (Howard 1979b). Pilger and Howard agreed to stay in touch, and in late October, when the *Year Zero* documentary aired, its last frame gave the address of an Oxfam-monitored postal box where people could send donations for food aid (Black 1992).

October of 1979 also marked a shift in the rapport between Oxfam and the Cambodian government. Oxfam gained more power and authority among other international charities and NGOS in an in-country meeting in early October when it agreed to coordinate and direct a consortium of eight separate charities who were interested in working in the RPK. However, after the consortium agreement was finalized, Oxfam's influence with the government of the PRK, specifically the Ministry of Foreign Affairs, headed by Hun Sen, and the Ministry of Commerce, under the direction of Taing Sarim, was sharply curtailed. The Ministry of Foreign Affairs moved to limit the influence and recognition of Oxfam throughout the country and limited the number of staff the consortium was allowed to have in the country to seven (four from Oxfam, three from other organizations), requiring them to remain in their rooms at the Samrika if they were not traveling under government escort. The Ministry of Commerce took charge of distributing and allocating donated food and other goods, tasks that had initially been delegated to the Ministry of Health. Perhaps not surprisingly after this change in oversight, some of the most valuable donations Oxfam brought into the country: Land Rovers, Leyland cargo trucks and commercial-grade sewing machines, were retained by the Ministry of Commerce for its own enterprises rather than being distributed to the Ministry of Health and occupational training programs, as Oxfam had planned (Thompson 1979a). Simply the fact that, at the start of what was perceived to be a famine-relief operation, Oxfam had been encouraged by the government of the PRK to fund infrastructure development programs such as occupational training schools, textile workshops and light goods factories is evidence of the government's focus on economic development above humanitarian basic support.

At the end of 1979, Oxfam officials and other participants in the consortium were surprised and alarmed to discover that the donations of food and equipment they delivered to the country were largely untraceable. Oxfam-sponsored goods had been arriving in the by sea and air since August of 1979, and upon arrival they were subjected to a thorough documentation system before they entered the Ministry of Commerce's extensive warehouses. However, once they arrived at the warehouses, their fate was less certain. Marcus Thompson, Oxfam's in-country director for 19,791–980, nicknamed these warehouses 'black holes' because it was nearly impossible to trace the movement of goods back out once they had entered (Thompson 1979b). In many cases, goods simply never left the warehouses; Thompson and other Oxfam employees noted that the Ministry of Commerce hoarded rice, medicine, soap, spare tires and other supplies. In other cases, goods were distributed in ways that were the most convenient and least taxing for the Ministry of Commerce; the relatively prosperous port town of Kompong Son received sewing machines, clothing and slightly damaged food supplies that had been earmarked for other locations, because the logistics of delivery surpassed the abilities of the Ministry of Commerce. On the other hand, villages, clinics and settlements far from main roads received fewer supplies, even though food, seed grain and other supplies were scarce in these areas as well.

2.4 1980

What about the famine that threatened the country so starkly in 1979, in the eyes of Jim Howard, John Pilger and others? An attempted review of where donations went in 1980 and a larger, retrospective review in 1983 showed most donations were untraceable (Stack 1983). In all probability Oxfam's donations, and those of the other consortium members in 1979 and 1980 had little to no impact on Cambodians outside the capital who needed them most. In 1979 the newly formed Ministry of Commerce was simply not skilled or powerful enough at the logistics of distribution to effectively distribute or keep track of the thousands of tons of rice, seed grain and other donated goods and tools coming into the country. Oxfam leaders documented the Ministry's poor distribution techniques as well as its tendency to hoard goods for personal use or resale.

In spite of this, when Oxfam's medical specialist, Dr. Henny Brown was allowed to travel to several smaller villages in Takeo, one of the poorest regions of the PRK in late January 1980 she and other officials expected to discover evidence of mass death due to malnutrition. However, happily, this was not the case. Dr. Brown and other representatives from Oxfam did not discover evidence of famine, Instead, just as they had already observed in hospitals and orphanages across Phnom Penh in November and December of 1979, in Takeo and elsewhere they encountered a weak and sickly population, but one that was now clearly recovering from the effects of several years of malnutrition and disease. Rates of tuberculosis, hookworm, and malaria were still alarmingly high across the country, and hospitals in Phnom Penh as well as rural medical outposts desperately needed supplies to help treat and cure these diseases, as well as staples like gauze, antibiotics, bedding, mosquito nets, needles and soap. However, the anticipated and much publicized potential deaths from starvation never materialized (Annual Meeting Report 1980).

While Oxfam had hoped to provide a lifeline in PRK, other inputs of food were probably more important, especially for rural populations. Foremost, early Vietnamese and Soviet shipments of grain, which were distributed by the Vietnamese military and the Soviet embassy respectively throughout the late summer and early fall of 1979 likely made a big impact on improving nutrition, especially near the Vietnamese border and near the port of Kompong Son. One of Oxfam's medical officers, Tim Lusty, perceived as much when he visited the country in 1980; "it is reasonable…to presume that (Oxfam's) relief food has only had a marginal effect on the overall improvement in rural areas…the source of all staples which we saw being distributed must be Russia or Vietnam" (Lusty 1980). More unexpectedly, the famous 'bicycle highway' between Thailand and Cambodia seems to have made a real difference in these months and beyond. Oxfam workers and other visitors to the country remarked on the incredible volume of goods (food, clothing, and cigarettes were the items most often mentioned) moving through the country along a dirt track that stretched from the Thai border, along Cambodia's massive inland water body; Tonle Sap.

While these two styles of distribution were very different, they were notable in that they both completely bypassed the new PRK government in order to get food and supplies out on the ground, relying instead on field-tested techniques of centralized distribution (in the case of Soviet and Vietnamese donations) and informal economic networks (in the instance of the bicycle highway). By May of 1980 Jim Howard noted in a second visit to the PRK that the threat of starvation had passed, and throughout Phnom Penh, in spite of the continued absence of currency, he noted "Several large markets have come into being selling rice, fish, vegetables, cloth, baskets, piles of fruit, pots and tools and great quantities of materials coming in from Thailand." (Howard 1980). Consumers used cans of rice and Thai cigarettes as currency in the markets (Harper 1979).

While the Ministry of Commerce had moved rapidly to try to prevent Oxfam and other aid agencies from establishing effective distribution networks within the country, none of the Ministries were able to offer competent leadership directed at averting the catastrophe of famine. These were skills that they would learn on-the-job, "they appear to be making it up as they go along," one Oxfam official noted in mid-1980 (Thompson 1980). The Consortium of aid agencies that Oxfam led dissolved at the end of 1980. While Oxfam moved on to focus on "development work" in the region, most significantly building and staffing community health centers in rural regions, the Kampuchea Team remained frustrated by the lack of cooperation and support they received from the PRK government (Warr 1981). By 1983, Oxfam had decided to phase out operations in Cambodia, focusing instead on its more productive regional partnerships with neighboring Vietnam and Thailand.

Oxfam may have left, but during the 1980s the International Committee of the Red Cross, UNICEF and the United Nations FAO all participated in hunger alleviation schemes to help improve nutrition levels and ensure famine conditions did not recur. In these years, the PRK needed between 900,000 and 1,500,000 tons of (milled) rice to meet basic subsistence needs. In most of the years of the 1980s, harvests fell short of these goals. In no year did the harvest exceed 1 million tons. In 1982, spring floods destroyed much of the young paddy rice, and the United Nations Relief Fund launched a massive campaign to purchase rice and seed for the country. Ultimately, they raised 27 million dollars. There were modest improvements in rice production in 1984, but droughts in 1985 meant that the PRK was once again asking for food aid to help cover harvest shortfalls. Although Cambodia received significant aid, agencies remained reluctant to provide it and each assistance contract was a painful negotiation between multiple parties, all of whom disliked Cambodia's client regime, but none of whom were eager to see the country lapse back into genocide.

2.5 The Longer Term

What can the relationship between the government of the People's Republic of Kampuchea and food security in 1979–1980 tell us about larger patterns of governance and a state's responsibility to its citizens in terms of finite resources? I am especially interested in analyzing this relationship in light of the concept of citizen *capabilities*, and the state's responsibilities vis a vis this category, as explored by Sen and Dreze.

Three impacts stand out. First, while the Khmer Rouge years were marked by an extreme level of terror and violence, there were some continuities in governance, especially with respect to the government's ability to capitalize on uncertainty. The opportunistic moves the Ministry of Commerce and the Ministry of Foreign Affairs made immediately after Oxfam signed on to a consortium agreement with the PRK shows a level of cynical calculation on the part of the government. The Ministry of Commerce's willingness to leave its storage sheds full in the face of countrywide food and goods shortages, and their takeover of vehicles intended for humanitarian purposes are further signs of this. It is interesting to note that these actions happened during a period when these ministries were not yet able to effectively distribute goods and services to the larger population. In place of this skill, their first actions were focused on consolidating power.

The second impact was that, even after the harvest and supply crises of 1979–1980 had passed, the PRK was forced to depend on outside aid for food, other consumer goods, and a variety of heavy industry, especially trucks and construction equipment. This was a situation Heng Samrin's government did not create, but also took few steps to reduce or eradicate. In part, this is surely a classic client-state mentality. Vietnam's occupation of the PRK did not include comprehensive support to rebuild or improve the country's infrastructure, yet it was in the best interest of both Vietnam and the PRK for this infrastructure to be improved as quickly as possible. The most expedient way to do this was to invite charities and international agencies to aid the new country with as much material assistance as possible. In fact, by August of 1981 Oxfam's in-country agricultural specialist was under the impression this was precisely what Oxfam was doing in the country. In is words, "there (was) no observable sign that the government regards Oxfam as more than a valued donor of expensive and otherwise difficult to obtain materials." (Anonymous 1981) This cynical relationship between charitable organizations and the PRK government continued steadily throughout the decade of the 1980s.

Finally, and perhaps most germane to a broader consideration of resources under regimes within Cambodia, one of the more unusual things this case study of famine prevention in the early days of the PRK demonstrates is the mismatch between what Heng Samrin's government defined as the most valuable resources its population required and those that a seasoned outside aid agency such as Oxfam judged necessary. This misreading cut both ways; Oxfam provided goods it believed to be essential that were clearly not highly valued by the Samrin government, such as soap (for hygiene), western pharmaceuticals (for medical triage) and dry skimmed milk powder (to feed babies and young children). These were the three items that Oxfam officials were most likely to discover months later, languishing in a storehouse or locked cupboard. On the other hand, the Samrin government urgently requested materials that seemed superfluous if not irresponsible to Oxfam and other aid agencies, including cheap plastic bags and baskets (to facilitate trade and distribution of all goods), vast amounts of sugar (to make the milk palatable), and cigarettes (to serve as a substitute currency). While neither group's list was "the wrong" set of essential requests, they reveal different definitions of "essential" and gesture at some of the fundamentally different priorities that public and private actors have when they identify and promote goods and other resources that are essential to life and well-being.

References

Anonymous. (1981). Unpublished Report, August 8, 1981. Oxfam Archive: MS Oxfam PRG/3/3.

Black, M. (1992). *A cause for our times: Oxfam the first 50 years.*

Chandler, D. (1998). *A history of Cambodia* (3rd ed.). Boulder: Westview Press.

Harper, M. (1979). *Audio tape number 2.* Nov 16, 1979 (Unpublished report from the Oxfam Archive: MS Oxfam PRG/3/3/2/13).

Howard, J. (1979a). Unpublished report MS Oxfam PRG/3/3/13 Team Reports Sep-Nov 1979 (a).

Howard, J. (1979b). Unpublished report MS Oxfam PRG/3/3/13 The Year of the Goat (b).

Howard. (1980). Unpublished Report MS Oxfam PRG 1/2/5 Asia Field Committee April–Jun 1980.

Kiernan, B. (1996). *The Pol Pot regime: Race, Power, and Genocide in Cambodia under the Khmer Rouge* (pp. 1975–1979). New Haven: Yale University Press.

Kiernan, B. (2003). The demography of genocide in Southeast Asia: The death tolls in Cambodia, 1975–79, and East Timor, 1975–80. *Critical Asian Studies, 35*, 585–597.

Lusty, T. (1980). Unpublished report, January 1980. MS Oxfam PRG/3/3/2/13 Team Reports Sep-Nov 1979.

Monro, D. (1979). *Year Zero.*

Murtha, A. (2014). *Brothers in arms: Chinese aid to the Khmer Rouge, 1975–1979* (p. 2014). Ithaca: Cornell University Press.

Pran, D. (1999). *Children of Cambodia's killing fields: Memoirs by survivors*, Yale University Press.

Sen, A., & Dreze, J. (1992). *Hunger and political action.*

Short, P. (2007). *Pol Pot: The history of a nightmare.* New York: Henry Holy & Company.

Stack, T. (1983, August). A report on the Kampuchea emergency programme. Commissioned by the Asia Field Committee (Restricted Circulation).

Thompson, M. (1980). Unpublished report from the Oxfam Archive: MS Oxfam PRG.

Tyner, J., & Rice, S. (2015). To live and let die: Food, famine and administrative violence in democratic Kampuchea, 1975–1979. *Political Geography, 48*, 1.

Warr, K. (1981). Report, January 7 1981 "the end of 1980". Oxfam Archive PRG/6/3/6/3 "Kampuchea Reports" February 1980–October 1983.

Widyono. (2008). *Dancing in shadows: Sihanouk, the Khmer Rouge, and the United Nations in Cambodia.* New York: Rowman & Littlefield Publishers, Inc.

Chapter 3
Rubbery Revolution: Plantations as Battlefields in the First Indochina War, 1945–1954

Michitake Aso

Abstract This chapter analyzes the First Indochina War as a watershed in Vietnamese environmental history. It focuses on southeastern Vietnam, a region whose strategic value largely resulted from its existence as a border area, not only geopolitically between Cambodia and Vietnam but also physically as a terrain marked by sharp divergence between a rubber monocrop and a forest environment. This chapter examines how the Việt Minh strategy evolved from one that sought to destroy imperial landscapes of labor to one that worked to co-opt the resources derived from rubber. At first, the Việt Minh accepted previously articulated definitions of plantations as places of rubber production and exploitation, and thus they sought to sabotage plantation operations. Only after 1950, when the Việt Minh imagined plantations as a potential source of supplies that could support their war effort, did they begin to tap into the food, money, and arms made available by rubber. Ironically, the regional characteristics that made this region useful for anticolonial resistance arose from the colonial refashioning of nature for rubber production.

Keywords Rubber plantations · First Indochina War · Việt Minh · Đông Nam Bộ · French Rubber Institute · Plantation landscapes

3.1 Introduction

The southeastern region of Vietnam, or Đông Nam Bộ, as it is commonly known as in Vietnam, forms a frontier zone where the central highlands meet the Mekong Delta, and is a coherent social and ecological unit. It has a distinct dry and wet season and is watered by three main rivers: the Sài Gòn, the Đồng Nai, and the Sông Bé. The vegetal cover on the undulating hills and sloping plateaus of the southeastern uplands include both dense and open forests, stands of bamboo, agricultural plots, and *tranh* (lalang) grass. The region's location between the plains of the delta

M. Aso (✉)
Department of History, State University of New York, Albany, NY, USA
e-mail: maso@albany.edu

M. A. Stewart, P. A. Coclanis (eds.), *Water and Power*, Advances in Global Change Research 64, https://doi.org/10.1007/978-3-319-90400-9_3

and the central highlands has made it an important trading point for delta and coastal products, such as rice and salt, and forest products, such as timber and wildlife. More recently, it has become a center of industrial production (Trịnh 1863; Sterling et al. 2006, p. 26; Baudrit 1936; Mạc et al. 1991; Taylor 2014). Đông Nam Bộ and the Mekong Delta have very different natures, cultures, and histories, yet these contiguous regions share many characteristics and have experienced parallel events. At the beginning of the twentieth century, both regions were sparsely populated and connected to international markets for agricultural products through Saigon (now Hồ Chí Minh City). Both regions were affected by colonialism and warfare. Because of its importance for anticolonial movements, the Mekong Delta and its history during the two Indochina Wars have received some scholarly attention (Brocheux 1995; Biggs 2010; Elliott 2000; Hunt 2009; Stewart and Coclanis 2011). This chapter explores Đông Nam Bộ and its relationship to the First Indochina War. It asks how this war shaped the environment and society of this region and how this region in turn shaped the war. These are broad questions that cannot be fully addressed in the space of one chapter, so I limit the scope of my investigation here to the rubber plantations, which played a prominent role in this region's twentieth-century history.

To explore the changing material and symbolic uses of rubber plantations between 1945 and 1954, this chapter examines the role of violence and revolution in Vietnam's environmental history. Violence has had mixed legacies for landscapes. On the one hand, warfare clearly has had destructive effects on Vietnamese landscapes, as Agent Orange, bombs and other weapons poisoned people and degraded environments (Biggs 2005; Martini 2012; Zierler 2011). On the other hand, war, and preparation for it, can have a conserving effect on certain landscapes (Cronon 1996, pp. 27–28; Russell 2001). In the case of rubber plantations, the First Indochina War both conserved and destroyed, and its effects were transformed by revolution. The result was a plantation industry that was heavily damaged by violence but ready, when the opportunity arose, to meet and exceed previous production levels as nationalism served to re-enchant plantation landscapes. As the anthropologist Philip Taylor has written, "The Vietnamese socialist state might be seen as thoroughly enchanted from its very inception" (2007, p. 50).

This chapter argues that the First Indochina War was crucial to the process of reconceptualizing rubber plantations not as evil colonial institutions but as centers of socialist production and beloved national landscapes. Three factors encouraged the regeneration of rubber landscapes still extant at the moment of communist victory in 1975. The first factor was the labor involved in rubber plantations and their economic value for nation-building programs of all ideological persuasions (Andrews 2008; Boomgaard et al. 1997; Latham 2000; Soluri 2005; White 1995). In addition to the workers' experiences on plantations, rubber represented a valuable export commodity in the socialist world, while multinational enterprises such as Michelin defended their interests against the assaults of various militaries. The second factor was the violence of the First Indochina War. Rubber plantations were strategically invaluable for the guerrilla efforts of the Việt Minh, while the French military and its Vietnamese allies sought to eliminate the sanctuaries of the plantations when not themselves co-opting these spaces. The third factor was the enchantment of nationalism. The events taking place between 1945 and 1954 refashioned rubber plantations from colonial landscapes of oppression to national landscapes of revolution, opening the way for their eventual incorporation into the Socialist Republic of Vietnam (SRV).

This chapter begins by examining the French and Vietnamese attitudes toward plantations immediately after World War II. At that time, the Việt Minh sought to destroy plantations and adopted scorched earth tactics aimed at destroying a hated colonial symbol and dissuading the return of colonial forces. Despite the Việt Minh's destruction of trees and threats against rubber workers, French interests successfully restarted rubber exports. Only in 1949 did the Việt Minh start to view plantations as sources of material, recruits, and sanctuary. Plantation landscapes were exploited as allies rather than foes of the guerilla movement, while the French started to view plantation landscapes with more ambivalence. As the war dragged on, plantations began to take on a revolutionary sheen and were re-enchanted by a spirit of nationalism. By the early 1950s, the Democratic Republic of Vietnam (DRV) prepared for the symbolic appropriation of southern landscapes, including rubber plantations, into a nationalist consciousness and socialist economy. DRV leaders aimed for an economic nationalism combined with complete control of domestic resources, not the rejection of science and technology. Eventually, plantations helped the communist leadership to establish a new economy and to cement ties to fellow socialist countries.

3.2 Scorched Earth

In 1945, the days of rubber plantations were seemingly numbered. Five years of Japanese military presence, culminating in the March 1945 coup d'état during which French officials, soldiers, and civilians were rounded up and imprisoned, had left the rubber industry in disarray. World War II caused other difficulties, as Allied bombs had destroyed the transport ships that had brought Vietnamese rubber workers from the north and center to the south. As a result, rubber production had fallen from a prewar high of more than 60,000 tons per year to near zero at the end of the war. Compounding the industry's internal difficulties, a legacy of colonial labor abuses meant that many Vietnamese violently opposed the continuation of plantation agriculture. As part of their "scorched earth" tactics, for example, the anticolonial Việt Minh initially attempted the complete destruction of the plantations.

The closing days of 1945 also marked the opening of the Cold War, and a complex negotiation of power among the French, the successive governments of South Vietnam, the Việt Minh, and other political and military forces led all sides to view rubber plantations as significant in the battle for the "hearts and minds" of the people. Although French reports after 1946 often emphasized heavy industry as necessary for Vietnam, many old colonial hands recommended modernizing agriculture through technical, social, political, and financial improvements. G. Wormser, for instance, a former plantation director, attempted to resuscitate late colonial projects of agricultural development. "Colonization," he argued, "was an extremely complicated problem. It not only dealt with several purely technical questions, but it also handled questions of labor, of subsistence, of populating, of finance and of policy." Furthermore, large organizations and institutions such as plantations, rather than smallholders, were at the heart of Wormser's project. "In general," he continued, "colonization can only be undertaken by the State, by Firms supported by the State, by wealthy owners possessing very great means, or by private individuals grouped together in cooperatives or in syndicates" (1946, p. 18). Other observers emphasized the role of rural engineers, trained in balancing human and technical factors, in such projects meant to bring about simultaneous improvements in rural living conditions and increases in agricultural exports (Malye 1949).

The harsh labor conditions of the rubber industry during the colonial era informed French and Vietnamese attitudes toward plantation landscapes at the beginning of the First Indochina War. To create and exploit plantations in the south, the French needed large numbers of unskilled workers, most of whom migrated from the north and center of Vietnam, known at the time as Tonkin and Annam, respectively. By the late 1920s, a combination of factors, including disease, malnutrition, and physical violence, led to high levels of morbidity and mortality among these plantation workers. Although conditions varied greatly from plantation to plantation and depended on a number of factors, including a plantation's proximity to forests, its laboring population, and its available medical treatment, death rates reached as high as 30% for certain months on certain plantations. By the 1930s, French and Vietnamese investigative journalism, colonial government reports, and scandalous exposés such as Andrée Viollis's book *Indochine S.O.S.* (1935) had created a degree of governmental and public awareness of the appalling plantation conditions (Boucheret 2008; Kalikiti 2000; Murray 1992).

Encouraged in part by these difficult labor conditions, rubber workers began to organize. The Indochinese Communist Party, formed in 1930, did not control these protests and sought instead to harness this labor activism under its banner of anticolonial politics. The successor to the Indochinese Communist Party, the communist-led popular front, the Việt Minh, emerged from among a number of contending successors to the colonial state after World War II in the best position to fill the power vacuum left by the defeat of the Japanese. Rubber workers contributed to the strength of the Việt Minh in several ways. First, they participated in the general uprisings that took place at the end of August 1945. Soon after, many rubber workers filled the ranks of the armed units of the Việt Minh, such as Chi đội (Battalion) 1, formed in 1945 and Chi đội 10, formed in 1946. These early efforts at resistance to the French reinvasion met with considerable difficulties, and by the middle of February 1946, the French and British forces had reoccupied most of the plantations (Đặng 2000, pp. 206–207, 210).

Memories of plantation landscapes played an important role in the Việt Minh's effort to frame the battle against the return of the French as an anticolonial war, and the Việt Minh coined several anti-plantation slogans and poems. Many of these sayings emphasized the connections between workers' bodies and rubber trees. For example, a section of one poem arising from events in the colonial era went: "Rubber grows well in this place; Every tree is fertilized with the body of a worker" (Huỳnh et al. 2003, p. 56). Another such slogan was: "Rubber, easy to go, hard to return." It is now commonly assumed that "hard to return" meant that most workers died on the plantations before they could return. This interpretation was indeed part of the understanding of the 1930s and 1940s; this slogan, however, was also understood to mean that workers often got married and settled down in the south rather than return to their natal villages in the center and north. Although Vietnamese communist-nationalist histories claim that all of these sayings arose spontaneously from the rubber workers' movement, it seems that some might have appeared after only after 1945. Whatever their merits as historical evidence, such characterizations of plantations have been able to survive almost uncontested in memory because of the hard life that most rubber workers experienced.[1]

As rubber workers sought better conditions and looked for ways to support the revolution, many attempted to destroy the equipment and the trees of plantation owners. According to one history of the rubber workers' movement, "All of the rubber workers' struggles during the period of resistance were aimed at the sole purpose of 'sabotage'" (Vũ 1950, p. 71). Later a journalist added that the Việt Minh too viewed the plantations as a key location, arguing that "the rubber plantations became the main target in the attack plans of the Việt Minh in the eastern region [southeast Vietnam] in order to weaken the French economy and weaken the vitality of the French Expedition" (Điệp 1965, p. 133). Attacks on plantations were part of the Việt Minh's broader strategy to wreak as much havoc in the colonial economy as possible (see Fig. 3.1). As Pierre Brocheux has pointed out, Việt Minh documents were full of the terms "scorched earth, blockade, and *guerre économique*" as "the economy … became an issue, a battlefield, and a weapon" (2002, p. 315).

In addition to memories of the colonial period and the practical motivation of disrupting French political and economic projects, the Việt Minh initially opposed plantation agriculture on ideological grounds. Throughout the colonial period, plantation owners and managers had emphasized the "scientific" nature of their agriculture. For rubber plantation owners and managers, the scientific approach helped justify their position as the "heads" who commanded the "hands" of Vietnamese labor. The ideological role of science during the colonial era prompted anticolonial leaders to portray scientific knowledge as the handmaiden of colonialism. This tendency was reinforced by a Marxist-Leninist tradition that viewed science as complicit in the control of society by the capitalist system. As with the elite of the People's Republic of China, Việt Minh leaders continued to assert "the class character of science" and to portray themselves as at the leading edge of a progressive movement that would overthrow "imperial" sciences and replace them with "socialist" sciences. Like Maoist political doctrine, the DRV "exalted everyday epistemology, and projected a utopian vision of scientific and political modernity" (Fan 2007,

[1]Phạm Ngộc Hồng, interview, transcript, 6. In 2008, Hồng was an 81-year-old former rubber worker from Nam Định. Many Vietnamese I talked with recited similar poems that emphasized the hard lives of rubber plantation workers under the French.

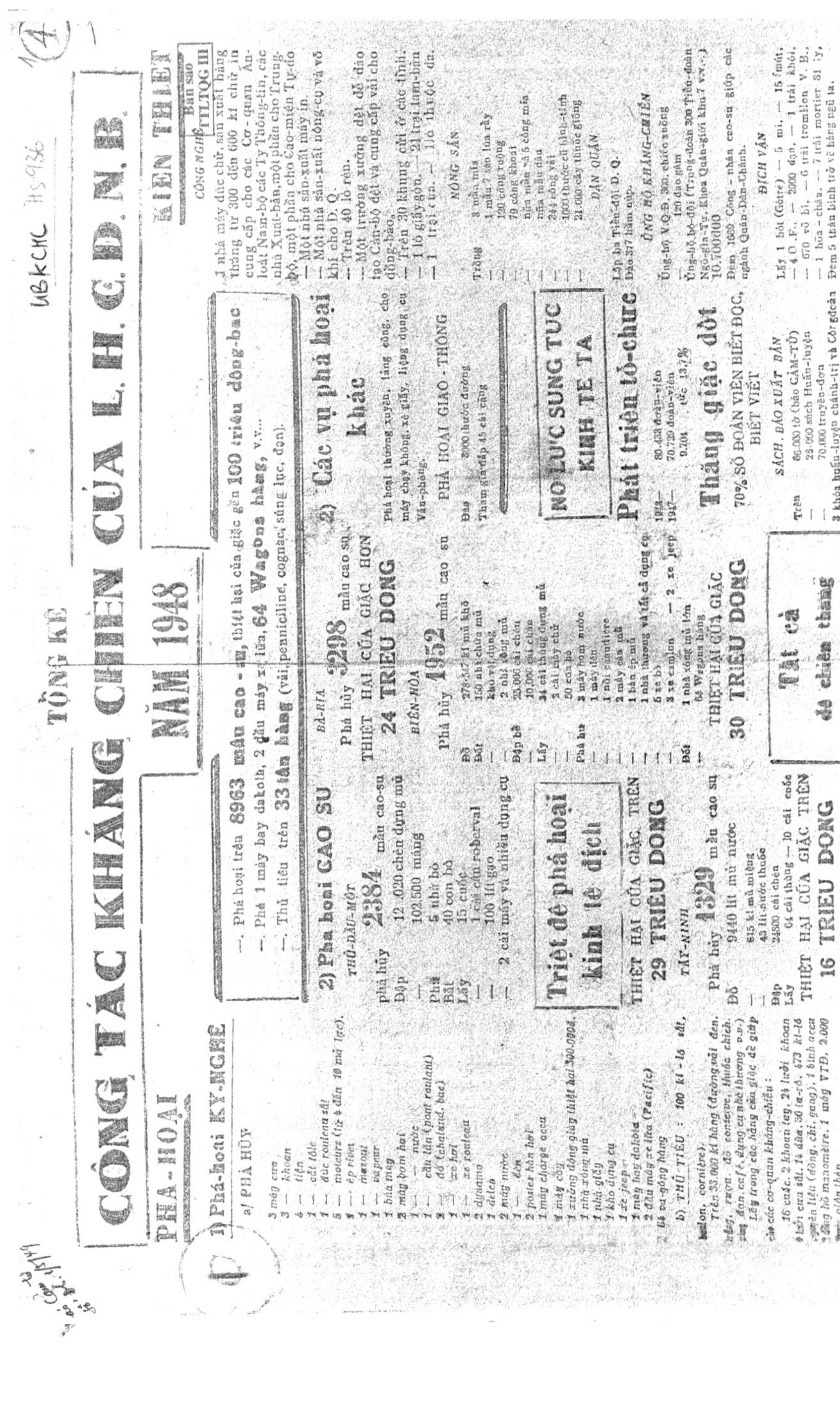

Fig. 3.1 Poster publicizing rubber sabotage. (From National Archives of Vietnam III, Hanoi)

p. 534). Yet Việt Minh leaders could never afford to be as dogmatic in their approach to science as their comrades to the north since the Việt Minh experienced great difficulties in procuring the goods needed for modern life and war, especially before the entry of the Chinese in 1949. With centers of science and technology quickly dispersed or destroyed by the French military, the Việt Minh resorted to low-level production in the countryside (Brocheux 2009, p. 189; Shapiro 2001).

The combination of practical and ideological motivations caused the Việt Minh to adopt a strategy of sabotaging trees and rubber producing equipment, an approach justified through memories of colonial landscapes of labor.[2] This collective effort was called the Rubber War Front (Mặt trận cao su chiến). In 1947, Việt Minh partisans founded the so-called Rubber Trade Union (Nghiệp đoàn cao su) at Long Khánh, the first such organization in the southeast region. The union's two main goals were to "turn the plantations into a battlefield in order to kill the enemy" and "destroy the enemy's economy." The rubber company (Đại đội cao su) belonging to Chi đội 10 had the responsibility of destroying rubber and helping the secret organizations on the plantations kill the militia members controlling the plantation space. In the campaign to sabotage French economic interests, the workers of Biên Hòa destroyed the astounding total of 300,000 rubber trees on 19 May 1947 alone, in honor of Hồ Chí Minh's birthday (Vủ 1950, p. 69). Overall, in 1947 the Việt Minh carried out 1600 incidents of rubber sabotage, with over seven million high-quality trees destroyed (a little less than 10% of the total number of *hévéa* trees in Indochina) covering 17,000 of 150,000 hectares dedicated to plantations. Consistent with slogans depicting trees as fertilized with the bodies of workers, these early campaigns focused on the trees themselves (Trần 2001, p. 33). According to Điệp Liên Anh, the mentality at the time held that "one rubber tree equals one enemy. To destroy one rubber tree is to kill one invader" (Điệp 1965, p. 133). Sabotage methods originally consisted of taking the bark off the trees, but plantation owners devised methods to quickly regenerate the lost bark, so starting in 1947 the Rubber Trade Union adopted the strategy of cutting a deep ring around the tree trunk that both prevented latex from traveling up from the roots and eventually killed the tree (Vủ 1950, pp. 72–73).

The defense of plantations by private militias, mostly composed of Nung and other Montagnard groups, curtailed the success of Việt Minh efforts to destroy plantations. These attacks were further limited by the organizational weakness of the Việt Minh in the south, "virtually a 'non-entity'" (Goscha 2011, p. 104). Historian of the rubber industry Đặng Văn Vinh agrees that the economic effects of Việt Minh sabotage were minimal. Yet psychologically, Vinh argues, the rubber workers' actions helped sustain the morale of the anti-French resistance, and rubber workers conducted the few Việt Minh activities that occurred in the region. To address their

[2] Archives Nationales d'Outre-mer, HCI CS 75: Main d'œuvre pour plantations caoutchouc, travailleurs du nord, terres rouges—Letters on the continuing policy of importing labor from north, including the idea of using prisoners of war. See *Report on Communist Organization of Rubber Workers*, including information on "Cao Su Chien" (Guerre du Caoutchouc [Rubber War]), a monthly starting in January 1950 and an "organe de propagande de l'Union Syndicale des Plantations d'Hévéas du nam-bo et du cambodge."

weaknesses, southern workers met in May 1948, where Việt Minh leaders sought to coordinate rubber sabotage with other anti-French activities (Đặng 2000, pp. 223–27). That year, Chi đội 1 became Trung Đoàn (Regiment) 301 and Chi đội 10 became Trung Đoàn 310 as Việt Minh leaders attempted to incorporated sporadic, local activism into larger structures. Plantations fed these military units, with rubber workers composing a majority of Trung Đoàn 310 and forming significant portions of Trung Đoàn 300, 301, 309, 311, and 312. Simply by drawing workers away from the industry, the resistance posed a real threat to the plantations, which faced a severe shortage of workers. In 1948, according to Việt Minh sources, 8000 cadres originated from the ranks of rubber workers, a number that compared favorably with the 21,000 rubber workers of the south (Vủ 1950, p. 77).

The consolidation of rubber sabotage activity was furthered discussed at a meeting on 11 January 1949, where 17 representatives of the southern forces of the Việt Minh reviewed the previous year's activities of the Rubber War Front. Overall, the meeting recorded many objectives met, such as the destruction of Michelin's Ven-Ven factory. Yet, the Việt Minh's success varied considerably by province. Cadres in Thủ Dầu Một, the province closest to Saigon, destroyed the most trees during the first half of the year and concentrated on other essential areas of the plantations, such as the factories. Yet they had failed to attract the support of ethnic minorities or specialized workers. After cadres in Biên Hòa organized a rubber company in May, the province recorded the highest destruction rates. Biên Hòa's rubber company also cooperated particularly well with Trung Đoàn 310, the provincial regiment. In Bà Rịa and Tây Ninh, numbers were low throughout 1948 because of only sporadic Rubber War Front activity. In Bà Rịa, Trung Đoàn 307 assisted in rubber-destroying activities, as did the ethnic minorities (*dân tộc thiểu số*). Tây Ninh, while having the spirit of sacrifice, had accomplished little. In Cambodia, the Việt Minh had been able to organize very few sabotage activities, and the French had successfully retaken control of the rubber industry. In all the regions, problems with lack of food, money, and other supplies hindered efforts.[3]

The delegates admitted that the plantations were still earning great profits for the colonizers, which given the limits on destruction efforts did not come as a surprise. More worrisome for the Việt Minh was the decrease in sabotage acts in 1948 compared with 1946 or 1947. Several reasons were put forward: (1) the enemy had prepared effective defenses; (2) organizational problems within the military had disrupted rubber-destroying activities since June; (3) a lack of food had forced soldiers to concentrate on growing crops rather than military activities; and (4) confusion among the soldiers meant that, when there was a rubber committee, the local cadre did not pay as much attention to sabotaging rubber. Such a trade-off pointed to tensions involving the question of who had the authority to decide strategy, how to divide out the war loot, and how to distribute the credit for actions. In the worst

[3] National Archives of Vietnam III, Hanoi (hereafter NAVN3), 2973 BNL: Tập tài liệu về hội nghị cao su năm 1948 do Bộ Canh Nông tổ chức ngày 20 đến 21-01-1949. "Ta chưa tổ chức phá-hoại cao-su được ở Cao Miên."

case, the secret activities of the Việt Minh cadre interfered with those of the Rubber Trade Union and vice versa.[4]

Not only the amount but also the type of activity is relevant when discussing attitudes towards plantation landscapes. During the first half of the year, little attention was dedicated to spilling latex and breaking collection bowls (the burning of dry rubber is more difficult to determine, since the numbers for January through October are aggregated into one category). These activities reduced rubber company profits without destroying the potential for future production. After June, these activities became more common, perhaps signaling the shift that was later announced by Lê Duẩn, the secretary of the southern region (*bí thư Xứ ủy Nam bộ*), at the end of 1949. Furthermore, a former rubber tapper and communist party member argued that this shift happened as early as 1947 (Lê 1979, p. 63). Yet the plans for 1949, which emphasized three main activities—organize, defend, and attack—reiterated the need to destroy the trees. This destruction involved both burning good trees and distributing knives for cutting trees.[5]

In August 1949, the Workers' Party decided to change military tactics in southern Vietnam. They sent instructions for Nguyễn Bình, the leader of the south, to prepare for more direct military confrontations with the French. On 18 August the Standing Committee of the Central Committee of the Party (Ban Thường vụ Trung ương Đảng) issued instructions to shift fighting from guerrilla warfare (*du kích chiến*) to mobile warfare (*vận động chiến*). In response to this directive, the Regional Party Committee of the South (Xứ ủy Nam Bộ) began preparations for a general counterattack (*tổng phản công*) (Hồ 2007, p. 237). Part of this shift entailed creating a more organized military structure, and on 18 November the Southern Command (Bộ Tư lệnh Nam Bộ) sent out instructions that created the inter-regiments and substructures. In this way, the Việt Minh formed the 301 and 310 inter-regiment (Liên Trung Đoàn 301–310), which concentrated its activities in the plantation-dominated southeastern region and attempted to bring order to a rather chaotic situation.

On 17 November, perhaps heartened by the communist victory in China, Lê Duẩn issued general directives for southern action. These directives included an order to destroy the profits of the French but to leave rubber trees in place. This decision may have been made to avoid alienating rubber workers from the Việt Minh, but Duẩn specifically mentioned the idea that rubber plantations were part of the future wealth of the country. "Rubber trees," Duẩn stated, "are a large source of profit of our Fatherland, we must take care of [*chăm sóc*], protect [*bảo vệ*], and not destroy [*chặt phá*] them" (Hồ 2007, p. 202; Việt Nam 1949, pp. 309–13). Sabotage activity apparently followed this shift in Việt Minh strategy from destruction to conservation. In the first 6 months of 1950, the workers of An Lộc plantation burned 2100 tons of rubber and destroyed six lorries, causing 2000 đồng of damage all the

[4] NAVN3 2012 PTT: Công văn, báo cáo của Bộ Canh nông, Liên đoàn Cao su Thủ Dầu Một, Ban Cao su Bà Rịa, Biên Hoà, Liên hiệp Công đoàn Nam bộ về tổ chức, phá hoại cơ sở cao su ở Nam bộ năm 1949, 1950, 1952.

[5] NAVN3 2973 BNL: Tập tài liệu về hội nghị cao su năm 1948 do Bộ Canh Nông tổ chức ngày 20 đến 21-01-1949.

while carefully avoiding the trees (Trần 2001).[6] From this point on, the Việt Minh leadership began to view rubber plantations as an immediate source of money and supplies and a natural resource that would remain after the departure of the French (Michon 2001, p. 46).

3.3 Material and Recruits

Events of the early 1950s encouraged the French and both Vietnamese governments to rethink their approaches to plantations. All sides viewed plantations as a key symbolic and material battleground during the First Indochinese War as they sought to transcend the colonial legacies of plantations and remake representations and material uses of these landscapes. For the French, since 1945, many plantation owners and workers had provided assistance to the French-led military to fight the Việt Minh. But not all plantations were eager to get involved with the violence, as the owners and managers faced a daunting task of restarting rubber production. In addition to the violent attacks by Việt Minh partisans, challenges included a lack of workers and growing nationalist scrutiny of plantation conditions (Điệp 1965, pp. 101–2). French plantation owners responded to both of these difficulties by increasing their efforts to portray plantations as inviting places to live and work. The Associated States of Vietnam, formed in 1949, attempted to navigate its way through muddy political waters by cautiously encouraging Vietnamese nationalism while allowing for foreign ownership of private businesses such as plantations. Finally, in official publications, the Việt Minh continued to condemn plantations as "hells on earth." Strategically, however, the Việt Minh viewed plantations as potential sources of money, goods, and people.

Vietnamese histories tend to portray the relationship between plantation owners and managers and the French-led military as close, but Marianne Boucheret has shown that tension existed between these factions. The military often complained that plantation owners were more concerned with profit than strategic interests, while plantation owners argued that the military disregarded the on the ground realities and left plantations to fend for themselves (Boucheret 2008). This relationship drifted further apart during the 1950s, as the French military became more reluctant to fight the plantation owners' battles against labor (Đặng 2000, pp. 215–16). More relevant to this chapter's question about the refashioning of plantation landscapes were the efforts of many plantations to adapt to the emerging national and Cold War situations. With the growing sense that an independent Vietnam, in whatever form, would be the outcome of the First Indochina War, Vietnamese were trying to hold plantation owners accountable for working conditions in a way than had not been possible during the colonial period. Unsurprisingly, plantation owners and managers were for the most part unable to make a sudden break with colonial mind-sets

[6] On 19 May 1950, the workers of Trảng Bom celebrated Hồ Chí Minh's birthday by burning 34,000 kg of rubber and 28 *xe* (vehicles), with a value of 600,000 đồng.

and simply recycled many of the social programs highlighted during the late colonial period.

Evidence of this recycling appears in a plantation-produced monthly journal called *Liên Lạc* (Connections), which highlighted the "heavenly" aspects of workers' lives. This journal ran for at least 30 issues between January 1952 and July 1954 and was produced by the official government printer, the Imprimerie Française d'Outre-mer in Saigon. The sub-header of the journal was "The Information Service between the Southern rubber plantations and Northern Vietnam," and true to this name, each of the journal's issues devoted extensive space to short letters from plantation workers to their relatives in the northern provinces. Reflecting the chaos of the war years, many letters are quite moving as they reassured fathers and mothers, aunts and uncles that the letter writers were still alive; these letters also inquired after the welfare of relatives. The voices in the letters range from fairly educated factor workers to barely literate rubber tappers. To have been at least a somewhat credible source of information, it stands to reason that the publishers would seek out real workers rather than simply invent them. Yet, none of the letters home complained about labor conditions on plantations even as workers' strikes continued to take place. In other words, the plantation-controlled journal was heavily censored, at the very least.

Rather than an indication of what workers thought, this journal reveals what the plantation owners thought would appeal to workers. For example, in an October 1952 article titled "The Price of Rice Has Risen," the piece's author, Đàn-Tâm (a literary term for loyalty), argued that while the rising price of rice hurt other consumers, especially the poor, rubber workers were unaffected. Đàn-Tâm stated that because of the rice provisioning system whereby each worker received a set amount of rice regardless of price, it was the plantations, rather than the workers, that would have to absorb the rising costs (see Fig. 3.2). In the highly inflationary environment between 1952 and 1954, food was a major concern, and the French believed that the provision of rice, medicine, and other basic necessities as part of workers' salaries would prove attractive, as they were in a way more resistant to inflation than cash payments.

Another common topic was the health conditions of the plantations. In the November 1952 issue of *Liên Lạc*, Cố-Nh discussed the reputation of the plantations as unhealthful places, acknowledging their bad reputation. While in the past plantations had been unhealthy, the author argued that this was no longer the case after the land had been cleared. One of *Liên Lạc*'s chief goals was to counteract Việt Minh efforts to dissuade northerners from working for plantations. In a November 1952 article called "Truly a Heaven?" Tran Quynh weighed in on the ongoing propaganda war between planters and the Việt Minh. Mentioning the attacks on plantations coming from the north, Tran argued that plantation living conditions were relatively better than almost anywhere else in Vietnam (Điệp 1965, pp. 111–16).[7]

[7] See Trân-Quynh, "Thiên-Đường Chăng?," *Liên Lạc*, November 1952. For a critique of this propaganda, see Điệp (1965).

Fig. 3.2 Rice distribution shack. (From Đàn-Tâm, "The Price of Rice Has Risen," *Liên Lạc*, June 1953)

The propaganda battle over plantations also took place in France. The French Rubber Institute (Institute Français du Caoutchouc), for example, held talks given by those with firsthand experience of the plantations. In 1949, the medical doctor Tran Dinh Que gave a historical overview of the medical system in Vietnam. Regarding plantations, he noted the existence of laws protecting workers on the books but argued that these laws needed to be adjusted to current realities.[8] By contrast, most French planters seemed unable to apprehend the realities of decolonization. The president of the technical commission, M. Bocquet, for example, expressed his views in a 1950 talk at the French Rubber Institute. During a meandering speech in which he recycled colonial-era stereotypes, Bocquet insisted that Vietnamese nationalism could not explain what was going on because the Vietnamese masses were incapable of acting on ideals. Instead, echoing the postwar French programs of Wormser and Malye, Bocquet argued that Vietnam suffered from material want due to a lack of a technical elite. According to this technocratic vision, agricultural engineers would bring prosperity to the countryside, thus eliminating the material motivations for the revolution (Bocquet 1950).

---

[8] Institut de Médecine Tropicale du Service de Santé des Armées 167, Affaires diverses. Documentation sanitaire: correspondance et texte de la conférence sur 'l'Organisation médico-sociale en Indochine et particulièrement au Vietnam', 1949, 19.

Owner and managers' views of plantations stand in stark contrast to Việt Minh understandings of the plantations and of the First Indochina War more generally. Paul Mus's *Sociologie d'une Guerre*, which appeared in 1952, highlights these conflicting views. In this book Mus, who had long lived in French Indochina, discussed the rubber plantations, which he called a "school of a new Vietnam." He wrote that, from the French perspective, these plantations were glittering visions of modernity, all straight lines and clean surfaces, with hospitals, schools, places of worship, and a progressive social vision. Mus warned, however, that what appeared to be modern and attractive improvements for Western observers might hold little attraction for the laboring population, who valued family above all else. For the Việt Minh, moreover, plantation landscapes were a sign of lack and represented all that could have been achieved during the colonial period under Vietnamese leadership, but was not. In *Sociologie d'une Guerre*, Mus points to an important component in what was later known as the battle for the "hearts and minds" of the people. Yet plantations were strategically important for other reasons as well. With the shift in strategy away from sabotage of trees, Việt Minh sought more finely tuned ways to take advantage of the industry. Plantation workers still served as potential recruits, and hostages, and above all the Việt Minh sought to redirect some of the money and supplies that were sent to the plantations in exchange for rubber. Strategically, too, plantations were located in a region that was a gateway to Saigon and the delta to the south. Ironically, the province of Biên Hòa, or "peaceful border," became a passage for war making (Đặng 2000, p. 218).

Overall, plantations became a source of money, goods, and people. Điệp Liên Anh noted that between 1945 and 1954 "the rubber workers became a reserve for the 2 sides. … The Việt Minh attempted to steer plantation workers into the struggle against the French" (Điệp 1965, p. 133). Along these lines, Regiment 303 in Thủ Biên Province formed in 1951 from Thủ Dầu Một and Biên Hòa comprised 95% rubber workers (Vủ 1950, p. 71). The bodies of French plantation personnel also became a valuable commodity, and strategies developed whereby plantation managers were taken hostage and held for large cash payments, which the plantation owners often paid out. Michel Michon, a former planter, recounts how he and other plantations managers were often captured and exchanged for cash. After the change in Việt Minh strategy from one of sabotage to one of extraction, the Việt Minh also provided a kind of protection for the plantations. Michon recalls how in the late 1950s the Việt Minh cell responsible for his plantation provided unsolicited help by searching for and punishing thieves who had supposedly stolen rubber coagulum (Michon 2001). Medicine was another invaluable commodity, and French and South Vietnamese authorities sought to stem the flow of medicines from plantations to the Việt Minh. Because of the dispersed nature of the Việt Minh, strategies varied from plantation to plantation, and there was no single way in which the Việt Minh sought to profit from plantation landscapes.

Geographically, plantations were well situated to play an important role in control of the border region of southern Indochina, and the French military decided to place many of its posts on plantations. For example, the Dầu Tiếng plantation had 8 posts and 300 soldiers, and the Quản Lợi plantation had 5 posts and 200 soldiers (Đặng 2000, pp. 212–13). From the point of view of the Việt Minh, the forests that bordered plantations provided a convenient place to evade detection, especially from French aircraft, and as a shield during both attack and retreat. Plantations provided many other conveniences for the guerrillas, including places to sleep, ready-made paths, wood and water, and places to hold meetings and to train (Đặng 2000, pp. 218–19). Plantations straddled both sides of the Vietnam and Cambodia border, and with Vietnamese working on plantations in Cambodia, rubber tapping served as a useful cover for travel between the two countries. Realizing the strategic uses of plantations, especially of small- and medium-size ones, the French also engaged in the destruction of rubber trees. Furthermore, the higher ground of the plantations, according to Đặng Văn Vinh, allowed some Việt Minh to escape the 1952 floods mentioned below, with plantation workers providing food and shelter (Đặng 2000, p. 234).

The establishment of the Associated States of Vietnam in 1949 resulted in a reorganization of the legal structure of associations in Vietnam and some reforms favoring Vietnamese control of governmental functions. On the one hand, French retained almost exclusive control of the plantation industry.[9] According to Điệp Liên Anh's *White Blood, Red Blood*, Vietnamese leaders of the Associated States continued to provide government support of the rubber industry. When Minister Trần Văn Hữu visited the plantations in 1951, for instance, he supposedly praised the plantation owners for their bravery in facing, among other threats, the "poison water" of the forests and Việt Minh ambushes. Anh wondered at the ability of a Vietnamese government official to ignore the suffering of Vietnamese labor and its vital role in making the plantations. He also noted the response of rubber workers themselves, who said, ironically, "That's really a case of the mother singing the praises of her children" (Điệp 1965, p. 119). While rubber remained under control of the French into the 1960s, the industry continued to provide important export earnings for the states of southern Vietnam, making it a difficult industry to reform. On the other hand, the government of Nguyễn Văn Tâm passed laws allowing rubber workers to form legal organizations. Worker organization happened only slowly before 1954, but between the end of 1954 and June 1956, the Plantation Workers' Union grew in membership from 2000 to 20,000 workers (Điệp 1965, pp. 139–41).

Even as Việt Minh strategies shifted, memories of colonial plantations continued to haunt nature. Lê Sắc Nghi, a former plantation worker and later general director of the Đồng Nai Rubber Company, recalled the predations of a man-eating tiger in 1950 that supposedly killed nearly 100 people and terrorized soldiers in Inter-regiment 301–310. This tiger had a particular relationship with plantations, as Việt Minh rumors held that Ba Móng, or Three-Claws, was the former pet of a plantation owner. Others believed that the tiger was a spirit of nature, something that the French military exploited for propaganda purposes (Hồ 2007, pp. 328–54; Lê 1979, p. 79; Trần 2001). The early 1950s were difficult years for the Việt Minh in other ways. While the frontal assaults on French military posts showed that the Việt Minh military in the south, including Inter-regiment 301–310, had come a long way, these assault resulted in heavy casualties as the French brought air and artillery power to bear in these battles. Then in 1951, Nguyễn Bình, commander of the southern forces, was ambushed and killed en route to northern Vietnam for a strategic meet-

[9] Ordonnance no. 10, 6 août 1950.

ing. In October of the following year, a terrible flood in the southeast region resulted in large-scale suffering and famine, in particular among the Việt Minh. Their job, as one partisan noted, was simply to stay alive.

Rubber production continued to increase steadily, reaching close to 73,000 tons by 1953, something of interest for the Việt Minh. A report from March 1954 summarized press from France, the United States, and South Vietnam about the rubber situation. In general, the rubber industry and plantation owners had been able to recover production, and the amount of rubber produced in 1953 exceeded the amount produced before World War II. The Việt Minh pointed out the major difficulties facing the industry, including the lack of workers and the profitability of rubber. The price of natural rubber was 12.5 times higher in 1953 than in 1939, partly due to the Korean War in 1950. There was, however, growing competition from synthetic rubber, especially as produced in the United States, and prices in Vietnam had risen steeply, rice in particular, which made the cost of labor relatively high. Insecurity too meant that plantations had to spend anywhere from 1 to 4 piastres per kilogram for security measures. The report concluded that peace would be helpful for the plantations, and reports of peace in Indochina caused increased buying of plantation shares on the stock market in Paris. This chapter now considers plantation landscapes from the viewpoint of a growing desire to produce socialist rubber.

3.4 Production and Propaganda

During the closing years of the First Indochinese War, Việt Minh leaders looked to start rubber production in northern Vietnam, and DRV leaders later prepared to take over rubber plantations throughout the country. Granted, the Việt Minh never accorded a high priority to growing rubber, but the fact that they devoted any of their scarce resources to discussing rubber raises the question of why. Attention to rubber made sense both domestically, within a newly envisioned socialist economy, and internationally, within a bloc of fellow socialist countries. The nation-states and empires forming this network, including the Soviet Union and the People's Republic of China, sought access to tropical raw materials, including rubber. The Việt Minh, and later the DRV, attempted to benefit from their climatic location to fill this need. Interest in rubber growing was also motivated by experiments conducted in southern China. In 1952, for example, Việt Minh leaders learned of groups of Chinese youth sent to replant the southern countryside in rubber trees. In a letter marked "secret" from 5 August 1952, Nguyen Van Luu reported on rubber growing in Guangxi. Đặng Văn Vinh contributed his comments on this question, writing a summary on the margins of the letter. Vinh noted that in general rubber could be grown below 23° latitude, which included all but the northern most points of Vietnam. The main limits to growth were temperature, rain, and especially wind conditions. With too strong a wind, *hévéa* trees and their shallow roots would be knocked over.[10]

[10] NAVN3 2012 PTT: Công văn, báo cáo của Bộ Canh nông, Liên đoàn Cao su Thủ Dầu Một, Ban Cao su Bà Rịa, Biên Hoà, Liên hiệp Công đoàn Nam bộ về tổ chức, phá hoại cơ sở cao su ở Nam bộ năm 1949, 1950, 1952, letter dated 5 August 1952.

To continue to develop rubber growth in the north, DRV leaders sought to appropriate the legacy of French colonial plant research. In a letter dated 14 April 1952, the Government Economic Board asked Nguyễn Xuân Cung, the minister of agriculture, to send information about those who had rubber-growing expertise in northern Vietnam and about any other material related to this question. Cung replied four months later that there were a handful of experts in the north, including Nguyễn Duy Cân, who had responsibilities at the Gia-rai experimental station from 1931 to 1937; Dào Thiên Tránh, who had worked at Bolovens; and the better known agricultural experts Đặng Văn Vinh, Vuong Gia Cân, and Nguyễn Khoa Chi.[11] The printed material, according to the minister, was even more dated with Henri Jumelle's 1903 *Les plantes à Coutcho* (*sic*), along with a handful of other material at experimental stations in the north serving as the only documented resources.

Even as Việt Minh attitudes toward the rubber industry evolved, narratives about rubber often remained harshly critical of plantation conditions. These strident critiques were aimed at achieving specific political goals. For example, as part of the Geneva Accords signed in July 1954, Việt Minh and South Vietnamese representatives agreed to allow people to move freely between the north and south for two years. During this period of "regrouping" (*di cư và tập kết*), there were intense efforts on both sides to influence population movements. With respect to the rubber industry, the DRV government depicted plantations as "hells on earth" in publications with such titles as *Rubber Hell*. These publications were aimed at countering the allure of plantations and a working life for farmers in the poverty-stricken north (Nguyễn 1955).

It is difficult to assess the success of these efforts, but there are some clues that suggest plantations continued to attract workers. For example, in 2008 a former director of a plantation area recalled from memory a recruitment poem from the early 1950s. It contained several markers of modernity, including an airplane ride. More important, it mentioned cash loans along with free rice, medicine, clothes, and even care for wives and children. The poem's form is "six-eight", where lines of six syllables alternate with those of eight syllables, a mnemonic device that has been used in Vietnamese epic poetry.

No money but you want to take a plane?
Come for the advance to go to Nam Kỳ
Your family can come too
Day and night come together, there is no need to be reluctant.
Old Thành Nam still has the recruiting company
Thái Bình is the street, the house is number three
After you sign the contract
A loan of one hundred piastres to spend
And you can come and go often
The office can accept all, according to common regulations
Medicine, rice, cloth for free

[11] Ibid., letter dated 14 August 1952, from Nguyên-xuân-Cung, the Minister of Agriculture, sent to Government Economic Board. Side note from Vinh reads "Too slow!" (*Cham qua*), since this letter was a response to request from 14 April 1952.

Sick child, wife in child birth will have care
Thirty months is the end of the contract
The money left over, in the ten thousands, you can bring it back
In South and North they come in crowds.

Yet, had recruits known the truth of plantation conditions, this former director said, they would not have traveled to the south. The real improvements in plantation conditions, he stated, happened only in 1958 due to Việt Minh pressure and the world market for rubber. What effect events subsequent to 1954 have had on these memories is difficult to say, but independence stands out as a convenient turning point for nationalist histories written from a variety of political viewpoints.[12]

Because of the work to make plantation landscapes Vietnamese, memories of French presence on plantations in Vietnam still rankled Vietnamese officials after 1975. When the French offered to help restart rubber production in the south, SRV officials hesitated to accept this aid. In 1980 during some of the most trying times of recent Vietnamese history, the SRV government provided substantial support for the rubber industry. Nguyễn Hữu Chất, onetime director of the Rubber Research Institute in Vietnam, wrote a letter arguing for the importance of Vietnam's contribution to the international rubber community that helped convince Prime Minister Tố Hữu to authorize a contribution of $10,000 to the International Rubber Research and Development Board at a time when foreign exchange was extremely scarce (Đặng 2000, p. 207).[13] Yet, until the economic renovation of *đổi mới* in 1986, rubber plantations had a difficult time rebuilding. Markets were limited, and those with the technical and managerial know-how were few and far between. Yet, leaders of the newly formed SRV held high hopes for the rubber industry in what was still essentially an agrarian country.

3.5 Conclusion

The continuing survival of plantation landscapes and the story of the tiger Three-Claws show that the decisions made to conserve some natures and destroy others resulted from both the geopolitical forces of the Cold War and local cultural contests. While revolutionary forces were able to turn plantation landscapes into national spaces, they were unable to tame Three-Claws. Not until the 1990s did southern rubber landscapes recover from the trauma of 30 years of war. This recovery was celebrated with a spate of histories of the rubber worker movements, along

[12] Phạm Ngọc Hồng, interview, transcript, 10-11.

[13] NAVN3 2216 PTT: Công văn của PTT và Bộ Nông nghiệp v/v Viện Cao su Việt Nam đóng góp vào kinh phí sưu tập giống cao su ở Nam Mỹ năm 1980. Letter, CN144s/O-GĐ và CN158s/O-GĐ, Phó Viện Trưởng, Nguyễn Hữu Chất gửi BNN và BTC, 20 June 1980, v/v xin ngoại tệ góp vào kinh phí sưu tập giống cao su Nam Mỹ; letter, 3080 V7, PTT gửi BNN, BNG, BTC, NHNN, 21 juil 1980, v/v đồng góp vào kinh phí sưu tập giống cao su ở Nam Mỹ. See also 135 BNN Tập QĐ nhân sự tháng 01.1977 của BNN.

with company museums. Yet rarely have these histories considered the role of violence in transforming these colonial landscapes into national ones.

In the 2000s, rubber production in Vietnam had rebounded to notable levels. Driven in part by China's accelerated growth, rubber exports in 2011 reached about 800,000 tons and brought in around US$3 billion of foreign exchange, employing thousands of Vietnamese.[14] As Vietnamese rubber companies expand into Laos and Cambodia, raising the specter of an expansionist Vietnamese nationalism, it is relevant to consider rubber once again as a potent tool for environmental change. This chapter's analysis of rubber plantations as sites of nationalist production is an attempt to shed light on other conflicts involving natural resources and nationalist sentiments. As outcries over Chinese-controlled bauxite mining operations in the central highlands shows, Vietnamese leadership has become caught between anti-foreign sentiments, specifically Chinese, and what is viewed as rational, efficient use of resources.

References

Andrews, T. G. (2008). *Killing for coal: America's deadliest labor war.* Cambridge, MA: Harvard University Press.

Baudrit, A. (1936). *Monographie d'une rivière cochinchinoise, Le fameux Song-Bé. Bulletin de la Société des études indochinoises* (Vol. 11, pp. 7–42).

Biggs, D. (2005). Managing a rebel landscape: Conservation, pioneers, and the revolutionary past in the U Minh forest, Vietnam. *Environmental History, 10*, 448–476.

Biggs, D. A. (2010). *Quagmire: Nation-building and nature in the Mekong Delta.* Seattle: University of Washington Press.

Bocquet, M. (1950). Visage du Vietnam vu par un français.

Boomgaard, P., Colombijn, F., & Henley, D. (1997). *Paper landscapes: Explorations in the environmental history of Indonesia.* Leiden: KITLV Press.

Boucheret, M. (2008). *Les plantations d'heveas en Indochine, 1897–1954.* Paris: Université Paris 1-Panthéon Sorbonne.

Brocheux, P. (1995). *The Mekong Delta: Ecology, economy, and revolution, 1860–1960.* Madison: Center for Southeast Asian Studies.

Brocheux, P. (2002). The economy of war as a prelude to a "Socialist Economy": The case of the Vietnamese resistance against the French, 1945–1954. In *Việt Nam exposé: French scholarship on twentieth-century Vietnamese Society* (pp. 313–330). Ann Arbor: University of Michigan Press.

Brocheux, P. (2009). *Une histoire économique du Viet Nam: 1850–2007: la palanche et le camion.* Paris: les Indes savantes.

Cronon, W. (1996). Introduction. In W. Cronon (Ed.), *Uncommon ground: Rethinking the human place in nature* (pp. 23–68). New York: W. W. Norton and Co..

Đặng, V. V. (2000). *100 năm cao su ở Việt Nam.* TP HCM: NXB Nông Nghiệp.

Điệp, L. A. (1965). *Máu trắng, máu đào: Đời sống đọa-đày của phu cao-su miền đất đỏ.* Sài Gòn: Lao-động mới.

Elliott, D. W. P. (2000). *The Vietnamese war: Revolution and social change in the Mekong Delta.* Armonk: M.E. Sharpe.

Fan, F. T. (2007). Redrawing the map: Science in twentieth-century China. *Isis, 98*, 524–538.

[14] http://www.asean-cn.org/Item/4351.aspx, last accessed 16 December 2012.

Goscha, C. E. (2011). *Vietnam: un Etat né de la guerre, 1945–1954*. Paris: A. Colin.
Hồ, S. Đ. (Ed.). (2007). *Lịch sử Liên Trung Đoàn 301–310 (1945–1950)*. Hà Nội: NXB Quân Đội Nhân Dân.
Hunt, D. (2009). *Vietnam's Southern revolution: From peasant insurrection to total war*. Amherst: University of Massachusetts Press.
Huỳnh, L., Hồ, S. Đ., Trần, Q. T., Hà, X. T., & Nguyễn, K. T. (2003). *Lịch sử Phong Trào Công Nhân Cao Su Việt Nam, 1906–2001*. Hà Nội: Nhà Xuất Bản Lao Động.
Kalikiti, W. S. (2000). *Rubber plantations and labour in colonial Indochina: Interests and conflicts, 1896–1942*. London.
Latham, M. E. (2000). *Modernization as ideology: American social science and "Nation Building" in the Kennedy era*. Chapel Hill: University of North Carolina Press.
Lê, S. N. (1979). *Đất Đỏ Miền Đông: Hồi Ký Cách Mạng*. Biên Hòa: Công ty Quốc doanh Cao su Đồng Nai.
Mạc, Đ., Trần, B. Đ., Công, K. L., & Nguyen, D. D. (1991). *Địa chí tỉnh Sông Bé*. Thủ Dầu Một: NXB Tổng Hợp Sông Bé.
Malye, M., 1949. Les problemes d'Indochine et le role des techniciens d'economie rurale.
Martini, E. A. (2012). *Agent orange: History, science, and the politics of uncertainty*. Amherst: University of Massachusetts Press.
Michon, M. M. (2001). *Indochina memoir, rubber, politics, and war in Vietnam and Cambodia 1955–1972*. Tempe: Program for Southeast Asian Studies.
Murray, M. J. (1992). "White Gold" or "White Blood"? The rubber plantations of colonial Indochina, 1910–40. In *Plantations, proletarians, and peasants in colonial Asia* (pp. 41–67). London: Frank Cass.
Nguyễn, H. T. (1955). *Địa Ngục Cao-Su*. Hà Nội: NXB Sự thật.
Russell, E. (2001). *War and nature: Fighting humans and insects with chemicals from World War I to silent spring*. New York: Cambridge University Press.
Shapiro, J. (2001). *Mao's war against nature: Politics and the environment in revolutionary China*. Cambridge: Cambridge University Press.
Soluri, J. (2005). *Banana cultures: Agriculture, consumption, and environmental change in Honduras and the United States*. Austin: University of Texas Press.
Sterling, E. J., Hurley, M. M., & Le, M. D. (2006). *Vietnam: A natural history*. New Haven: Yale University Press.
Stewart, M. A., & Coclanis, P. A. (Eds.). (2011). *Environmental change and agricultural sustainability in the Mekong Delta*. New York: Springer.
Taylor, P. (2007). *Modernity and re-enchantment: Religion in post-revolutionary Vietnam*. Singapore: Institute of Southeast Asian Studies.
Taylor, P. (2014). *The Khmer lands of Vietnam: Environment, cosmology, and sovereignty*. Honolulu: Asian Studies Association of Australia in association with University of Hawai'i Press.
Trần, T. M. H. (2001). *Địa Chí Đồng Nai*. Biên Hòa: Nhà xuất bản tổng hợp Đồng Nai.
Trịnh, H. Đ. (1863). *Histoire et description de la Basse Cochinchine (pays de Gia-dinh)*. Paris: Imprimerie impériale.
Việt Nam, Đ. C. S. (1949). *Văn Kiện Đảng*.
Viollis, A. (1935). *Indochine S.O.S.* Paris.
Vũ, T. (1950). *Công Nhân Cao Su Chiến Đấu, Thành tich tranh đấu của công nhân cao-su tù 1920 đến 1948*. Hanoi: NXB Lao Động.
White, R. (1995). Are you an environmentalist or do you work for a living? Work and nature. In W. Cronon (Ed.), *Uncommon ground: Rethinking the human place in nature* (pp. 171–185). New York: WW Norton.
Wormser, G. (1946). *De quelques facteurs de la situation economique et politique du sud-indochinois*. Saigon: Impr. de l'Union Nguyen-Van-Cua.
Zierler, D. (2011). *The invention of ecocide: Agent orange, Vietnam, and the scientists who changed the way we think about the environment*. Athens: University of Georgia Press.

Chapter 4
Remembering Lost Landscapes in Cambodia

W. Nathan Green

Abstract This chapter narrates a short history of agricultural and environmental change in the village of East Big Lake in southern Cambodia. Primarily based upon villagers' memories of place, it explores the changing relationship between people and their local environment over the second half of the twentieth century, during which the country gained independence from France and then later became embroiled in civil war, genocide, and national reconstruction. People's memories of the environmental past are retold through the landscape, drawing upon their personal experiences. The information retold in this historical narrative was learned through a yearlong ethnographic research project, in which the author carried out participant observation, interviews, and structured surveys with a group of 26 families. This chapter provides a local, village-based perspective on Cambodia's agricultural and environmental history, about which relatively little has been written. It also demonstrates how place, memory, and landscape can contribute to environmental histories.

Keywords Cambodia · Environmental history · Landscape · Memory · Rice agriculture

The village of East Big Lake once had an abundance of fish, recalls Sok, a wiry and well-tanned farmer who has lived in the village most of his 63 years. I stand next to him and watch his cattle graze in the harvested rice fields now dried out by the November winds. He tells me that in the old days, there were so many fish in the village's big lake that a person could catch enough in one afternoon to feed their family for days. But that was before the Pol Pot time, when the villagers cut down all the trees around the lake and dug canals to drain its water. After Pol Pot, they split up the land and planted rice where the lake had once been. Fish still swim into the fields, but not like before, and there are fewer every year now that farmers use chemicals to kill the crabs and the grasses. These days if people want fish, they have

W. N. Green (✉)
Department of Geography, University of Wisconsin-Madison, Madison, WI, USA
e-mail: wgreen@wisc.edu

M. A. Stewart, P. A. Coclanis (eds.), *Water and Power*, Advances in Global Change Research 64, https://doi.org/10.1007/978-3-319-90400-9_4

Photo 4.1 Farmers transplant rice together in East Big Lake

to go to the market or dig a fish pond. Sok jokes that in the past, the fish raised the people, but now people raise the fish.

Like many place names in Cambodia, the name of the village East Big Lake suggests a story of a different environmental past. The clues are inscribed on the landscape (see Photo 4.1). The village is in southeast Kampot Province where most of the land is flat, but in the orange light of the early evening Sok and I can see several karst mountains standing tall on the western horizon. They are the calcified remnants of the ancient sea that covered much of the lower Mekong region during the Paleozoic era some 260 million years ago. Closer to us, we can spot pond herons flying over an irrigation canal that forms a border around the rice fields of Sok's village. During the wet season the canal allows fish to swim from a different lake by the mountains into the paddies. Small paperbark trees that used to line the entire lake shore now grow on top of the dikes between the rice fields. Behind us is Sok's house, only a short walk along a small dirt path shaded by sugar palms, coconut, and mango trees. There are 25 other houses in Sok's hamlet, approximately one tenth of all the families in the village. Most houses in the village have been built along the national highway a kilometer to the east, but no cars can be heard in his small hamlet.

If we look critically, these physical and built features of the landscape—mountains, canals, dikes, and fruit trees—help us to piece together what might be called the agro-environmental history of East Big Lake (Gregg 2010). But they are not enough. We also have to understand the ways that people organize themselves to produce their sustenance and the cultural value that they place upon this food. Nor can a place like East Big Lake be thought of as isolated from the wider world. How villagers like Sok become dependent upon money to exchange goods and subject to the enforcement of policy and law all bear directly on people's relation to nature.

We must follow these connections out of the village (Cronon 1992). In Cambodia, these connections lead to Cold War geopolitics, brutal genocide, and the long socialist reconstruction efforts that eventually gave way to a market economy governed by a limited democracy.

When paired with people's memories of the past, the history of this landscape becomes animated by stories of joy and suffering, abundance and loss. For so many Cambodians, war, genocide, and socialist reconstruction in the second half of the twentieth century re-defined the relation between people and nature. Beginning in September 2016, I lived in East Big Lake for a year to learn how these tumultuous times changed people's values, meanings, and practices of living on the land. I conducted oral history interviews in Khmer with men and women from each of the original families in the hamlet as well as the main local authorities. I surveyed all but one of the 26 families living in the hamlet about their family relations, history of land ownership, rice agriculture, and livelihood activities. Many of the details of the agro-environmental history in this essay have also come through conversations during participant observation of farming tasks, religious and spirit ceremonies, and daily activities within the hamlet. In my efforts to get at people's memory of the past, in my participant observations I heeded the insight of anthropologist Anne Guillou who argues that in Cambodia, "individual acts of communication that produce and transmit collective memory are mediated by places" (Guillou 2016). I therefore paid special attention to place and landscape to learn about the collective memories of agro-environmental change that inform this historical narrative.

Inevitably, individuals sometimes remember past events differently. After all, two people may have lived in the same time and place, but remember that period quite differently based upon their individual identity and experience. And in a place of war like East Big Lake, where neighbors once killed neighbors and many present-day authorities came to power under the Khmer Rouge, collective memories are fraught with not only anger and trauma that shape neighborly relations, but they are also colored by the politics of the nation (Baird and Le Billon 2012; Nguyen 2016). How people retell their stories—what is remembered and forgotten—is always affected by the interests of those in power, especially when the leaders of a country have ruled for so long. I can neither overlook nor sidestep these politics of memory. Instead I have read people's stories alongside each other, and seek to contextualize conflicting memories within local and national contexts. Ultimately, in retelling the story of East Big Lake, it is my aim to illustrate how landscape, memory, and place can teach us about local agro-environmental change, and the lessons such histories offer for the future.

On the eastern edge of Sok's hamlet, there are a group of 20 graves marked by large conical mounds of sand. In the middle towers a grave for Sok's grandfather, who immigrated from southern China in the 1920s. He came as part of a large wave of Chinese migrants to Cambodia during the boom years of the global economy. In that decade, more than 5000 Chinese arrived in Cambodia each year. Sok's grandfather came to try his luck growing pepper in Kampot Province, where Chinese traders from Hainan had settled in the eighteenth century. East Big Lake was ideal for growing pepper, because the crop grew well in the sandy soil. It was also just a day's walk south to the prominent Vietnamese seaport of Hatien, where pepper merchants shipped the spice to global markets (Willmott 1966).

After marrying a local Khmer woman, Sok's grandfather cleared land with axe and oxen and built a small home with wood from an old resin tree. When Sok's mother was born several years later, there were only four other families living in their hamlet. The houses were far apart from each other, separated by field and forest. The forest was not big like the deep forests in the Vine Mountains a day's walk to the west where tigers and elephants lived. Most of the local forests grew up in small groves around large termite mounds and home to local spirits. Sok's mother remembers being afraid of the forest beyond her village. At night she could hear the high-pitched whistle of dhole, a kind of wild dog, that would roam into the village to kill people's chickens and ducks. Some men dared to venture into the bigger forests near the mountains to capture quail with nets and dig traps for deer. Eventually the forests in East Big Lake began to diminish, however, as families had children and more land was put under the plow. Such was the pattern elsewhere in Cambodia during the early twentieth century, when rice farmland expanded fourfold and the population grew from two to five million people (Nesbitt 1997).

In the days when Sok's mother was still young, when she was not helping at home preparing the rice, she always enjoyed going with the other young people to catch fish in the lake just west of her home. She would go with the other young boys and girls, because it was more fun and almost impossible to catch fish alone. She remembers chasing the fish into her nets by pounding the water with a stick, or sometimes when the water was shallow, she would run around thrusting her *rut* trap over the slippery snakehead fish. Some of the older men used larger traps made out of bamboo and netting that looked like large hands rising out of the water. In those days, when people fished together, they could catch enough to fill two or even three large basins that could feed the family for weeks.

In the late 1940s, after France retook control of Cambodia from the Japanese at the end of World War II, Sok's mother and aunts recall a brief period when French soldiers roamed the countryside. They were stationed in the district town 6 km to the west of their hamlet, and would drive their cars along the highway during the day time in search of a revolutionary group known as the Khmer Issarak. She did not see any fighting herself, but she remembers the young Khmer men who sometimes slipped in and out of her village at night to attack the French. Those days are a faded memory now, because there have been too many other wars since then to remember those times clearly.

Sok was born in 1953, the year that King Norodom Sihanouk helped Cambodia gain independence from France. For most of the older generation in East Big Lake, the nearly two decades of Sihanouk's reign are recalled with nostalgia. They remember that era with a sense of lost connection to the animals and forests that were once more abundant. In those times, clouds of bats stretching for kilometers would fly from the Golden Boat Mountains in the north to come and eat the insects plaguing people's fields. Wherever these nightly creatures flew, the rice fields grew greener in their wake from their guano.

Photo 4.2 A farmer uses oxen to plow a raised rice field near his home

It was also a time when people still farmed naturally. There were few chemicals, so villagers made their fertilizer from cow manure that they would mix with ashes, animal bones, and the soil from the large termite mounds. Sok would help make the fertilizer in the afternoons, after he returned from studying at the local primary school. He was one of the fortunate few in that area who lived close enough to school to be able to attend. He remembers how much he enjoyed gathering with other children to watch the men race their cow carts around the fields after a day of hard work. Back then, no one had tractors like they do now. The diesel powered machines of today may be swift and powerful, but they are replacing the cattle that replenish the sandy rice fields with manure (see Photo 4.2). Today less and less rice is grown naturally, and so most villagers feel that it is fit only to be sold in the market rather than eaten at home.

Not everyone remembers the post-independence period fondly, however. For some, Sihanouk's reign in the 1960s was a time filled with poverty and inequality. In the north end of Sok's village lives an older woman named Pisal whose family has been in East Big Lake for generations. For Pisal, farming has always entailed bitter hardship. When she was growing up, her father had to supplement their vegetable garden with cash from work on nearby pepper plantations owned by Sino-Khmer businessmen. He never earned enough money, and once he was forced to sell off part of the family's land to pay back a loan that he had used to purchase a small motorcycle trailer to take people to market. Growing up was difficult for Pisal. She never went to school to learn to read and write, and had to work hard on and off the farm to help her parents raise her younger siblings.

Pisal was hardly alone. Rural poverty and inequality marred Sihanouk's reign throughout the 1960s and fueled the flames of civil war in the 1970s (Osborne 1994). Sihanouk had sought to remain neutral throughout the 1960s as war engulfed neighboring Vietnam and Laos. However, as war spilled over into Cambodia by the end of the decade, and the country's economic and security conditions collapsed, both the political left and right came to oppose Sihanouk. The first communist-led revolts against the government were led by thousands of tenant farmers in the far northwest of the country in 1967 and 1968. They rose up against landlords and government officials to protest against land inequalities and a draconian state rice buying scheme (Kiernan 1982). Then in 1970, the US-backed army general Lon Nol staged a coup d'état against Prince Sihanouk in an effort to fight the communists mobilizing a guerilla force in the remote reaches of the country. Many rural Cambodians, who had long felt exploited by the more affluent cities, joined the communist insurgency when Sihanouk himself allied with the Khmer Rouge after his fall from power. Led by Pol Pot, and other French-educated communists, the Khmer Rouge fought a bloody civil war against the Lon Nol government for 5 years from 1970 to 1975 (Chandler 1991).

When the war made its way to East Big Lake it came on the wings of America's B-52s. From 1969 to 1973, the United States bombarded eastern Cambodia to destroy the supply line of the North Vietnamese Army, and to dislodge the Khmer Rouge guerrillas from their mountain hideouts. Like the thunder and wind that shakes the coconut trees above people's homes at night, US bombs rained down upon Sok's family and their neighbors on several occasions, leaving behind cratered roads, destroyed homes, and dead loved ones.

One of Sok's neighbors had been a butcher during the war. He is a short and stocky man with a chiseled face that can still stop people in their tracks. He recalls the Lon Nol government soldiers coming by day, and the Khmer Rouge guerillas by night, to steal peoples' pigs and chickens and stored rice. He recounts how one day five heavily armed Lon Nol soldiers tried to steal from him, but he refused to give them his livestock, instead demanding to know how much they would pay him per kilogram for his pig. After his house was bombed, however, he fled to the woods and joined the Khmer Rouge. Not everyone was fierce like Sok's neighbor though. Most people chose instead to abandon their homes and fields, fleeing to Phnom Penh and other villages where there was less fighting. By the end of the 1970s, more than half of the land cultivated with rice in 1968 was lying fallow, home only to wild pigs and landmines (Nesbitt 1997).

I first heard Sok recount those hard years of civil war one evening as we led his cattle back from the fields. We could see pond herons circle upwards and fly towards his house. Their white underbellies were lit up red from the setting sun. They came to roost above his home in the lone tall resin tree left in the hamlet. When Sok saw the birds, he told me about the sarus cranes, whistling ducks, and cormorants that once lived here. But like humans who flee when their homes are destroyed, these birds too left and never came back. During the Pol Pot time, Sok told me, he was ordered to go to the marshes to collect baby egrets and carry them back to the communal kitchen to supplement the meager food stocks. People had always eaten the waterfowl in the nearby lakes. But that time was different, because they took them all.

Like a loadstone, older villagers in East Big Lake are drawn to the Pol Pot time when describing the environmental past. In April 1975, Pol Pot's forces gained control of the country and established the new state of Democratic Kampuchea. While people rarely use this official name anymore, the 3 years, 8 months, and 20 days under Khmer Rouge rule have been collectively burned into people's memories and the landscape.

Sok was amongst the many who worked night and day on the infrastructure projects that would launch Cambodia's "Super Great Leap Forward." Apart from enforcing a radical Maoist ideology that abolished nearly all social institutions at the foundation of Cambodian society, the Pol Pot regime sought to create a modern socialist society through the physical labor of transforming nature. At the heart of the regime's strategy was a technical plan to free Cambodian farmers from their dependence upon rain-fed agriculture. The government initiated a nationwide program to construct irrigation systems and modernize production in order to produce three tons of rice per hectare (Tyner and Will 2015). Sok and his siblings were forced to dig canals with pick axe and hoe. They also drained the big lake, chopped down the trees in the fields, leveled the dirt, and straightened the dikes to make fields no smaller than one hectare. In spite of record yields in some parts of the country, the regime strictly rationed food in the communal dining halls in order feed its soldiers and trade rice for Chinese weapons. With only rice porridge and water lily to eat, laboring as many as 20 hours per day without medical care, many of the people in Sok's village—including his younger sister—were worked to death.

Like other youth, Sok's sister had been forced to work in a mobile work group. She carried dirt and broke rocks to build the Koh Sla dam near the Elephant Mountains some 60 km northwest of East Big Lake. The dam was designed to provide a water source for the massive irrigation canals that would transport water to the dry fields in southern Kampot. The mobile youth groups slept outside at night underneath flimsy bamboo and forest leaf shelters located in a densely forested region infested with malarial mosquitos. In those conditions, Sok's sister, who had always been slight of build and prone to illness, soon contracted malaria and died. Her death, like millions of others, was the result of what James Tyner and Rachel Will (2015) call "letting die," in which violence is perpetrated not by a singular act, but is built into the physical and administrative structures that value some lives more than others. Far from an egalitarian society, the Pol Pot regime created a hierarchical system that granted special privileges and authority to the "old people" who had always lived in the countryside, ensuring that the "new people" from the cities and upper classes, like Sok and his sister, would bear the burden of the violent labor required to conquer nature. Sok still remembers seeing his sister on the other side of the dam where the girls worked just before she died: her hair had fallen out and she was as emaciated as a skeleton. It was a savage time, Sok says, when Khmer killed Khmer and there was no Dharma.

One of the tragic ironies of the Pol Pot regime is that while people starved to death, life teemed all around them. The canals, rice fields, and lakes were abundant

Photo 4.3 Two women sort tamarind pods picked in their village forest

with food, but those caught hiding food would swiftly have their hands tied behind their backs and marched out into the fields never to return.

It was this natural abundance that allowed villagers to survive after the Vietnamese army and a small group of former Khmer Rouge leaders liberated Cambodia at the beginning of 1979. In East Big Lake, the year's rice harvest was burned by the fleeing Khmer Rouge soldiers, and so villagers turned to the crabs, snails, frogs, and fish in the fields to feed themselves. Others went to the woods to trap rabbits, quail, and fruit bats (see Photo 4.3). Some food aid from the USSR was distributed in the village, mostly powdered milk and wheat, but it was not enough to feed everyone. And more often than not it was sold only to the rich and well-connected. That first year, in need of rice, Sok traveled with other families from his hamlet to the nearby province of Takeo. They brought with them two ox-carts of sea-salt that had been gathered in the salt flats near the ocean 15 km south of the village. They traded one basket of salt for each basket of rice, enough to supplement what they could forage to feed their families.

Once the villagers of East Big Lake found food, they also had to create shelter. People's houses had been destroyed or fallen into disrepair during the years of fighting and genocide. Making do with the building materials at hand, most families constructed small, single room houses out of sand mixed with hay and the dirt from termite mounds. They cut palm leaves to make low-angled roofs. Sok remembers how these were difficult homes in which to raise a family. Inside, the cooking fire filled the home with an acrid smoke. Parents lost sleep worrying about the safety of their babies as scorpions and centipedes took up residence with impunity. During the wet monsoons, the roof leaked and at night water puddled under people as they tossed and turned listening to the trees crash above their heads. In spite of this hardship, Sok fondly remembers re-establishing his life, marrying, and having children in those difficult years. When he and his wife first built their home, like their neighbors, they planted every kind of fruit tree for which he could find seeds—from mangos and coconuts to kaffir lime and milk fruit. With the Khmer Rouge gone, people planted the trees because they felt that they had a future in which to invest.

In the rainy season of 1979, with the country on the brink of famine, the new Cambodian government known as the People's Republic of Kampuchea had to turn its attention to agricultural production. The government that had ousted the Pol Pot regime was led by Vietnamese-backed communists, many of whom had defected from their posts as Khmer Rouge just a year or two before. In the first few years after overthrowing Pol Pot, there was almost no functioning state. The government lacked the resources and trained people needed to administer and rebuild the country. Moreover, much of the new government's resources remained tied up fighting with the remaining Khmer Rouge forces that had fled to Cambodia's western border (Gottesman 2003).

Faced with these challenges, and encouraged by their Vietnamese patrons, the new socialist government decided to confront the food problem in the country by collectivizing agricultural production. Starting in 1979 they organized people into solidarity groups of around 10–20 families that would farm state land together (Frings 1994). East Big Lake was divided into four such groups, with Sok's father taking command of one group. Every household was responsible for contributing to rice farming, with land, equipment, and harvest split equally within their solidarity group. After each of the harvests in 1980 and 1981, Sok's father oversaw the distribution of un-milled rice to members within the group. In contrast to many groups throughout the country, which distributed the harvest based upon a system of labor points (Diepart and Sem 2015), in Sok's group each person received one basket of rice equivalent to 12 kg until it was all gone.

Yet agricultural collectivization was doomed nearly from the start. Most villagers rejected the government's socialist policies that were so reminiscent of the genocidal Pol Pot regime. Aside from the village chief who was himself a young 30-year-old farmer, there were no state officials to assist in the day to day affairs of rebuilding East Big Lake village. And like in many places throughout the country during the 1980s and early 1990s, Khmer Rouge guerillas continued to wage war from their camps in the nearby mountain forests.

Like much of the country, moreover, people in East Big Lake lacked essentially everything needed to start farming again. Few agricultural tools remained from before. Most plows, harrows, and sickles had been lost or destroyed as the Khmer Rouge fled the area, and there was little seed left to replant. A few cattle had survived the war years, but it would be a long time before people could breed enough to meet the needs of the village. Even though the agricultural equipment and oxen were supposed to be split up equitably amongst the families in each solidarity group, the distribution of tools and animals tended to favor the families that had already been wealthy prior to the war years.

Confronted with these challenges of collectivization, in 1983 the local authorities in East Big Lake set about re-distributing land for families to farm on their own. Legally land was still owned by the state but this distribution of land effectively reestablished traditional forms of family-owned private property. Given the limited ability of the central government to oversee the process of dividing up land, considerable discretion was given to group and village leaders. Some villagers were able to take advantage of kinship and patronage ties to gain access to the fields with the best soil fertility and location. Many of the Sino-Khmer merchant families that had little interest in returning to the farm took less agricultural land in exchange for receiving residential plots next to the main road. Because families received land based upon the total number of people in their family, the youth who had not yet married only received one plot, which made it nearly impossible to raise a family on farming alone when they finally did marry and start to have children in later years. While the re-distribution of land was widely welcomed by villagers in East Big Lake, these small variations also sowed the seeds for land differentiation that would become more significant in determining land inequalities within the village in later years.

East Big Lake was not alone in changing the structure of collectivization. There was a general movement towards de facto privatization of land throughout the country, even as the government continued to tout the importance of the solidarity groups for building a socialist state. At the end of 1983, minister of planning Chea Soth presented a national progress report about the solidarity groups to the top leaders in the government. The report concluded that:

> We should educate and guide the people to understand deeply and clearly that only when there is organization to increase the harvest as a collective can we eradicate the shoots of oppression and make all the people equal, happily supporting each other in increasing the harvest and the livelihood of all the people. Then all families will eat to their fill and have happiness. (Slocomb 2003, 107)

What the new government leaders and their Vietnamese patrons did not fully appreciate is that while they were scrambling to figure out how to build a socialist state, the villagers in East Big Lake, as elsewhere, had already re-adopted a communal form of mutual aid that Cambodian farmers had practiced for centuries known as *brovas dai*.

Under *brovas dai*, villagers formed reciprocal work groups during the plowing, transplanting, and harvesting periods of rice production. Neighbors kept track of the labor hours and tasks that they owed each other; owners of oxen would plow neighbor's fields, who in turn would join together to transplant the ox owner's field and so on. Working in groups of 20–30 people, farmers would make short work of the relatively small plots owned by individual families before moving onto the next person's field. In this way, farmers balanced those peak moments in the rice production process when time was limited and labor needs were high (see Photo 4.4).

For many villagers, this traditional form of mutual aid exemplified an important relation with the natural world. On a late November afternoon, I was walking through the golden rice fields with Sok's wife Sothy, when the rice was beginning

Photo 4.4 Four neighbors thresh rice together as a form of mutual aid

to fall over from the weight of the heavy grain. In her late 50s, Sothy is a tiny woman who walks with a stiff gait from years of squatting to do household chores. I asked her an open question about what the environment used to be like in the village. She responded that previously people would do *brovas dai* to transplant and harvest the rice. I was confused by Sothy's answer because I was expecting something about the forest, fish, or fields—what I associate with the word environment. Instead she went on to explain how much she used to enjoy helping her neighbors to farm rice. They would laugh and tell jokes to each other, and sometimes they would break into song to make the work go quicker. She would spend all day outside near her fields during the harvest because she wanted to be ready to join a group at any time so that others would come to her field when it was ready. Sothy always enjoyed sitting in the shade, eating fruit and feeling the breeze on her back while she waited to join a work group. She smiled wistfully thinking of those years when villagers cultivated the land together.

Like most people in the village, Sothy does not remember a specific year when villagers quit doing *brovas dai* together. Even today some poorer families still engage in mutual aid. Yet the vast majority of people now hire workers or machines to farm their fields. Some people say that they began to hire workers after the first elections were organized by the United Nations Transitional Authority of Cambodia (UNTAC). First outlined in the 1991 Paris Peace Agreement, UNTAC was a multilateral effort designed to assist the Cambodian government hold national elections in 1993 and finally bring peace to a war-torn country. UNTAC employed 20,600 foreign personnel, including peacekeeping forces and advisors, and more than three times that number of local Khmer staff, making it the largest nation-building mission ever carried out by the UN at the time (Strangio 2014).

UNTAC transformed both the political landscape and the national economy. By the time that it was over, UNTAC had injected more than US$2 billion into the economy, significantly increasing foreign currency circulating in the country. Moreover, when Cambodia adopted a liberal democratic constitution and free-market economic reforms, many of which had been underway since the late 1980s, the country opened itself up to international trade, investment, and development aid. Between 1992 and 1995, donor countries like Japan, France, and the United States disbursed approximately US$1.4 billion to facilitate investment projects, improve technical capacity, and bolster the national budget (Slocomb 2010).

These changes in the country's governance and economy had a profound impact on life in East Big Lake. Simply put, money became more important in people's daily lives. That is not to say villagers had not used money before the 1990s. After the Pol Pot time, some villagers had managed to reclaim family valuables such as gold that they had buried during the war. Currency had also begun to circulate within the village once the government re-introduced the riel in 1980 and trade resumed along the nearby Vietnamese border. However, trade remained limited throughout most of the 1980s due to ongoing conflict with local Khmer Rouge guerillas and poor infrastructure. As such, there was little need for money in most people's daily affairs during the decade after 1979.

One morning while visiting a neighbor just south of Sok's home through the village forest, Sok and I stood and watched local carpenters renovate his neighbor's two-story wooden home. I listened to Sok chat with his neighbor, a friendly, retired wedding musician in his late 50s. They discussed the old days when people could call their neighbors for help if they wanted to build a home. Now people like Sok's neighbor borrow from the bank to hire construction workers. The two of them went on to list several other ways that money has changed how people in the village relate to each other. They said that the village used to have a number of communal ponds that people dug out and maintained together. No one paid to dig private wells like they do now (see Photo 4.5). Weddings and funerals were small, local affairs in which neighbors helped each other by cooking food together, in contrast to today's lavish events that involve hiring professional caterers and photographers and have hundreds of guests. There were few doctors, and so people had to treat illnesses with medicinal plants and traditional spirit doctors rather than paying for trips to local clinics. With limited infrastructure or trade, there were few goods to buy at the local stores along the highway that ran through the village to the east of Sok's hamlet. These days, even within the hamlet, people can purchase many of their daily goods.

Neither Sok nor his neighbor wanted to return to those days without money. The distance between the past and the present is often measured by the conveniences, festivities, and luxury goods that can be bought with money today. But they agreed that people in their hamlet used to help each other more, because they were all poor together.

For most families in the hamlet, money became more significant only when youth began to migrate to find jobs in the city and abroad. After 1993, there was a

Photo 4.5 A local family hires an excavator to dig a private pond for small-scale irrigation

large increase in foreign investment into labor intensive and export-oriented industries for garment, textiles, and shoe production (Slocomb 2010). These jobs attracted predominately younger people who had little interest in farming, or who did not have any village land to farm because it had all been claimed by the older generation. As young adults migrated out of the village, however, the family members who stayed behind had little choice but to hire people to help them transplant and harvest their rice fields. Families thus became more dependent upon remittances sent home from those members working in the cities. This trend has only increased in the last two decades with foreign investment, particularly from China, into the country's manufacturing and construction industries. Rural jobs in contrast have languished (Hughes and Un 2011). Of the 26 families in Sok's hamlet, more than 80% now receive remittances from family members working outside of the province.

Rising household debts have also contributed to the migration out of East Big Lake since the 1990s. Part of the development aid to Cambodia during the UNTAC period was channeled into small microfinance institutions in order to build a formal banking system in the Cambodian countryside (Norman 2011). The first microfinance institutions arrived in East Big Lake in 1996, when their sharply dressed staff showed up and began to offer villagers loans and training courses in animal husbandry. Grandfather Dee who lives just north of Sok was one of the first to borrow from these rural banks. Dee is in his late 60s and only recently returned from Thailand, where he had been working for the past 20 years. Dee had first borrowed money to expand his pig operation, and then later borrowed from a different organization to finance his children's education. However, when his mother became fatally ill and a thief stole his motorcycle at the same time, he felt compelled to sell his land to repay his debts. He joined many others from his village on a voyage to Thailand, where he found work near the Cambodian border picking fruit and farming shrimp. Eventually Dee brought his entire family with him to Thailand, where he could help look after the children while he and his wife worked. Although he has since returned to East Big Lake and purchased new land for farming, without anyone in his family to help him, Dee must hire workers from outside of the village for most of his agricultural labor needs.

Dee's experience of debt-driven migration has become common amongst families in the village, as foreign money has flooded into Cambodia's microfinance industry since the 2008 global economic recession. More than two dozen microfinance institutions and banks now service East Big Lake, providing families the credit needed to purchase everything from farming equipment to luxury timber for new homes. Yet these debts are rarely repaid from within the village. Rather, predominately wives and grandparents stay at home to raise children, and rely upon money from family members working in the cities and abroad.

While some indebted families like Dee's have sold their farmland in the last 25 years, other families have accumulated more land beyond what they could farm alone. Such patterns of land ownership have been made possible by both the rise of money within the village as well as changes in property law. Initially, the families that had received land from the state during redistribution in 1983 had enough land to support their needs. According to the village chief who has led East Big Lake since 1980, when the government redistributed land, it had sought to provide people with "a pot of rice" that with proper care could always sustain a family. As children were born, and parents sought to secure their future, however, some neighbors and relatives began to buy and sell land using the gold that they had hidden away during the Pol Pot time. These initial land transactions were either entirely informal or brokered by the village chief who provided only a hand written note that certified ownership. The government did not formally legalize private property ownership until 1989. Even after land privatization, throughout the 1990s, the government lacked the technical and financial means to develop a functioning land registry system (Guillou 2006).

For these reasons, during much of the 1980s and 1990s, land transactions were rare and small within East Big Lake. Then in 2001, the government enacted sweeping land reforms that set the legal and technical stage for systematic, national land titling and registration. East Big Lake villagers received formal recognition of their private property in 2004. With these new reforms, in tandem with the growing local economy, some villagers were able to expand their agricultural production beyond household needs. In particular, as the price for land along the national highway has skyrocketed in the last 10 years, the local land market has contributed greatly to rising inequality within the village. Many of the merchant families that live along the highway have greater livelihood security, and migrate less than the families living in Sok's small hamlet.

Walking around East Big Lake today, it is possible to read this shifting relationship between money, labor, and land on the landscape. Between Sok's hamlet in the west and the highway to the east stretches a complex quilt of rice fields in which a dual system of rice agriculture is practiced. On the one hand, families maintain small plots of land that they transplant by hand and that receive only natural fertilizer. These plots are often close to home and on land that does not flood easily. When the monsoon rains come in June, they are full of activity as older men and women join together in work groups to farm together. From these fields, families harvest traditional varieties of rice that they will store and consume over the year, if they can afford to. On the other hand, there are also larger, more uniform rice fields to the west of Sok's hamlet where the lake used to be as well as adjacent to the canal built during the Pol Pot time. Most of these larger fields are owned by just a handful of families in the hamlet. They now hire agricultural laborers and machines in order to practice high-input, broadcast rice agriculture. The families that broadcast rice in these larger fields sell their surplus to Vietnam where it is transported to global markets in Europe, China, and America.

As some farmers in East Big Lake transition to broadcast rice agriculture, traditional methods to manage pests and weeds are no longer effective. When rice is transplanted, the fields are plowed and harrowed twice to kill the weeds, and rice seedlings are able to out-grow whatever remains. Wading through a transplanted field to pull or cut grass is also much less tedious than in a field that has been broadcast. But as more farmers switch to broadcast rice farming, they increasingly rely upon herbicides and pesticides purchased along the main highway, where several merchants sell farm supplies and provide short-term loans. These chemicals, villagers lament, have begun to kill the animals in the field. Fish, frogs, and crabs are harder to find than they once were. And like these animals, people too have begun to suffer from the rise of chemicals in the fields and canals. Most of the older generation attribute the modern ailments of today, from stomach pain to cancer, to the loss of natural farming.

The sun has gone down as I accompany Sok from the fields back to his house where his grandchildren are busy playing marbles. Already Sok has taught them how to take the cattle out to the fields, and how to tell when the milk fruit is ripe (see Photo 4.6). But he does not want his grandchildren to be farmers. Farming is hard work, and these days the prices of crops only seem to go down, never up. The fact is that even if Sok's grandchildren wanted to be farmers, there is not enough land for all of the new generation in this village. And these days his children working in the city can earn in 2 months what he earns the whole year from farming. Sok is happy for the young generation—they have so many more opportunities for advancement than his generation ever did. It is a familiar sentiment that Sok and the other villagers have expressed often.

Nonetheless, as we listen to the soft clicking of bats chasing night time bugs, Sok expresses regret that this new generation has never seen or even known about the big

Photo 4.6 A grandfather teaches his grandson how to collect snails and crabs in their rice field

lake or the animals that once lived in the village. The crows and the deer are gone. The wild dogs that once lived in the mountain forests are silent. Like the rain that replenishes the ponds and the cattle that fertilize the fields, Sok hopes that people will take care of the land and water that remains.

References

Baird, I. G., & Le Billon, P. (2012). Landscapes of political memories: War legacies and land negotiations in Laos. *Political Geography, 31*, 290–300.

Chandler, D. P. (1991). *The tragedy of Cambodian history*. New Haven/London: Yale University Press.

Cronon, W. (1992). Kennecott journey: The paths out of town. In W. Cronon, G. Miles, & J. Gitlin (Eds.), *Under an open sky: Rethinking America's Western past* (pp. 28–51). New York/London: W.W. Norton and Company.

Diepart, J.-C., & Sem, T. (2015). *The Cambodian peasantry and the formalisation of land rights: Historical overview and current issues*. Gembloux: GRAESE: Groupe de Recherches Asie de l'Est et du Sud Est.

Frings, V. (1994). Cambodia after decollectivization. *Journal of Contemporary Asia, 24*, 49–66.

Gottesman, E. R. (2003). *Cambodia after the Khmer Rouge: Inside the politics of nation building*. New Haven: Yale University Press.

Gregg, S. M. (2010). Cultivating an agro-environmental history. In D. C. Sackman (Ed.), *A companion to American environmental history* (pp. 425–441). Chichester/Malden: Wiley-Blackwell.

Guillou, A. Y. (2006). The question of land in Cambodia: Perceptions, access, and use since de-collectivization. *Moussons Recherche en sciences humaines sur l'Asie Sud-Est, 10*, 299–324. https://doi.org/10.4000/moussons.2060.

Guillou, A.Y. (2016). Potent places as embodied memory in Cambodia. *Kyoto Review of Southeast Asia, 20*. https://kyotoreview.org/issue-20/embodied-memory-cambodia/

Hughes, C., & Un, K. (Eds.). (2011). *Cambodia's economic transition*. Copenhagen: Nias Press.

Kiernan, B. (1982). The Samlaut rebellion, 1967–1968. In B. Kiernan & C. Boua (Eds.), *Peasants and politics in Kampuchea, 1942–1981* (pp. 166–205). London/Armonk: Zed Press/M.E. Sharpe.

Nesbitt, H. J. (Ed.). (1997). *Rice ecosystems and varieties*. Phnom Penh: Cambodia-IRRI-Australia-Project.

Nguyen, V. T. (2016). *Nothing ever dies: Vietnam and the memory of war*. Cambridge, MA/London: Harvard University Press.

Norman, D. J. (2011). Neoliberal strategies of poverty reduction in Cambodia: The case of micro-finance. In C. Hughes & K. Un (Eds.), *Cambodia's economic transformation* (pp. 161–181). Copenhagen: Nias Press.

Osborne, M. E. (1994). *Sihanouk: Prince of light, prince of darkness*. Honolulu: University of Hawaii Press.

Slocomb, M. (2003). *The People's Republic of Kampuchea, 1979–1989: The revolution after Pol Pot*. Chiang Mai: Silkworm Books.

Slocomb, M. (2010). *An economic history of Cambodia in the twentieth century*. Singapore: NUS Press.

Strangio, S. (2014). *Hun Sen's Cambodia*. New Haven: Yale University Press.

Tyner, J. A., & Will, R. (2015). Nature and post-conflict violence: Water management under the Communist Party of Kampuchea, 1975–1979. *Transactions of the Institute of British Geographers, 40*, 362–374. https://doi.org/10.1111/tran.12080.

Willmott, W. E. (1966). History and sociology of the Chinese in Cambodia prior to the French protectorate. *Journal of Southeast Asian History, 7*, 15–38. https://doi.org/10.2307/20172832.

Chapter 5
A Brief History of Mangrove Distribution and Coastline Development in Soc Trang Province, Vietnam, to Address Coastal Management Strategies

Olivier M. Joffre and Klaus Schmitt

Abstract Coastlines and their mangrove forests change over time under the influence of human and natural drivers. To design appropriate mangrove reforestation interventions, we use the Vietnamese province of Soc Trang, at the Bassac River mouth, as a case study to understand coastal zone changes. Our research, covering 1904–2007, is based on historical material from the French colonial period (topographic maps, reports), satellite images, and onsite interviews with key informants.

Since 1904, the coastline and mangrove forests have changed significantly, including a sequence of deforestation and reforestation in some areas, changes in tree species composition, transformation of the coastline landscape from sand dunes to mangrove forests, and large-scale accretion at the river mouth. The natural processes of accretion and erosion have changed over time for the same area in Vin Chau District, thus influencing mangrove cover and reforestation programs. Damage to the mangrove forest during the Vietnam War due to defoliants was localised to specific areas along the coastline, and damaged trees were later cut for local use. Deforestation for fuelwood, expansion of farming areas, access rights, and usage of the mud flats during the French colonial period, followed by reforestation that modified the original species composition, are the main drivers of coastline changes.

These drivers influence the coastline and mangrove cover dynamics in different ways. The knowledge of historical processes and coastal dynamics is important in developing climate change adaptation strategies, which combine such site-specific

O. M. Joffre (✉)
WorldFish, Mekong Regional Office, Phnom Penh, Cambodia

Knowledge, Technology and Innovation Group, Wageningen University, Wageningen, The Netherlands
e-mail: o.joffre@cgiar.org

K. Schmitt
Deutsche Gesellschaft für Internationale Zusammenarbeit (GIZ) GmbH, Manilla, Philippines

M. A. Stewart, P. A. Coclanis (eds.), *Water and Power*, Advances in Global Change Research 64, https://doi.org/10.1007/978-3-319-90400-9_5

measures as effective mangrove protection and management, mangrove rehabilitation, and engineering measures.

Keywords Mangrove · Land cover change · Shrimp farming · Vietnam · Mekong Delta · Coastal erosion · Mangrove rehabilitation · Coastline change

5.1 Introduction

Intensification of human-induced climate change is expected to increase coastal hazards in tropical countries both in frequency and in magnitude, and sea level rise and climate-related events are likely to increase the number of people affected by floods in coastal areas (UNDP 2007).

In the coastal area of the Mekong Delta, rising sea levels and increases in the occurrence and severity of storm surges, flooding, and drought are some of the expected impacts of climate change. Sea levels could rise up to 1 m by 2100, rainfall during summer monsoon is predicted to be more intense, and together with intensification of tropical storms, floods will affect larger areas of the delta. About 35% of the population of Soc Trang Province (see Fig. 5.1) would be affected by inundation if the sea levels were to rise by 1 m today (Carew-Reid 2007); the most affected would be poor communities dependent on natural resources such as coastal fisheries.

One of the measures to adapt to climate change in coastal areas is to improve the management of the mangrove belt for increased resilience (McLeod and Salm 2006; Schmitt and Duke 2016). Mangroves maintain water quality, play a role in carbon dioxide sequestration and climate regulation, and provide nursery grounds, habitat, and food for many aquatic species, as well as habitat for numerous forms of animal life. Mangroves also provide fuelwood and construction material, as well as cultural services, and have potential for tourism development (Millennium Ecosystem Assessment 2005; Alongi 2014; Lee et al. 2014).

One key function for mangroves in climate change adaptation is their role as physical barriers against waves, floods, and storms. Mangroves contribute to stabilizing the coastline by trapping sediments and protecting beaches and coastlines against storm surges and waves (Barbier et al. 2008, 2011; Chatenoux and Peduzzi 2007; Danielsen et al. 2005; Das and Vincent 2009; Horstman et al. 2014; Kathiresan and Rajendran 2005; Mazda et al. 1997; McIvor et al. 2012a, b). Recent studies show that mangroves can act as a bioshield against tropical storms (Das and Vincent 2009; Chatenoux and Peduzzi 2007) by attenuating wave energy and stabilizing the substrate against erosion. Their effectiveness as bioshield has been critically discussed by several authors (for an overview, see Feagin et al. 2010); however, their ability to attenuate short-period wave energy and reduce erosion (e.g., Mazda et al. 1997) is not disputed and can also generate financial benefits. In northern Vietnam US$1.1 million invested in mangrove rehabilitation resulted in a savings of US$7.3 million annually from reduced costs in dyke maintenance (Brown et al. 2006). Hence, effective mangrove management is a key challenge for climate change adaptation measures in the coastal area of the Mekong Delta.

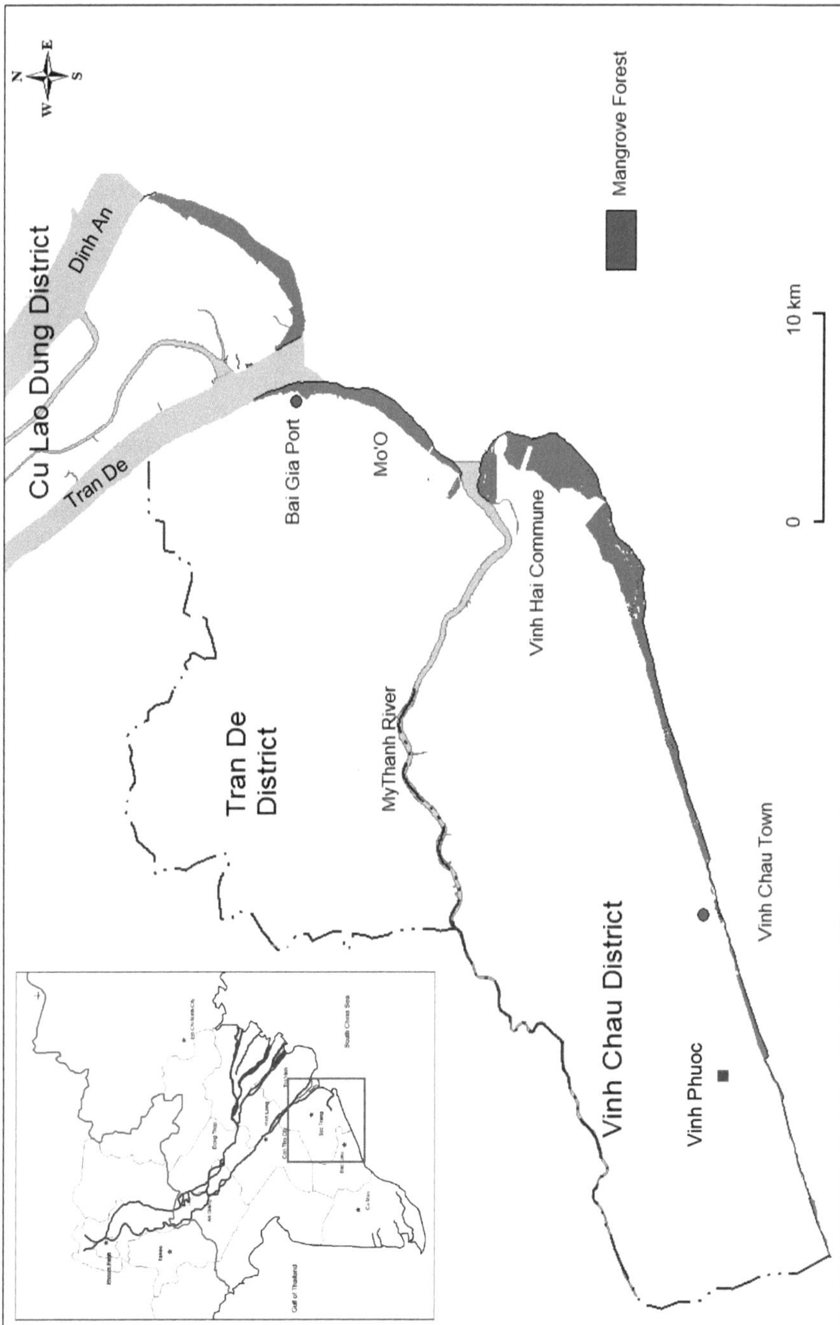

Fig. 5.1 Study area, coastal zone of Soc Trang Province, Vietnam

In Soc Trang Province the mangrove belt has suffered from anthropogenic activities such as defoliants sprayed during the American War (Vietnam War), exploitation for firewood, and conversion into agriculture, salt, and shrimp farms. Anthropogenic activities, combined with erosion, have resulted in a discontinuous mangrove belt along the coastline, ranging from absent to about 1200 m in width.

Locally, a common perception is that the mangrove belt was continuous all along the coast before the Vietnam War, protecting the coast from erosion, and that these mangroves were mainly destroyed by defoliants used during the American War. However, documented facts do not support this statement. Tran et al. (2004) show a map of herbicide spray missions over the Mekong Delta during the Vietnam War (1965–1971), which clearly indicates that herbicide spraying along the coast of Soc Trang Province was restricted to Cu Lao Dung Island and the eastern part of Vinh Chau District, near the My Thanh River mouth (a large-scale version of this map has been published by the International Crane Foundation). This is also supported by maps shown in Stellman et al. (2003).

The statement about the continuous mangrove belt also does not explain the fact that severe erosion is now destroying mangrove forests along parts of the coast of Vinh Chau District. Why did the mangroves withstand erosion before the use of defoliants (1965–1971) but not afterward? Why are forests planted in the 1990s being destroyed by erosion while forests in the same place before 1965 were allegedly not affected by erosion? Those questions are based on the hypothesis that the mangrove belt was continuous along the coast. To validate this hypothesis and answer those questions, we investigated the history of mangrove cover and coastline development from 1904 to 2007 to understand its coastal dynamics and changes and to differentiate natural and anthropogenic factors involved in those changes.

By using an historical approach in a field that usually applies hydrological modelling and GIS analysis, we aimed to illustrate how historical material can add to spatial analysis in general and how it can help reconnect with the studied local environment (Stewart 2011). We have combined historical maps with other sources such as interviews to understand not only how the coastline changed but also how its resources were used by local people, to gain a better sense of anthropogenic factors and provide a comprehensive picture which can be used to reach nonscientific audiences such as policy makers.

5.2 Study Site and Methodology

5.2.1 Study Site

The coastal zone of Soc Trang Province includes three districts, Vinh Chau,[1] Tran De, and Cu Lao Dung. It is 72 km in length and comprises more than 10,000 ha of mud flats, which are mainly located in the Cu Lao Dung and Vinh Chau Districts

[1] During the French colonial period (1862–1954), the administrative divisions were different: Vinh Chau District was part of Bac Lieu Province, which also included present-day Ca Mau Province.

(Fig. 5.1). Cu Lao Dung District is the largest island in the province, surrounded by the Tran De and Dinh An branches of the Bassac River.

Both Tran De and Cu Lao Dung Districts have mangrove forests composed of similar species (mainly *Sonneratia* spp.), whereas Vinh Chau District's mangrove forest is mainly composed of *Avicennia* and *Rhizophora* spp. Vinh Chau District can be divided into two main agro-ecological areas, with an accretion area in the eastern part, including a well-developed mangrove belt, mud flats, and sandbank, and a western part with erosion along the coastline. In this area, the mangrove forest is not well established, and the mud flats are less developed.

The flow regime of the Mekong River, the tidal regime of the South China Sea, and coastal longshore currents driven by monsoon winds create a dynamic process of accretion and erosion along the coast. From November to March strong winds and currents create erosion up to 30 m per year. In the rainy season, when the Mekong River carries a high sediment load, accretion occurs, expanding mud flats and sandbanks by up to 64 m per year (Joffre 2010; Pham 2011).

5.2.2 Historical Material and Field Research

The evolution of mangrove forests in Soc Trang Province was documented using historical material from 1904 to 1965, information for 1965–2007 provided in Pham (2011), and satellite images from 2006 to 2007. The historical material includes topographic maps and aerial photos, as well as documents related to the mangrove forest from Archives Nationaux d'Outre Mer (Aix en Provence, France), Bibliothèque Nationale de France, Départements des Cartes et Plans (Site Richelieu, Paris), and Services Historique de l'Armée de l'Air (Château de Vincennes, Paris); for aerial photos of 1953, Service Historique de l'Armée de Terre and Service Historique de la Marine (Château de Vincennes, Paris) and Institut Géographique National (Saint-Mandé, Paris); and for aerial photos of the study area in 1950, National Archives II in Ho Chi Minh City (Vietnam).

The maps produced during French administration of Vietnam used the Système de Référence Géodésique de Hanoï as the geographic coordinate system. It expresses coordinates in "grades" (100 grades are equivalent to 90°), and the zero-reference line for longitude is the Paris meridian, which is 2°20′13.95″ west of the Greenwich meridian. Therefore, a transformation of the coordinates used in the maps from 1904, 1933, and 1952 was necessary to enable comparison with the 1965 maps and 2006/2007 satellite images.

Semi-structured interviews were conducted with key informants in five locations along the coastal zone (in Vinh Chau District, east, west, and north; in Tran De District; and in Cu Lao Dung District) to corroborate the information from maps and aerial photos on mangrove forest and land use changes along the coast.

5.3 Results

5.3.1 Accretion and Erosion in Vinh Chau District

Significant changes have occurred along the coastline of Vinh Chau District from 1904 to 2007, with the degree of change varying by both place and time.

5.3.1.1 Western Vinh Chau

In the western part of the district, the original vegetation cover, composed of *Avicennia* spp., gradually disappeared until the 1950s (Fig. 5.2). The present-day forest cover of *Rhizophora apiculata* is largely the result of reforestation programmes, which started in 1993.

The 1904 map (Fig. 5.2a) shows the presence of mangroves between the sand dunes and the coastline. According to local villagers, the plant cover was not wide and was mainly composed of small *Avicennia* spp., *Thespesia populnea*, and herbaceous species such as *Acanthus* spp. The 1952 map (Fig. 5.2b) shows no forest cover along the entire Vinh Chau coastline.

This change in forest cover was due to cutting mangroves for local fuelwood consumption and clearing mangroves to use the resulting mud flats to trap fish and shrimp. The local administration had been renting out plots on the mud flats through auction since before the 1950s. These plots, usually rented by merchants from Vinh Chau Town, Bac Lieu Town, or better-off households from Vinh Phuoc Commune, were used to trap fish (mullets, goby) and shrimp (*Penaeus* and *Metapenaeus* spp.) at high tide when the mud flats were submerged (Fig. 5.3a). While this system was in place, the mud flats were not part of an open access system, and local villagers were allowed only to collect sesarmid crabs, mud crabs, and other low-value crustaceans.

A 2006 satellite image (Fig. 5.3b) shows that the mud flats were converted into shrimp and *Artemia* ponds following the development of a sea dyke in the 1990s. On the seaside of the dyke *Rhizophora* spp. were planted, but they have been partially destroyed due to strong erosion.

Comparison of the 1904 and 1952 maps (Fig. 5.2) shows that the western part of the district was an accretion site during this period. If we compare the distance from the Vinh Chau canal to the seashore in 1904 and 1952, there is an increase of approximately 1.7 km in 48 years (about 35 m per year). The accretion area started at the western side of Vinh Chau Town and continued up to Bac Lieu Province. The location of the coastline in 1952 is shown as a red dotted line on the 1904 map (Fig. 5.2a). By 1965 accretion at the same location appears to have stopped, with a distance of about 3.4 km between the Vinh Chau canal and the seashore.

a

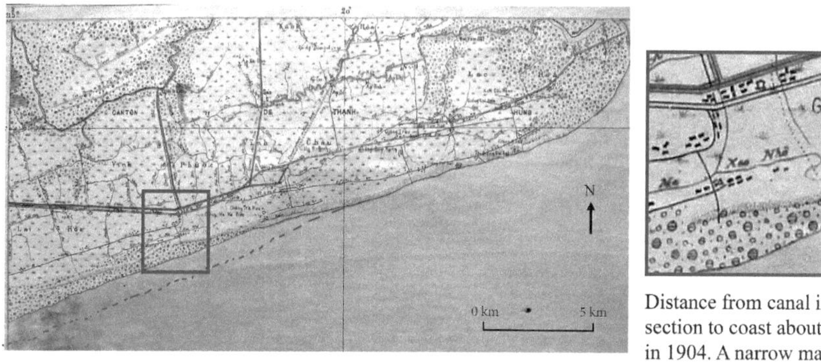

Location of the coastline in 1952 in red dots

Distance from canal intersection to coast about 2 km in 1904. A narrow mangrove belt is present

b

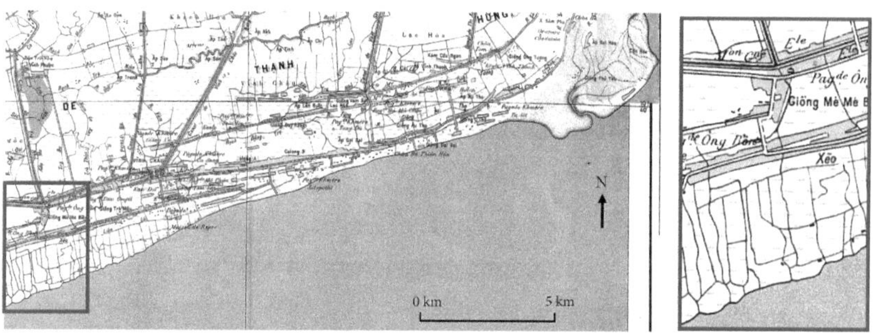

Distance from canal intersection to coast about 3.7 km in 1952. No mangrove forest left

Fig. 5.2 Maps of Vinh Chau District from 1904 (**a**) and 1952 (**b**) and details of the accretion area. (**a**) Map of 1904. Green dots on the coast represent mangrove according to the original legend. The location of the coastline in 1952 is represented by the red dotted line, showing the accretion in this part of the district between 1904 and 1952. Scale, 1:100,000. (Service Géographique; scale added by O. Joffre). (**b**) Map of 1952 (edited by Service Géographique d'Indochine in 1928 and updated in 1952)

5.3.1.2 Eastern Vinh Chau

In the eastern part of the district, no mangrove belt was shown from Vinh Chau Town to Vinh Hai Commune on the 1904, 1952, and 1965 maps (Figs. 5.2 and 5.4). The absence of mangroves in this part of the coastline was confirmed by villagers. The present-day *Avicennia* spp. forest is the result of reforestation programmes in the late 1990s.

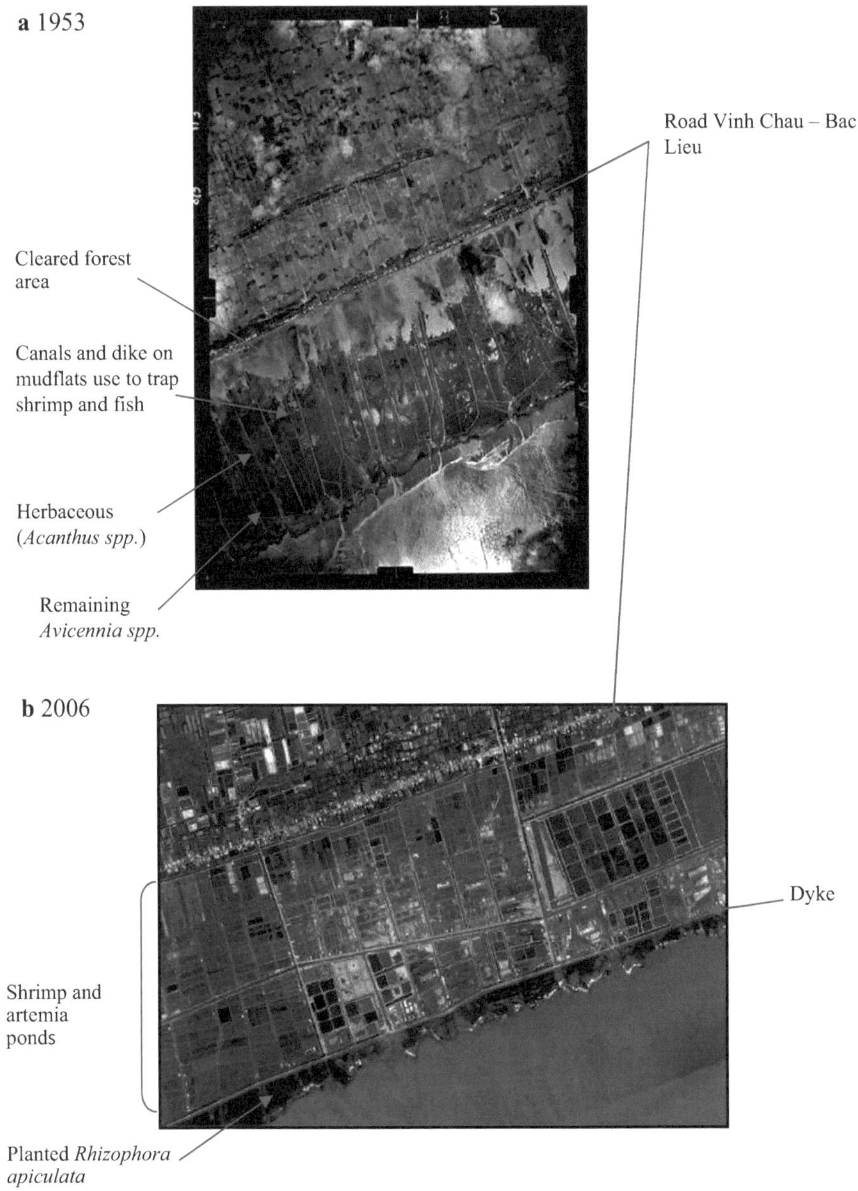

Fig. 5.3 West Vinh Chau District in 1953 (**a**) and 2006 (**b**). (**a**) Aerial photo, 1953. Altitude, 5200 m; scale, 1:42,000. (Institut Géographique National, mission 89, focale 125 mm). (**b**) Satellite image, 2006 (QuickBird image 04/12/2006)

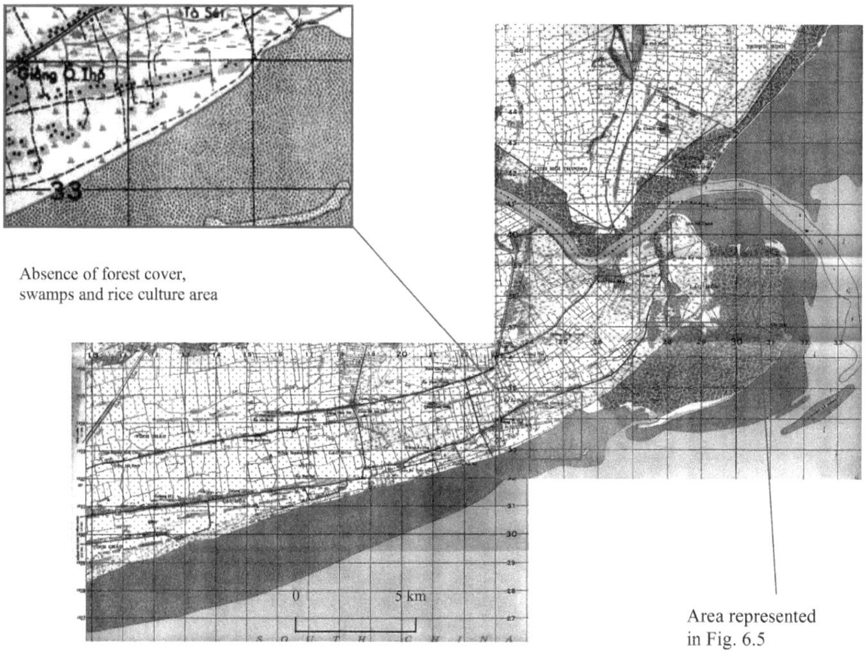

Fig. 5.4 Map of east Vinh Chau and Vinh Hai Commune, 1965. Mangroves are absent along the Vinh Chau coast (see Fig. 2 for 1904 and 1952) but present in Vinh Hai Commune. Scale, 1:50,000. (Prepared by the US Army in 1966, based on 1965 aerial photos; scale on the map added by O. Joffre)

Forest cover changes near the My Thanh River mouth in Vinh Hai Commune can be observed during the twentieth century. By 1952, some forested areas observable in 1904 maps had been converted into settlements and rice fields (Figs. 5.2 and 5.5a). The 1953 photo shows a sand dune covering the eastern side of Vinh Hai, alongside an area composed of forest, *Nypa* palms, and dense forest. At this time, the forest was mainly composed of *Sonneratia* and *Avicennia* species with scattered *Ceriops* spp., *Bruguiera* spp., *Excoecaria agallocha*, and *Lumnitzera racemosa*. By 1965, Vinh Hai forest had expanded to the southwest (compared to 1952), and settlements and swamps covered the northern part adjacent to the My Thanh River mouth (Figs. 5.4 and 5.5a). In 1965, the shape of the coast was still comparable to its appearance in 1952, and it was only later that accretion modified the coastline, leading to the disappearance of the sandbank surrounding the forest (Figs. 5.4 and 5.5b).

With the presence of permanent settlements of guerrillas during the Vietnam War, Vinh Hai forest suffered from the use of herbicides by the American Army between 1965 and 1971 (Tran 2006). After the war, the Vietnamese government started reforestation programmes, using *Rhizophora apiculata*, which explains the change in species composition. According to Tran et al. (2004), this species, with high commercial value, was chosen for economic considerations.

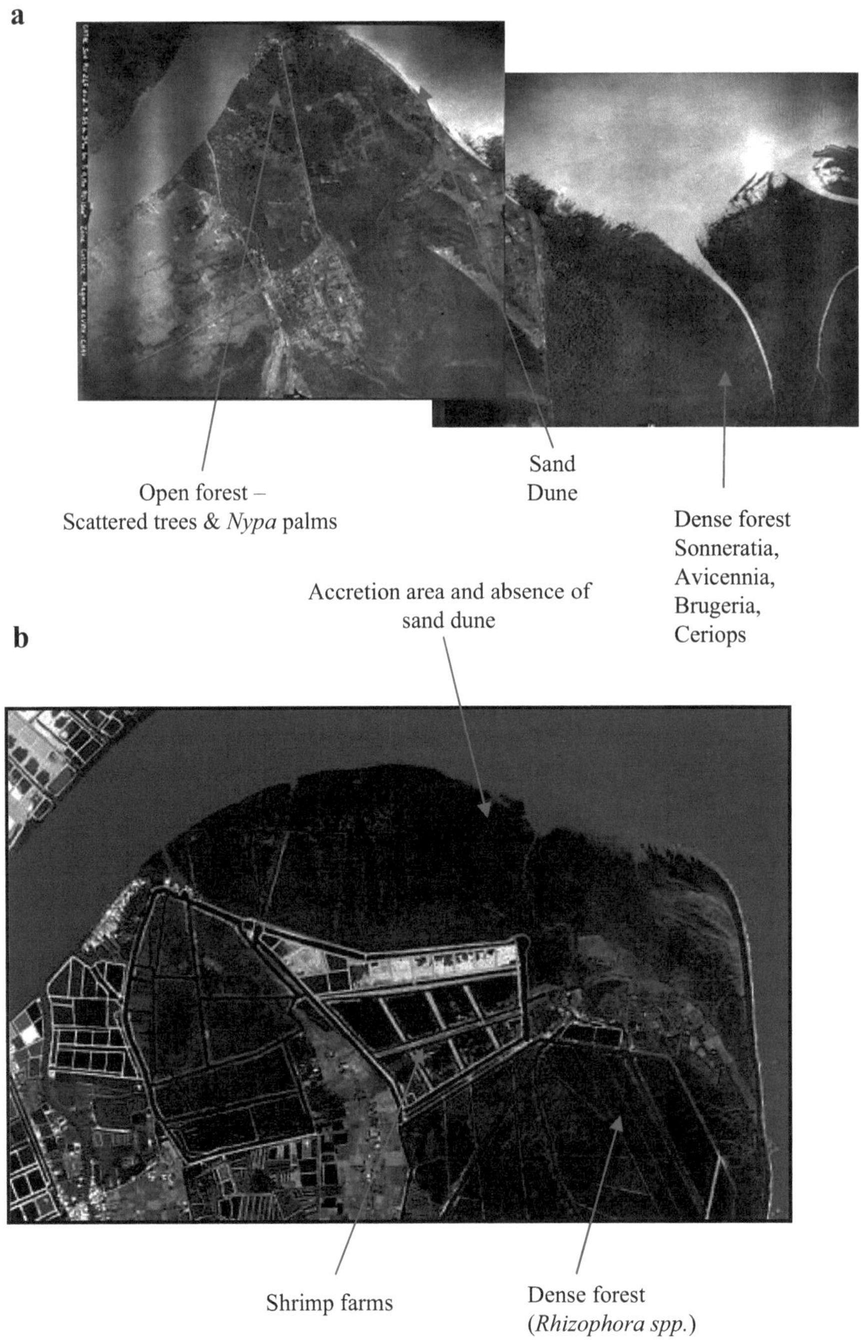

Fig. 5.5 North of Vinh Hai Commune, 1953 (**a**) and 2006 (**b**). (**a**) Aerial photo of Vinh Hai forest and My Thanh River mouth, 1953. Altitude, 1500 m; scale, 1:10,000. (Armée de l'Air Française mission RV 245, focale 5 mm). (**b**) Satellite image of Vinh Hai forest, 2006 (QuickBird satellite image 06/02/2006)

5.3.2 Changes in Cu Lao Dung and Tran De Districts

Comparison of 1904 and 1933 maps (Fig. 5.6) shows accretion sites in both districts. During this period Tran De District gained land toward the east, with an accretion both north and south of Bai Gia port. In 1904, only a strip of land 0.7 km wide (maximum width) and 3.7 km long (north to south) was located east of the longitude 115°40 E (106°11′50″ E). In 1933, the same strip of land was 7.8 km long (north to south) and 1 km at its maximum width. The difference in land area represents the accretion between the 1904 and 1933. In 1965, the same strip of land was estimated (based on the 1965 map) to be around 11 km long, with a maximum width of more than 2 km. Since 1965 the accretion rate has decreased, and in 2006 the strip of land east of 115°40 E was 2.9 km wide and 11.3 km long (including the mangrove belt). This represents an average accretion of 20.7 m per year from 1904 to 2006.

According to local people in the southern part of Tran De District, the forest of *Nypa* palm, *Sonneratia* spp., and *Excoecaria agallocha* suffered from exploitation by villagers mainly for fuelwood and palm roofing for home use. During the Vietnam War, inhabitants from Mo'O and the My Thanh River mouth area were relocated for security reasons ("Strategic hamlet" policy) to near Bai Gia port, which increased the need for fuelwood in this area. It was only in the late 1970s and 1980s that the government started reforestation with *Sonneratia*. The present-day mangrove belt in Tran De District is the result of these reforestation programmes.

A comparison of maps from 1904 to1933 shows that Cu Lao Dung and Cu Lao Tron Islands both increased in size toward the southwest (Fig. 5.6). The southernmost point of Cu Lao Dung Island was located 1.6 km north of the latitude 10°60 N (9°32′24″ N) in 1904, while in 1933 the island had already "crossed" this latitude. Cu Lao Tron Island had also crossed this latitude by 1933, whereas in 1904 it was located 3 km farther north. Further growth of Cu Lao Dung Island can clearly be seen in 1965 (Fig. 5.7b), with its edge estimated to be 2.1 km south of latitude 9°32′24″ N, where the edge of island is shown on the 1933 map. By 2006, the edge of Cu Lao Dung Island had almost reached latitude 9°29′24″ N, 4.96 km south of latitude 9°32′24″ N. From 1904 to 2006, the average accretion of the island was 64.3 m per year.

The shape of the islands also changed. In 1904 Cu Lao Dung Island had a width of 3.2 km, while in 1933 it was approximately 5.4 km at its maximum width. Aerial photos of Cu Lao Dung and Cu Lao Tron from 1953 show that the islands were clearly separated (Fig. 5.7). The 1965 map shows the emergence of new land covered with mangroves between Cu Lao Dung and Cu Lao Tron Islands. During that period the gap between the two islands narrowed due to continuous accretion. Subsequent to 1965, this newly created area (which was covered with mangroves) was converted into farmland.

On Cu Lao Dung Island, before 1953 the mangrove belt was continuous along the southern part of the island facing the South China Sea. This forest was composed mainly of *Avicennia* and *Sonneratia* species, while land was flooded as far as Rach Trang Village (Fig. 5.7). According to villagers, agriculture expansion started

a 1904

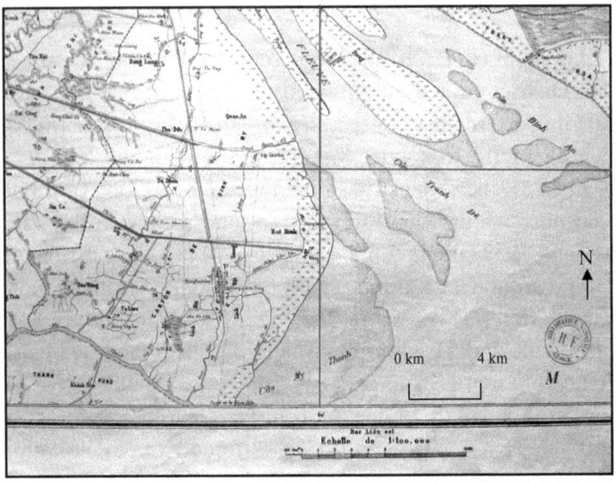

b 1933

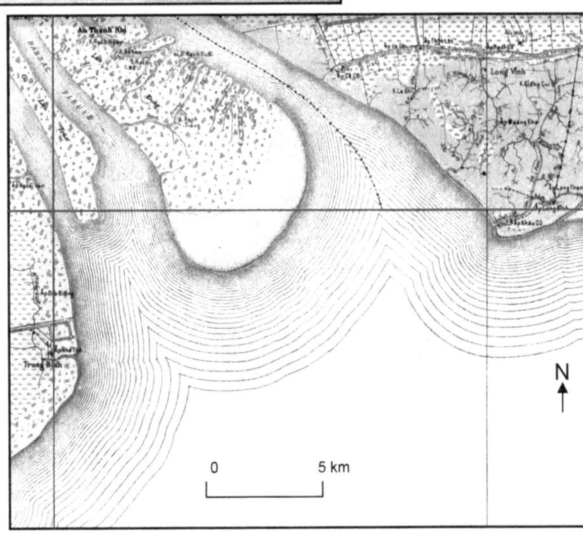

Fig. 5.6 Map of Tran De and Cu Lao Dung Districts, 1904 (**a**) and 1933 (**b**). Green dots on the coast represent mangrove according to the original legend. Red lines represent 115°40 E longitude and 10°60 N latitude (Hanoi system), or 106°11′50″ E and 9°32′24″ N (sexagesimal system). Scale, 1:100,000. (Service du Cadastre et de la Topographie de la Cochinchine; scale on the map added by O. Joffre)

from there, with rice fields and orchards around the village. Behind the mangrove belt along the coast the forest was more open, and in 1953 agriculture expansion had already encroached on this forest. After the 1960s and 1970s, deforestation started, with people from Tra Vinh Province (northeast of Soc Trang Province) specializing in forest clearing for agriculture, while inhabitants from Tran De used the forests mainly for fuelwood. During the Vietnam War the forests were affected by herbi-

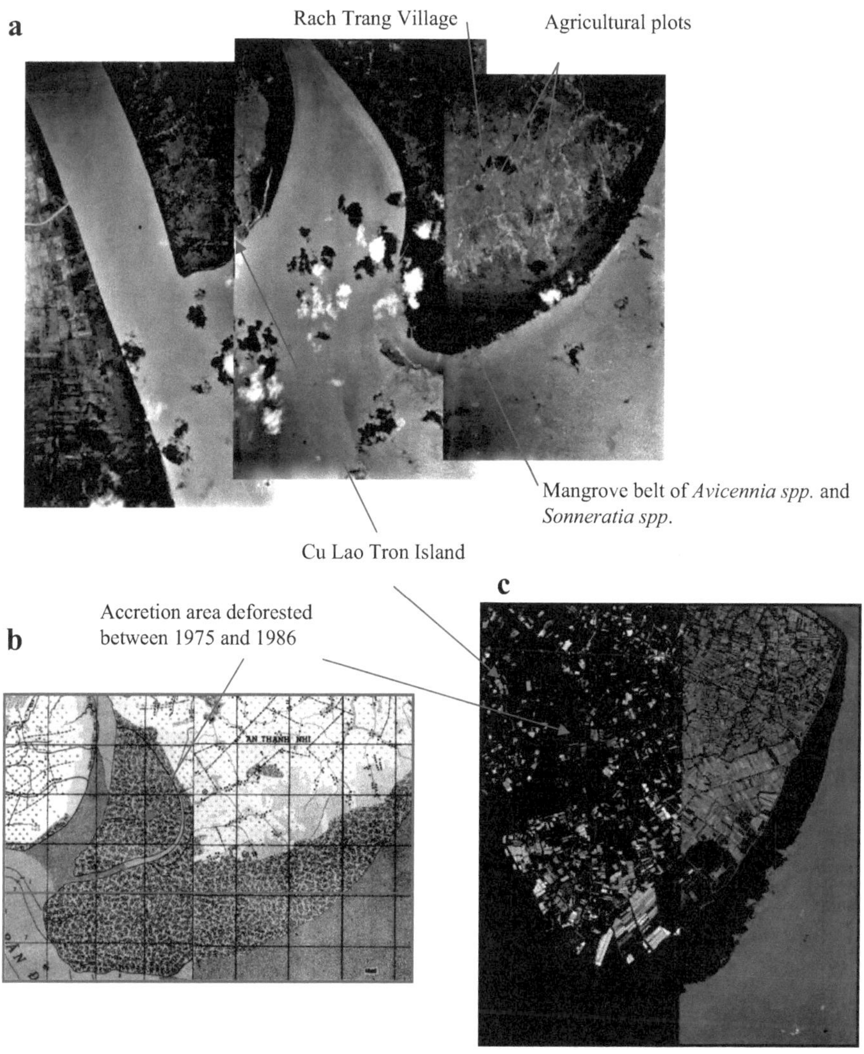

Fig. 5.7 Changes in Cu Lao Dung District, 1953 (**a**), 1965 (**b**), and 2006 (**c**). (**a**) Aerial photo of Cu Lao Dung and Cu Lao Tron Island, 1953. Altitude, 5200 m; scale, 1:42,000. (Institut Géographique National, mission 89, focale 125 mm). (**b**) Detail of 1965 map, representing the south part of Cu Lao Dung Island. Approximate location: 10°60 N latitude (Hanoi system) or 9°32′24″ N (sexagesimal system). Scale, 1:50,000. (US Army 1966; scale latitude 9°32′24″ N on the map added by O. Joffre). (**c**) Cu Lao Dung Island in 2006 (QuickBird satellite image 06/02/2006)

cides, and according to villagers, *Sonneratia* suffered most. Villagers emphasized that the main impact on the mangrove forest was not the herbicides sprayed during the war but intensive deforestation in the southwest part of the island for fuelwood and land reclamation. Deforestation continued at a high rate on Cu Lao Dung Island until the development of the State Forest Farm in 1986.

5.4 Discussion

A comparison of maps (1904–1965), aerial photos (1953), and satellite images (2006, 2007) shows rapid changes in the coastal area of Soc Trang. The mangrove belt in the province was not continuous until the 1990s, when planting programmes started. Before that, parts of Vinh Chau District had only sand dunes along its coastline or a thin belt of *Avicennia* spp. Therefore, the hypothesis stated in the introduction, that the mangrove belt was continuous along the coast, is rejected, and the related research questions can be answered: Why did the mangroves withstand erosion before the use of defoliants (1965–1971) but not afterward? Why are forests planted in the 1990s being destroyed by erosion while forests in the same place before 1965 were allegedly not affected by erosion? First, the thin mangrove belt in West Vin Chau was cleared by human activities in areas currently subject to erosion, and those areas never suffered from herbicide spraying. Second, our analysis shows that erosion and accretion patterns are changing during the twentieth century. Former accretion sites in west Vinh Chau District during the early twentieth century are now sites subject to strong erosion.

Along this coastline, mud flats were traditionally used by local people until 1975 to trap fish and shrimp in plots delimited by dykes and canals, which were rented annually from the local administration. Forest composition changed over the century, due to deforestation from herbicides used during the Vietnam War, and later from reforestation programmes initiated by the government. Reforestation in eastern Vinh Chau changed the original species composition, most likely for practical reasons: *Rhizophora* is easier to plant than *Avicennia* or *Sonneratia*.

In Tran De and Cu Lao Dung Districts, accretion occurred throughout 1904–2007. In west Vinh Chau District accretion occurred only up until the 1950s, while more recently the area has been subject to severe erosion. When comparing this analysis of historical material with the common perception that a continuous mangrove belt protected the coast until the American War (Vietnam War), it becomes obvious that coastline development and mangrove history are more complex and dynamic. Shoreline changes induced by erosion and accretion are natural processes, and this kind of change happened continuously during the twentieth century. It was not just the use of defoliants that had a significant impact on the mangroves of Soc Trang Province. Human pressure for firewood, exploitation of the mud flats, conversion of mangroves into agriculture and aquaculture, and reforestation programmes contributed to changes in mangrove cover, species composition, and the coastal landscape.

5.4.1 Accretion and Erosion Dynamics

Accretion and erosion patterns have continually changed during the twentieth century. Some former accretion areas are now subject to severe erosion, resulting in a narrow and patchy mangrove belt west of Vinh Chau Town (Fig. 5.1). These changing erosion patterns in turn modified the development of mangrove forests and determined the success or failure of reforestation programmes. Reforestation in west Vinh Chau has not been successful, while reforestation in east Vinh Chau, Tran De, and Cu Lao Dung has largely been successful.

Dynamics and changes in erosion/accretion need to be taken into account when planning mangrove reforestation and afforestation. In sites where the mud flats are severely eroded, planting mangroves will not be successful without additional engineering measures such as breakwaters to reduce wave energy and restore the eroded foreshore. In addition, in the specific case of Soc Trang Province, the influence of the sandbank in front of Cu Lao Dung Island on the longshore currents needs to be better understood. The sandbank's role in modifying current and erosion patterns needs to be modelled to predict future changes and to plan site-specific and appropriate erosion protection measures as part of an integrated approach to coastal area management. This will be an important contribution to developing climate change adaptation measures that consider changes in space and time, as well as improving coastline protection through mangroves.

The role of mangroves as a bioshield was already known by foresters during the French administration. Already in 1909, Ducamp explained the importance of mangrove forests for protection against tidal waves: "*Using mangrove trees, with a minimum cost, like a glaze of living matter capable of protecting dykes on the sea shore and the local population, probably not against water intrusion but against the massive hit of a tidal wave rushing onto an uncovered beach.*"[2] In 1918, Couffinhal described the root system of mangrove trees as a web or a tight net useful in protecting and fixing soils and sediments, with trees shielding river embankments and mangroves producing land while the forest progressed toward the sea.

5.4.2 Access Rights to the Mud Flats

In addition to knowledge of coastline changes and a better understanding of accretion and erosion patterns, the combination of historical materials and information from local people provides interesting insight into a very specific land use that affected the mangrove belt. Canals located on the mud flats of the west Vinh Chau coast can clearly be seen on aerial photos from 1953 (Fig. 5.3a). However, the reasons for their presence and their use were not known, and only information collected during interviews provided an answer.

Mangroves and mud flats were already exploited systematically for their fisheries resources before 1975. Usually considered common-pool resources, mangroves and mud flats were leased out by the French local administration in the western part of Vinh Chau District for trapping fish and crustaceans with tidal movement.

[2] "De faire par le palétuvier, à peu de frais, comme glacis de matière vivante et agissante capable de protéger les digues maritimes et leurs populations, non pas peut-être contre l'invasion des eaux, mais contre l'énorme coup de bélier qu'estun raz de marée se ruant sur une plage découverte." English translation by O. Joffre.

Villagers were employed by the tenant of the plot to dig canals and operate sluice gates to trap shrimp and fish. Under this system, locals were allowed to collect only certain types of resources for their own use, while the most valuable ones were collected for the tenant. This type of land-use practice, where wild juveniles are trapped in ponds in the salinity-affected areas in the Mekong Delta, is similar to the traditional rice-shrimp system (Xuan and Matsui 1998). This traditional system was established farther inland in the 1980s but has not been described for the coastal shore before.

Even if this practice and access regime disappeared after the Vietnam War, it is interesting to note that, before the boom of the shrimp industry in the region, specific areas of the mud flats and mangroves, which are now commonly considered as open access, were traditionally managed through restricted-access regimes. This finding shows that mud flats and mangroves were not always open-access resources before the development of the shrimp industry, as stated in the literature (Lutrell 2006; Barracoulgh and Finger-Stich 1996; Primavera 1997, 2006).

5.5 Conclusion

This study is based on collection and comparison of historical archive material with up-to-date satellite images and information from people with long-term knowledge of the area. No digital analysis was carried out, but simple map measurements and comparisons clearly show that in the western parts of Vinh Chau District accretion occurred between 1904 and 1953 and stopped sometime between 1953 and 1965 and that this area is now an erosion zone.

The study also proves that, contrary to common belief, mangroves were not present in certain parts of Vinh Chau District before 1965. The presence and absence of mangroves in Soc Trang over time has been influenced by both human actions and the natural process of erosion and accretion driven by sediments from the Mekong River, the tidal regime of the South China Sea, and coastal longshore currents.

Sediments from the Mekong River have led to accretion in front of Cu Lao Dung Island. Changes in sediment transport will modify the accretion/erosion dynamics. Recent studies on the sediment loads of the main branches of the Mekong River do not include the coast of the delta (Walling 2008; Kummu et al. 2008). However, a strategic environmental assessment of hydropower on the Mekong mainstream (ICEM 2010) predicts that the dams in the Mekong River catchment will destabilise the coastline of the Mekong Delta.

Historical maps contributed greatly to our understanding of dynamic processes along the coastline. The indication of vegetation (even if not detailed), sand dunes, and mud flats with the clear location of canals helps us draw a comprehensive picture of the history of the coastal zone in Soc Trang Province. However, historical maps alone cannot provide all answers. First, we needed to complement our understanding of the dynamic process by adding recent satellite images, to estimate accretion and erosion rates over a longer period and also understand the current dynamic in the early twenty-first century. We triangulated historical information from maps and aerial photos with results from interviews in the field. This source of information is important, as it gives detailed information missing on the maps or aerial photos (such as species composition) and provides essential information about anthropogenic factors influencing deforestation (fuelwood collection and mud flat exploitation). Interviews in the field with elderly people provided a clear understanding of different factors—herbicide spray and fuelwood collection—contributing to deforestation. Finally, additional information collected in archives provided a general context regarding forest and coastal zone management during the French colonial era and complemented our understanding of how the mangrove forest was perceived and managed.

This study shows the important contribution of historical maps for understanding coastal dynamics. It also shows that maps alone can show only spatial changes—to understand reasons for changes, other available materials (aerial photos, reports in archives) are needed to complete the picture, together with information from field interviews.

The importance of the understanding of mangrove history to improve its management has been highlighted by Saenger (2002, p. 229): "*Mangrove systems ... are changeable, they are dynamic, they are unpredictable, they are subject to aperiodic and periodic fluctuations of the extreme kind, and ... each mangrove community has a history. Reading that history from the tell-tale signs of today, is the artful skill of the silviculturalist or restoration ecologist who is likely to succeed.*" The knowledge of historical processes and coastal dynamics is an important input in the development of climate change adaptation strategies that use a combination of site-specific measures, including effective mangrove protection and management, mangrove rehabilitation, and engineering measures. Other inputs, such as site selection based on suitability for the growth of mangrove forests, appropriate species selection, numeric modelling that simulates hydrodynamics and shoreline development, and an assessment of the impact of upstream damming on the sediment flow of the Mekong River, must also be considered when developing a climate change adaptation strategy. All these must be used as part of an integrated approach to coastal area management to develop strategies that consider different options depending on site conditions and will not lead to path dependencies and the possible reduction of adaptive capacities (Smith et al. 2013).

Acknowledgment This study forms part of the Coastal Zone Management Project, Soc Trang, implemented by Deutsche Gesellschaft für Internationale Zusammenarbeit (GIZ) GmbH, financed by the German Federal Ministry for Economic Cooperation and Development (BMZ). This work was undertaken as part of the CGIAR Research Program on Fish Agrifood Systems (FISH).

References

Alongi, D. M. (2014). Carbon cycling and storage in mangrove forests. *Annual Review of Marine Science, 6*, 195–219.

Barbier, E. B., Koch, E. W., Silliman, B. R., Hacker, S. D., Wolanski, E., Primavera, J., Granek, E. F., & Polasky, A. S. (2008). Coastal ecosystem-based management with nonlinear ecological functions and values. *Science, 319*, 321–323.

Barbier, E. B., Sally, H. D., Kennedy, C., Koch, E. W., Stier, A. C., & Silliman, B. R. (2011). The value of estuarine and coastal ecosystem services. *Ecological Monographs, 81*(2), 169–193.

Barracoulgh, S., & Finger-Stich, A. (1996). *Some ecological and social implications of commercial shrimp farming in Asia, UNRSID Discussion Paper 74*. Switzerland: The United Nations Research Institute for Social Development and World Wide Fund for Nature.

Brown, O., Crawford, A., & Hammill, A. (2006). *Natural disasters and resource rights: Building resilience, rebuilding lives*. Manitoba: International Institute for Sustainable Development.

Carew-Reid, J. (2007). *Rapid assessment of the extent and impact of sea level rise in Viet Nam* (Climate Change Discussion Paper 1). Brisbane: International Centre for Environmental Management.

Chatenoux, B., & Peduzzi, P. (2007). Impacts from the 2004 Indian Ocean tsunami: Analyzing the potential protecting role of environmental features. *Natural Hazards, 40*, 289–304.

Couffinhal, J. F. (1918). *La Situation des Forets en Cochinchine*. Rapport du Chef du Service Forestier de la Cochinchine.

Danielsen, F., Sørensen, M. K., Olwig, M. F., Selvam, V., Parish, F., Burgess, N. D., Hiraishi, T., Karunagaran, V. K., Rasmussen, M. S., Hansen, L. B., Quarto, A., & Suryadiputra, N. (2005). The Asian tsunami: A protective role for coastal vegetation. *Science, 310*, 643.

Das, S., & Vincent, J. R. (2009). Mangroves protected villages and the reduced death toll during Indian super cyclone. *Proceedings of the National Academy of Sciences USA, 106*(18), 7357–7360.

Ducamp, R. (1909). L'arbre et l'eau en Indochine. *Revue Indochinoise, 11*, 18–24.

Feagin, R. A., Mukherjee, N., Shanker, K., et al. (2010). Shelter from the storm? Use and misuse of coastal vegetation bioshields for managing natural disasters. *Conservation Letters, 3*, 1–11. https://doi.org/10.1111/j.1755-263X.2009.00087.x.

Horstman, E. M., Dohmen-Janssen, C. M., Narra, P. M. F., van den Berg, N. J. F., Siemerink, M., & Hulscher, S. J. M. H. (2014). Wave attenuation in mangroves: A quantitative approach to field observations. *Coastal Engineering, 94*, 47–62.

ICEM. (2010). *MRC Strategic Environmental Assessment (SEA) of hydropower on the Mekong mainstream: Summary of the final report*. Hanoi.

Joffre, O. (2010). *Mangrove dynamics in Soc Trang Province 1889–1965*. Deutsche Gesellschaft für Internationale Zusammenarbeit (GIZ) GmbH, Management of Natural Resources in the Coastal Zone of Soc Trang Province, Vietnam. Available at: http://daln.gov.vn/r/files/ICMP-CCCEP//tai_lieu/Document/Soc-trang/2010/Mangrove-history-1889-1965-EN.pdf. Accessed Nov 2016.

Kathiresan, K., & Rajendran, N. (2005). Coastal mangrove forests mitigated tsunami. *Estuarine, Coastal and Shelf Science, 65*, 601–606.

Kummu, M., Penny, D., Sarkkula, J., & Koponen, J. (2008). Sediment: Curse or blessing for Tonle Sap? *Ambio, 37*(3), 158–163.

Lee, S. Y., Primavera, J. H., Dahdouh-Guebas, F., McKee, K., Bosire, J. O., Cannicci, S., Diele, K., Fromard, F., Koedam, N., Marchand, C., Mendelssohn, I., Mukherjee, N., Record, S. (2014). Ecological role and services of tropical mangrove ecosystems: A reassessment. *Global Ecology and Biogeography, 23*(7), 726–743.

Luttrell, C. (2006). Adapting to aquaculture in Vietnam: Securing livelihoods in a context of change in two coastal communities. In C. T. Hoanh, T. P. Tuong, J. W. Gowing, & B. Hardy (Eds.), *Environment and livelihood in tropical coastal zones* (pp. 17–29). Wallington: Cab International.

Mazda, Y., Magi, M., Kogo, M., & Hong, P. N. (1997). Mangroves as a coastal protection from waves in the Tong King Delta, Vietnam. *Mangroves and Salt Marshes, 1*, 127–135.
McIvor, A., Möller, I., Spencer, T., Spalding, M. (2012a). *Reduction of wind and swell waves by mangroves*. Report 1. Cambridge Coastal Research Unit Working Paper 40 (p. 27). The Nature Conservancy and Wetlands International. https://www.conservationgateway.org/ConservationPractices/Marine/crr/library/Documents/wind-and-swell-wave-reduction-by-mangroves.pdf. Accessed July 2017.
McIvor, A., Spencer, T., Möller, I., Spalding, M. (2012b). *Storm surge reduction by mangroves* (Natural Coastal Protection Series: Report 2. Cambridge Coastal Research Unit Working Paper 41, p. 35). The Nature Conservancy and Wetlands International. https://repository.tudelft.nl/islandora/object/uuid%3A79fe752e-ce52-4bf6-a45c-2847bead07ab. Accessed July 2017.
McLeod, E., & Salm, R. V. (2006). *Managing mangroves for resilience to climate change*. Gland: IUCN.
Millennium Ecosystem Assessment. (2005). *Ecosystems and human well-being: Wetlands and water synthesis*. Washington, DC: World Resources Institute.
Pham, T. T. (2011). *Mangroves of Soc Trang 1965–2007*. Deutsche Gesellschaft für Internationale Zusammenarbeit (GIZ) GmbH, Management of Natural Resources in the Coastal Zone of Soc Trang Province, Vietnam.
Primavera, J. H. (1997). Socio-economic impacts of shrimp culture. *Aquaculture Research, 28*, 815–827.
Primavera, J. H. (2006). Overcoming the impacts of aquaculture on the coastal zone. *Ocean and Coastal Management, 49*, 531–545.
Saenger, P. (2002). *Mangrove ecology, silviculture and conservation*. Dordrecht: Kluwer Academic Publishers.
Schmitt, K., & Duke, N. C. (2016). Mangrove management, asssessment and monitoring. In L. Pancel & M. Köhl (Eds.), *Tropical forestry handbook* (2nd ed., pp. 1725–1759). Berlin: Springer.
Smith, T. F., Thomsen, D. C., Gould, S., Schmitt, K., & Schlegel, B. (2013). Cumulative pressures on sustainable livelihoods: Coastal adaptation in the Mekong Delta. *Sustainability, 5*, 228–241.
Stellman, J. M., Stellman, S. D., Christian, R., Weber, T., & Tomasallo, C. (2003). The extent and patterns of usage of agent orange and other herbicides in Việt Nam. *Nature, 42*, 681–687.
Stewart, M. A. (2011). William Gerard de Brahm's 1757 map of South Carolina and Georgia. *Environmental History, 16*((July), 524–535. https://doi.org/10.1093/envhis/emr079.
Tran, T. (2006). *Les perturbations anthropiques contemporaines dans les mangroves du Sud Viet-Nam entre nature, civilizations et histories*. PhD thesis. Universite de Paris IV.
Tran, T., Barzen, J. A., Cong Kiet, L. E., & Moore, D. (2004). Wartime herbicide in the Mekong Delta and their implications for the post – war wetland conservation. In H. Furukawa, M. Nishibuchi, Y. Kono, & Y. Kaida (Eds.), *Ecological destruction health, and development* (pp. 198–211). Victoria: Kyoto University and Trans Pacific University Press.
UNDP. (2007). *Human development report 2007/2008. Fighting climate change: Human solidarity in a divided world*. Available http://hdr.undp.org/en/. Accessed Nov 2016.
Walling, D. E. (2008). The changing sediment load of the Mekong River. *Ambio, 37*(3), 150–157.
Xuan, V. T., & Matsui, S. (1998). *Development of farming systems in the Mekong Delta of Vietnam*. Ho Chi Minh City: HCMC Publishing House.

Part II
Farmer Livelihoods, Strategies for Sustainability, and Adaptation to Environmental Change

Chapter 6
Trade-offs Between Ecosystem Services and Opportunity Costs in Maintaining the Tonle Sap Lake Agro-ecosystem (Cambodia)

Malyne Neang, Philippe Méral, Olivier Aznar, and Christophe Déprés

Abstract The usefulness of the ecosystem services framework (ESF) to emphasize relationships between agriculture and ecosystems has received little attention, and studies applying ESF to understand links between ecosystem services and rice production systems are lacking. This chapter tries to fill this gap by combining the ecosystem services (ES) and dis-services (EDS) approach suggested in 2007 by Zhang, Ricketts, Kremen, Carney, and Swinton and with agrarian system analysis and diagnosis methodology to identify ES and EDS provided by rice production systems adopted by farmers on the agro-ecosystem of Tonle Sap Lake (TSL) floodplain. Our findings show that organic rice production systems do not perform well economically or ecologically in ES provisions. In contrast, rainy-season rice, floating rice in particular, has the best performance for ES provision. This study proposes three choices to reconcile economic and ecological performance: (1) to promote production systems with medium performance for ES but low opportunity cost, promote adoption of rainy-season rice, excluding floating rice, in combination with short-term rice; (2) for medium performance for ES and medium opportunity cost, promote adoption of rainy-season rice, including floating rice, in combination with short-term rice; and (3) for high performance for ES and high opportunity cost, promote adoption of floating rice alone.

Keywords Ecosystem services · Ecosystem disservices · Rice cropping system · Organic rice · Floating rice · Trade-off · Tonle Sap lake · Production cost · Opportunity cost

---

Malyne Neang (✉)
ECOLAND Research Center, Royal University of Agriculture, Phnom Penh, Cambodia
e-mail: nmalyne@rua.edu.kh

P. Méral
IRD, UMR GRED, Montpellier, France

O. Aznar · C. Déprés
Clermont Auvergne University, AgroParisTech, INRA, Irstea, VetAgro Sup, Territoires, Clermont-Ferrand, France

M. A. Stewart, P. A. Coclanis (eds.), *Water and Power*, Advances in Global Change Research 64, https://doi.org/10.1007/978-3-319-90400-9_6

6.1 Introduction

The Millennium Ecosystem Assessment (2005) has provided a new framework based on the ecosystem services (ES) concept in order to stress the need for ecosystem conservation. In tropical literature, this ecosystem services framework (ESF) has been used mainly to provide economic and ecological arguments for protected areas, mainly in forest ecosystems (e.g., regulation services through hydrological function or carbon sequestration).

However, as explained by Zhang et al. (2007), the usefulness of ESF to emphasize relationships between agriculture and ecosystems has received much less attention, except for specific value chains such as coffee and cocoa (Rapidel et al. 2011). Yet, in tropical developing countries, this issue is particularly relevant. Agriculture is the main form of land management in these countries, in which food security and food sovereignty are key matters for farmers and policy makers. Several recent studies have shown the importance of agro-ecosystems in terms of sustainable development in rural areas. Most of these publications discuss the links and feedback loops between ES and agricultural activities (Dale and Polasky 2007; Power 2010; Swinton et al. 2006, 2007; Zhang et al. 2007). Based on a case study of rice production on the floodplain of Tonle Sap Lake (TSL) in Cambodia, this chapter looks further into this issue.

Cambodia provides a good illustration of this topic for several reasons. This country, which in 2014 was ranked by the United Nations Development Programme in the medium human development category (137th among 187 countries), is mainly a poor and rural country. According to Mund (2010), about 80% of Cambodian people live in rural areas, and 85% of them are rice producers. Moreover, rural areas account for 90% of the poor. The main drivers of rural development are dedicated to the agricultural system of lowland rice production (Mund 2010). According to the World Bank (2013), the drivers of poverty reduction between 2004 and 2011 were increased rice production (23%) and rice prices (24%), far ahead of other factors: farm wages (16%), nonfarm business (19%), urban salaries (4%), and unexplained reasons (14%). In this context, the government tries to increase rice productivity in different ways, such as machinery and agricultural technology (new varieties, fertilizer, cultivation techniques). Whatever the policy promoted, the key factor in the adoption of any rice production system by these small farmers is the availability and controllability of water in the rice terrace agro-ecosystem of the TSL. This floodplain is the most suitable for rice production thanks to increased soil fertility through sedimentation and abundant water for this crop. But the agro-ecosystem also has increased risk of yield loss caused by floods. Rice production terraces also serve as flood control solutions through dikes built between rice fields (Dan et al. 2005; Ly et al. 2012; Masumoto et al. 2008; Someth et al. 2009; Tsubo et al. 2007). Thus, farmers face both positive and negative interactions with the TSL ecosystem. The implementation of rural development policies based on rice productivity in this critical ecosystem provides a good illustration of trade-offs between provisioning services and regulation services.

Little research has been conducted on this issue. Although literature related to the rice sector is abundant, and the functioning of the TSL is well known, studies applying ESF to understand links between ecosystem services and rice production systems are still lacking. The objective of this study is to try to fill this gap.

The next section describes the case study and the functioning of the TSL ecosystem. We present our methodology, based on the general structure of ecosystem services (ES) and dis-services (EDS) suggested by Zhang et al. (2007). To study interrelations between agricultural activities and the ecosystem, we apply agrarian system analysis and diagnosis. In the third section, we present our results, focusing on ES and EDS provided by the agro-ecosystem in the different rice cropping and production systems adopted by farmers. In doing so, we show the different trade-offs and opportunity costs for these systems. The last section presents a general discussion about the usefulness of ESF based on this work.

6.2 Materials and Methods

6.2.1 Study Site

The Tonle Sap Lake (TSL), in the Mekong River Basin, is the largest freshwater lake in Southeast Asia. In the rainy season (May to October), this large lake receives and stores the water flowing back from the Mekong River and its tributaries and expands until it covers up to 15,000 km^2. In contrast, in dry season (November to April), water flows into the Mekong River downstream from late October or early November, and the lake shrinks down to 2500 km^2 (Arias et al. 2012; Brooks et al. 2007; Varis and Keskinen 2006). This natural mechanism ensures the flow of the Mekong River, protects the agricultural land of the Mekong Delta in Vietnam from saltwater (Pham et al. 2008), and ensures water availability for dry-season and receding-rice irrigation in Cambodia and Vietnam (Dan et al. 2005).

This lake, the first Biosphere Reserve of Cambodia, is also classified as one of the world's most productive wetland ecosystems (UNESCO 2012; Varis et al. 2006). Research confirms a highly productive fishery in the TSL, ranging from 179,500 to 246,000 t/year (TKK et al. 2008) to 289,000–431,000 t/year (Van Zalinge et al. 2000). The lake is the fourth most productive captive fishery in the world, representing 16% of the Mekong River fish capture. It provides 60% of the protein intake of the entire Cambodian population, which consumes 20–60 kg of freshwater fish per capita per year (TKK et al. 2008; Van Zalinge et al. 2000).

The flood pulse creates vast areas of seasonal floodplain habitats for birds and fishes, as well as a rich plain for agriculture, which ensures local livelihood from rice production (see Fig. 6.1), fish and nonfish aquatic species, timber, and nontimber products (Lamberts 2006; MacAlister and Mahaxay 2006). Thus, this is the world's highest biodiversity and the most productive ecosystem for inland fish in Southeast Asia (Brooks et al. 2007; Yen et al. 2007). The whole ecosystem of the

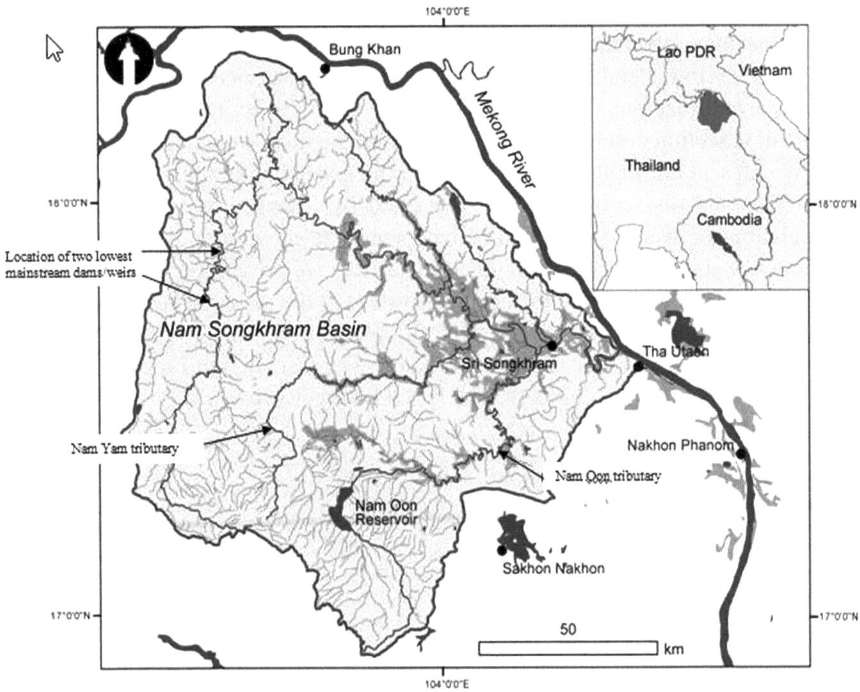

Fig. 6.1 Rice fields in Cambodia. (Source: Open Development Cambodia)

lake, floodplain, and riparian flooded forest and shrublands provides an ideal wetland habitat for Mekong fish species to feed, breed, and rear young (Kummu et al. 2006). Varis and Keskinen (2006) show that the TSL ecosystem also plays an important role of flood regulation by preventing and mitigating floods in the lower floodplains. This floodplain provides a large seasonal reproductive grassland habitat to two-thirds of the world's bird population, particularly the threatened Bengal florican (*Houbaropsis bengalensis*).

This entire ecosystem has also supported the livelihoods of local people for more than 1000 years, with grazing and traditional land use for wet-season rice growing and dry-season fallowing. Many places in this ecosystem are used for floating and flood-recession rice cultivation, which has low productivity. These paddies play an important role in regulating floods and fostering groundwater. Their dike systems use water harmoniously by storing it for irrigation and helping to reduce the risk of flooding for the local towns. The excess water is stored and discharged slowly into the lake and then down the Mekong (Masumoto et al. 2008; Pham et al. 2008). Every year, 1.6 million tons of sediment is stored in the lake and floodplain, making the soil naturally fertile with young alluvial deposits (Gray et al. 2007; Kummu and Sarkkula 2008), with a long-term sedimentation rate of 0.75 mm per year (Dan et al. 2005).

In summary, this ecosystem provides several essential services for local people, including supporting, provisioning, regulating, and cultural services (Millennium Ecosystem Assessment 2005, 2007):

- Supporting services
  - Soil formation and fertility (sedimentation, biomass from forest)
  - Nutrient cycling
  - Primary production
- Provisioning services
  - Fish and other aquatic species, including plants
  - Nontimber forest products (wild foods, honey)
  - Rice
  - Grass for grazing
  - Timber for firewood, house construction, and equipment for agriculture and fisheries
- Regulating services
  - Carbon sequestration by flooded forests
  - Regional and local water regulation
  - Natural habitat/biodiversity
  - Nursery
  - Waterways for transportation
- Cultural services
  - Ecotourism (floating villages, birds, Tonle Sap trips)
  - Cultural heritage (floating villages)
  - Sense of place in cultural practices (Water Festival)
  - Spiritual services (Arak Teuk, "Water Guardian")
  - Cambodian culture (adapted from Millennium Ecosystem Assessment 2005)

6.2.2 Methodology

Following Zhang et al. (2007), ecosystems and agriculture are embedded in a complex relationship based on positive and negative interrelations and feedback loops. Ecosystems provide not only supporting and regulating services but also disservices, such as pest damage and flood disasters (see Fig. 6.2). Thus, these marketed and nonmarketed services are the two main outputs of agro-ecosystems.

To analyze these different flows of services and disservices, we adopted the field methodology from agrarian system analysis and diagnosis (Cochet 2012; Cochet et al. 2007; Cochet and Devienne 2006; Dufumier 2006). We conducted the survey

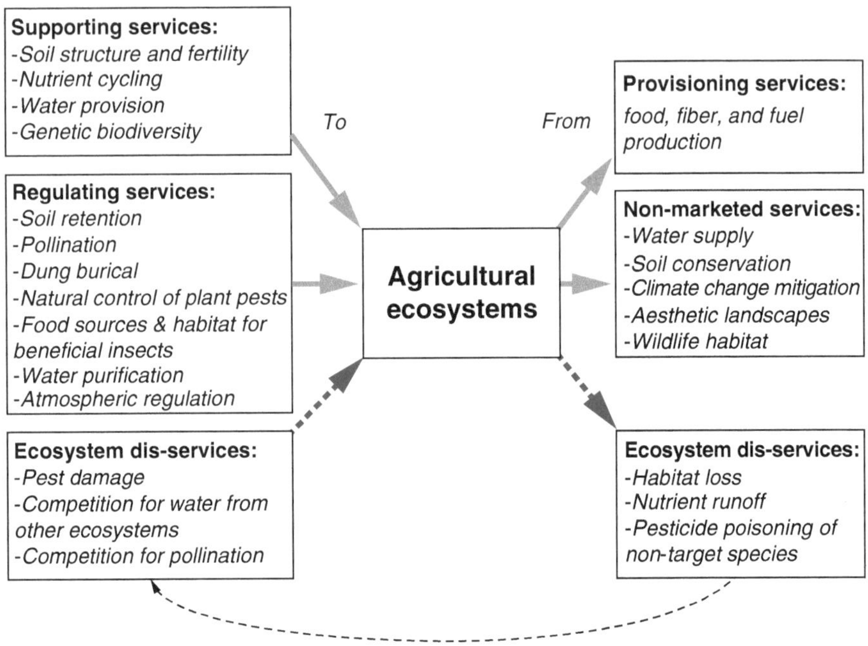

Fig. 6.2 Ecosystem service/disservice framework suggested by Zhang et al. (2007)

in three stages in order to understand farmers' choices under social, economic, and political conditions:

- *Landscape reading:* This allows us to understand the various zones in the agro-ecosystem. Started by observation the agro-ecosystem and vegetation, the question "why" guides us to meet the elder and local people for a better understanding of land use change in the study area.
- *Historical study:* The current agricultural situation is the result of long- or medium-term evolution. Historical study helps us identify the key factors of change, which create the actual agricultural practices.
- *Production system modeling and performance economic calculation:* This stage involves comparison of economic performance (value added and agricultural revenue per family) of production systems to clarify and explain why farmers practice different production systems in the same region (Neang et al. 2017).

We used the following economics calculation formulas (Neang et al. 2017):

$$GO_i\ /\ \text{ha} = Q_i\ /\ \text{ha} \times P_i,$$

where GO_i = gross output for each cropping system i, Q_i = rice yield (family consumption + sold production), and P_i = average selling price on the local market;

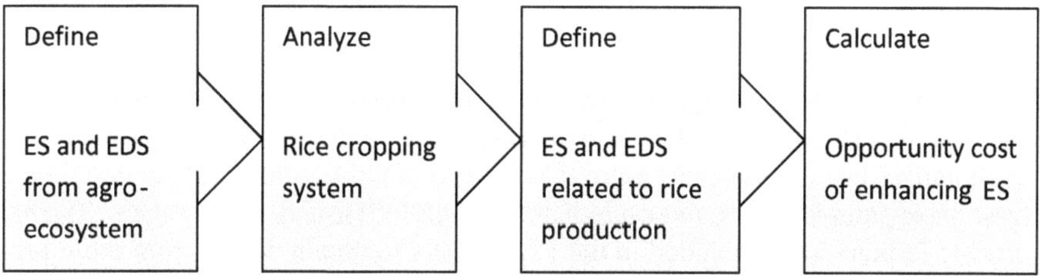

Fig. 6.3 Steps to identify ecosystem services (ES) and dis-services (EDS)

$$II_i \,/\, \text{ha} = \Sigma Q_{\text{inputs used/ha}} \times P_{\text{inputs}} + \Sigma Q_{\text{services used/ha}} \times P_{\text{services}},$$

where II_i = monetary value inputs, such as seeds, chemical inputs, and services used (plowing, transplanting, weeding, harvest, transport), during 1 year of production;

$$GVA_i \,/\, \text{ha} = GO_i \,/\, \text{ha} - II_i \,/\, \text{ha},$$

where GVA_i = gross value added; and

$$GR_j = \Sigma GVA_i \times \frac{S_i}{fl},$$

where GR_j = gross remuneration for the family's labor (*fl*) in the family's production system ($j = 1 - \infty$), and S_i = total surface area of the production system.

In summary, agrarian system analysis and diagnosis gather specific field data on agro-ecosystem management and agricultural practices. As illustrate in Fig. 6.3, our approach can be divided into four steps that involve defining ES and EDS related to the agro-ecosystem in general and to rice cropping in particular, and then using these data to calculate costs related to enhancing ES.

Combining agrarian systems analysis and diagnosis with the ES and EDS framework allows us link the economic performance of production systems that contribute to farmers' livelihood, on one hand, and the ecological efficiency of ES provision for sustainable agro-ecosystem use, on the other. Comparison of value-added and ES provided across different production system typologies shows the trade-offs among them. Our purpose is to identify production systems that are most effective and efficient, while being operational, productive, and feasible for farmers. In other words, we are looking for production systems that maintain ES with low opportunity costs.

We interviewed farmers living in two districts: Steung Sen (Srayove commune: Srayov Tbong, Roka, and Rolous villages) and Santuk (TbPhanhagy, Ompus, and Porkhav villages). We chose our samples randomly in these villages along the floodplain of the TSL (flooded grassland, flooded shrubland, and clear flooded forest).

Data collection has to be done with an understanding of the agrarian system and economic calculations in order to explain diverse situations and trajectories of a production system. For this study, 208 farmers and key informants were interviewed during 2010 and 2012. Qualitative data helped delimit the study zone and understand its history and change in agriculture. To obtain qualitative data, 36 individual interviews were carried out, as well as one group discussion of 12 elderly farmers. For quantitative data, 172 farmers were interviewed. Sample selection was based on a reasonable sample choice to ensure heterogeneity of farmers in the region. Twenty organic farmers were included in the 172 farmers to obtain details from these production systems. After data cleaning, only 167 farmers remained for analysis. We decided to choose 50% of them (87 farmers) for deep interviews on their thinking about the impact of agricultural inputs and agro-ecosystem change on their health, the rice field ecosystem, and the fishery sector.

6.3 Results

6.3.1 Ecosystem Services and Dis-services Provided by the TSL Ecosystem

This lacustrine active floodplain has brown or gray clayey or loamy topsoil classified as the Toul Somroung soil type by the Cambodian Agricultural Research and Development Institute, which has slow drainage, cracks into hard blocks when dry, and is well suited to irrigation. This soil is classified by Crocker (1962) in the brown, gray, or cultural hydromorphic soil units. It would be Luvisol or Vertisol using the FAO/UNESCO soil classification system (White et al. 1997, 2000; Shimizu et al. 2006). Based on recommendations of White et al. (1997), irrigation systems are needed to increase its potential for rice production. To maintain the field, this soil needs 62–100 kg N and 40–52 kg P_2O_5 per hectare.

Even though this soil is naturally fertile thanks to alluvial deposits, sedimentation does not reach the middle and upper terraces due to low flow speeds caused by vegetation (flooded forest, flooded shrubland, and flooded grassland) (Dan et al. 2005; Kummu et al. 2008). Without proper water management systems, rice yield is low because of poor soil, as well as floods and droughts (Fujisaka 1991; Mitchell et al. 2013; Nguyen et al. 2011). Local farmers call their agricultural situation "tveu sre rompeung mak," which means "producing rice by counting on the sky," This local saying illustrates their vulnerability to floods and droughts during the production season. Their harvest is hazardous.

Following Keske and Huon (2002), this agro-ecosystem landscape can be divided into three different zones based on different elevations that receive different effects of the flood pulse, each with different ES and EDS (see Figs. 6.4 and 6.5). We also give the local terms to refer to those zones, corresponding to the elevation measures by Keske and Huon (2002).

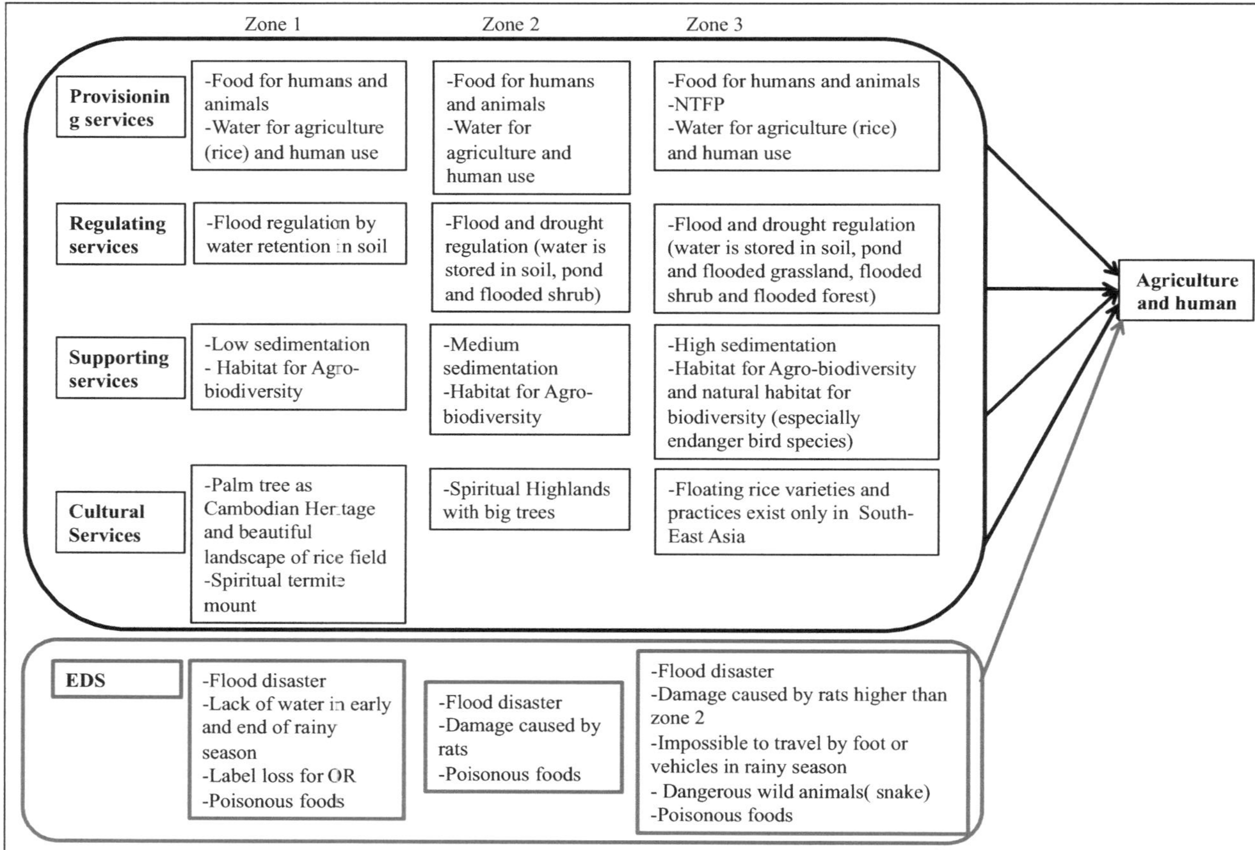

Fig. 6.4 Ecosystem services (ES) and dis-services (EDS) related to the three TSL floodplain agro-ecosystem zones. *NTFP* nontimber forest products, *OR* organic rice. (Adapted from Zhang et al. (2007))

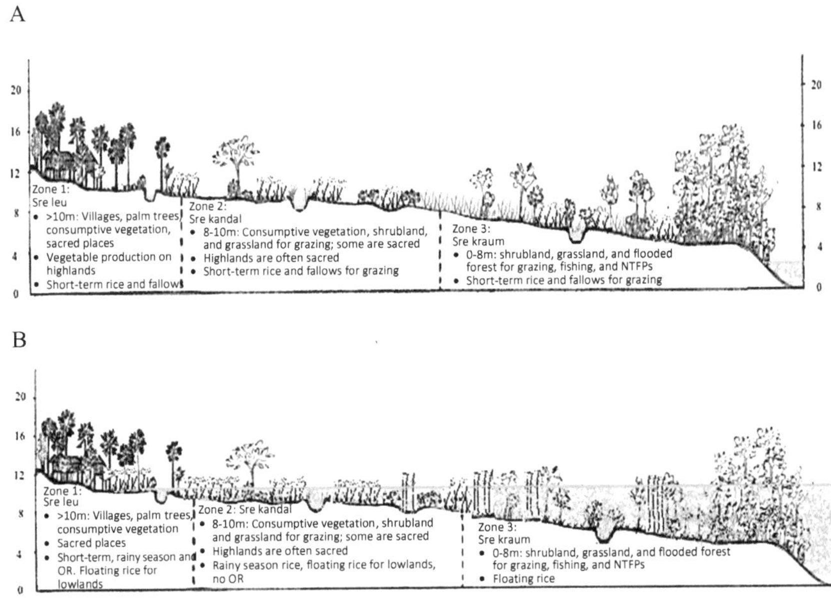

Fig. 6.5 Current land use in dry season and early rainy season (Dec–Jun; **a**) and middle of rainy season (July–Nov; **b**). *OR* organic rice, *NTFP* nontimber forest products

6.3.1.1 Zone 1: Sre Leu

"Sre leu" means "upper rice terraces" (approximately 10 m of elevation; Keske and Huon 2002). This agro-ecosystem, including the villages, often has ring dikes up to 30 cm high to retain water as long as possible after the rain. This zone is flooded last by the flood pulse, with 10–30 cm of water during August and September. The zone provides several provisioning services. First, water provision enables rice production in the rainy season (rainy-season rice and floating rice on lowlands, as well as short-term rice—only one cycle). Second, the floodplain absorbs water in the rainy season, which prevents floods in the villages. Third, the ecosystem is a source of food (fish and other aquatic animal and plant species) in the rice fields. Additionally, palm trees (*Borassus flabellifer*) and other trees on ring dikes complete the provisioning services (fruits, leaf, wood). The particularities of palm trees make them important in Khmer society by ensuring different ES. They provide habitat for bats, which eat insects and contribute the most fertile excrement to the soil, ensuring both regulating and supporting services. These trees also represent Khmer identity, ensuring cultural services, and provide sweet juice for producing sugar. Ring dikes also have a lot of termite nests that local people believe host guardian spirits. In terms of EDS, the risk of lack of water for rainy-season rice and floating rice in the beginning of the rainy season is a real problem for farmers. Conversely, flood pulses of the TSL can contaminate organic rice paddies, thus causing loss of organic labeling, and the high risk of floods decreases the potential to produce short-term rice from August to October.

6.3.1.2 Zone 2: Sre Kandal

"Sre kandal" means "middle rice terraces" (8–10 m of elevation; Keske and Huon 2002). The agro-ecosystem is characterized by rice fields with low ring dikes, around 10 cm high. There are fewer domestic perennial trees because of long and high inundation from the natural flood pulse. Farmers graze their animals in the dry season and fish in natural ponds and waterways surrounded by flooded shrublands. This zone is flooded before zone 1, with 15–40 cm of water from mid-July until the end of November. Water provision is the main ES. It enables rice production, except for organic rice due to flood contamination. Floating rice is more important in zone 2 than in zone 1. In contrast, zone 2 also provides different sources of aquatic food in rice fields and natural ponds or small rivers, and firewood from flooded shrublands. This zone, with its highly nutritious grass, provides ideal conditions for animal grazing, particularly buffaloes, which like the ponds in this zone. Some areas have highlands with big trees, which local people believe host the guardian spirits that protect them from natural hazards. They usually graze their animals on the highlands in the rainy season. In times of flood disaster, those highlands become safe places to keep their animals. On the other hand, rat hunting during the dry season and at the beginning and end of the rainy season is an important source of income for farmers, who sell the rats to Vietnam. In terms of EDS, the natural flood pulse of the TSL makes producing organic rice impossible, and each year farmers face a low risk of yield loss from rats for rice fields close to flooded shrub areas.

6.3.1.3 Zone 3: Sre Kraum

"Sre kraum" means "lower rice terraces" (0–8 m of elevation; Keske and Huon 2002). In this agro-ecosystem, rice fields are associated with clear flooded forest, which is called "prey kraum" or "roneam." This ecosystem is the richest zone, made up of flooded forest, shrubland, and grassland. It is permanently flooded in the rainy season, from June to the end of November, with more than 1.5 m of water. This ecosystem provides enormous ES to agriculture and farmers' livelihoods. Despite the high fertility of the soil, thanks to alluvial sedimentation, this zone faces a high flood risk, which has led farmers there to adopt floating rice. Nowadays most farmers have converted their floating rice fields to produce two cycles of short-term rice. This zone is also an important source of firewood, some strong wood useful for agricultural tools and for house construction, nontimber forest products (honey and medicinal plants), fishing for family consumption and sale, and nutritious grass for animal feed (grazing during the dry season and grass collecting during the rainy season). Farmers have traditional practices of grazing associated with fishing by organizing work sharing between farmers to graze their animals in zone 3 (two persons in charge of 15–30 animals for 10 days to 1 month at a time), which lies 20–40 km from their village. They also fish and collect nontimber forest products, which provide financial revenues for their family. Rat hunting is also an important occupation during the dry season and at the beginning and end of the rainy season, for sale to Vietnam. These rats, called "rice rats," are in the most demand because consumers believe that they are healthy and clean by eating only rice in the purified ecosystem. Regarding EDS, risks of floods and rat damage are still high for floating rice because it is grown during high water levels, so it is not possible to use rodenticide. Rats climb on trees and eat rice panicles.

6.3.2 Services and Dis-services Provided by Rice Cropping Systems

In our case study, we classify rice cropping systems into three main categories, short-term rice, rainy-season rice, and floating rice, comprising nine cropping systems based on varieties and practices farmers use in plots located in different flood pulse agro-ecosystems of the TSL (Neang et al. 2017). The relationships between rice technique practices and ES and EDS are detailed in Table 6.1.

6.3.2.1 Short-Term Rice

Short-term rice systems have been recently adopted (2000–2002) to better adapt to the frequent flood disasters during the rainy season. Short-term rice cropping systems are called dry-season rice by farmers. These nonseasonal and nonphotosensitive varieties enable farmers to produce outside of the flood period by using three different cropping calendars: (1) early-season rice, which they can start at the end of February and harvest in May or start in May and harvest in August, with a productivity around 4.9 t/ha; (2) receding rice, for which they wait until the water recedes, starting in December and harvesting in January, for 4.57 t/ha, a lower yield because draining flood water is difficult; and (3) combined early-season and receding rice in the same rice field, which enables farmers to double their productivity to almost 10 t/ha. Water management is important: early-season rice needs irrigation, while receding rice needs drainage most importantly.

All kinds of short-term rice require agrochemical inputs, such as fertilizer and pesticides. Our short-term rice respondents asserted that, based on inputs sellers' advice, they use chemical pesticides preventively and mix two or three pesticides together as a cocktail in case of pest attacks. In addition, herbicide use is becoming a common practice for short-term rice to reduce plowing. Farmers perceive that technical practices of these rice cropping systems could degrade the agro-ecosystem and agro-biodiversity, and may produce poisonous food for local farmers: 75% of respondents believed that fish and other aquatic species from their short-term rice fields are poisonous because of chemicals and don't consume them anymore for fear of chemical residues. Regrettably, the poorest farmers still continue to eat this food for lack of an alternative.

Table 6.1 Ecosystem services and dis-services by rice type

Rice cropping systems	Practices and land use	Ecosystem services (nonmarketed) from agro-ecosystem	Ecosystem dis-services from agro-ecosystem
Short-term rice Early-scason rice Zone 1: 21% Zone 2: 34% Zone 3: 45% Receding rice Zone 1: 17% Zone 2: 36%	Using the existent rice field in zone 1	Cultural services: preserve spiritual practices and beauty of agricultural landscape, such as rice fields with palm trees Provisioning services: high-yield rice; leaves, trunks, fruit, and juice from palm trees for farmers' basic needs	
Zone 3: 47% Double cycle of early + receding rice	New "high-yield" variety		Reduce the genetic resources in daily food consumption
Zone 1: 24% Zone 2: 40% Zone 3: 36%	Deforestation of flooded clear forest, shrubland, and grassland in zone 3 and high use of pesticide		Degrade soil and water quality, habitat, biodiversity, and flood regulation capacity of this ecosystem
Rainy-season rice Medium-term rice: direct seeded or transplanted	Dependence on water regime from flood pulses of TSL	Regulating services: respect water regime and alluvial deposit	
Zone 1: 90% Zone 2: 10% Long-term rice: direct seeded or transplanted Zone 1: 38% Zone 2: 62%	Maintenance of existing highlands, spiritual places, and palm trees. Palm trees are replanted every year in zone 1	Cultural services: preserve spiritual practices and beauty of agricultural landscape, such as rice fields with palm trees Provisioning services: rice; leaves, trunks, fruit, and juice from palm trees for farmers' basic needs and sale	
	Absence of chemical use or small amounts of fertilizer and pesticides used if needed	Regulating services: preserve agro-biodiversity—fauna, flora, and amphibians of rice fields—and water quality	
	Use of hybrid medium-term fragrant rice varieties to adapt better climate irregultation and market		Degrade natural varieties in genetic bank

(continued)

Table 6.1 (continued)

Rice cropping systems	Practices and land use	Ecosystem services (nonmarketed) from agro-ecosystem	Ecosystem dis-services from agro-ecosystem
Organic rice	Use of only existing rice fields, thus absence of new deforestation of flooded forest, shrub, or grassland	Regulating services: preserve indirectly flooded clear forest for habitat and biodiversity	
Zone 1: 100%	Highland and spiritual places, palm trees are kept. Palm trees are replanted every year in zone 1	Cultural services: preserve spiritual practices and beauty of agricultural landscape (rice fields with palm trees)	
		Provisioning services: organic rice; leaf, trunk, fruits, and juice for farmers' basic needs and sale	
	Restrain from use of chemicals	Regulating services: preserve agro-biodiversity (fauna, flora, and amphibians of rice fields) and water quality	
	Use new varieties fragrant rice		Reduce the genetic resources in daily food consumption
	Rice fields are protected from floods by ring dikes around 40 cm high to avoid chemical contamination for preserving organic label		Degrade regulating services: soil formation from deposit[a]
Floating rice	Use of only existing rice fields with many trees on it (flooded clear forest in zone 3)	Conserve directly flooded clear forest	
Zone 2: 64%		Regulating service: flood regulation, habitat and biodiversity	
Zone 3: 36%		Provisioning services: firewood, nontimber forest products, and inland fish	
	Dependence on water regime from flood pulses of TSL	Respect water regime and alluvial deposit	
		Regulating services: soil formation from deposit[a]	

(continued)

Table 6.1 (continued)

Rice cropping systems	Practices and land use	Ecosystem services (nonmarketed) from agro-ecosystem	Ecosystem dis-services from agro-ecosystem
	Existed highlands for spiritual places and palm tree are kept in Zone 1	Cultural services: preserve indirectly spiritual places and beauty of agricultural landscape (rice fields with palm trees)	
		Provisioning services: leaves, trunk, fruits, and juice for farmers' basic need and sale	
		Provisioning services: material and food from flooded forest	
	Absence of chemical use or use of small amount of fertilizer and pesticide if needed	Preserve fauna, flora, and amphibians of rice fields; no chemical residue leaching into water	
		Regulating service: soil biodiversity and water quality	
	Use natural and local varieties	Preserve natural varieties for genetic bank and daily consumption	
		Cultural Services: Preserve the identity of lowland flooded agro-ecosystem of floating rice	

[a]Soil deposit (sedimentation) on floodplain is very low because of the low speed of water flow from flood pulse (Dan et al. 2005; Kummu et al. 2008)

6.3.2.2 Rainy-Season Rice

These cropping systems are farmers' traditional practices, with seeds selected naturally and locally by them and their ancestors. Medium-term rice, with 120–150 days to maturity, starts in May. It can be transplanted if water is too high or directly seeded if they start early enough, if water is low enough. They often grow fragrant rice, which farmers use to produce "ambok" (rice grilled and flattened by crushing) to sell in Phnom Penh at the national Water Festival in November. This rice has a medium duration of maturity, and its flowering time is between 10 and 15 October (CARDI 2007). This rice is mostly cultivated in zone 1 (90%) because it cannot survive deep and long floods. Sometimes, this rice can also be found in zone 2 where land is not flooded and unsuitable for long-term rice.

Directly seeded long-term rice is cultivated predominantly in zone 1 and in some highlands in zone 2, where there is less water in the early rainy season, enabling farmers to sow on muddy land. In contrast, transplanting is practiced in zone 2 and some lowlands in zone 1 where the water is 20–30 cm high. Because labor has become rare in the region, farmers prefer direct seeding. According to farmers, long-term rice can survive very well in floodwater until 60–70 cm. These varieties have a 6-month life cycle, starting in April and ending in December.

For rainy-season rice, farmers who have money will use pesticides against crabs and rats, and some low amounts of fertilizer (50–100 kg/ha diammonium phosphate). On average they get from 1.3 t/ha for direct seeding to 2.2 t/ha for transplanting.

Organic rice is cultivated only in zone 1, particularly on highlands, to avoid floods even from the natural flood pulse. It has the same life cycle as other medium-term rice systems. This rice is transplanted only one stem at a time, because farmers have received training from some nongovernmental organizations about SRI (System of Rice Intensification) to improve their productivity by increasing organic fertilizer use (Ly et al. 2012). Unfortunately, it was adopted by only a small number of farmers because in that region it is very hard to manage water in order to transplant in muddy soil with single young stems, as well as the lack of organic matter for compost. The organic label came later, in 2003, to improve farmers' practices and increase their revenues. Among farmers in Raksmey Steung Sen Association, 55% produce organic rice on 100% of their land, with an average of 0.77 ha per household. Because of the ecological risk of floods, others use only the suitable part of their land for organic rice and continue to produce floating rice associated with long-term or medium-term rice on the rest of their land. Farmers said organic rice needs 2–3 t of compost per hectare, but they can find only one to two oxcarts of compost (around 35 kg/oxcart) per year. This is the main constraint and factor limiting their yield to 2.2 t/ha on average.

6.3.2.3 Floating Rice

Floating rice is normally a cropping system for rainy-season rice, but here we address it separately because it can grow very tall in flood conditions and thus offers good protection against the risk of flooding. Since 2002, many floating rice fields in zone 3 were converted to short-term rice; thus, only 36% of floating rice fields are found in zone 3, compared to 64% in the lowlands of zone 2. In the rainy season, predominantly in September and October, overflow from the lake floods the paddy fields with up to 4 m of water, creating conditions that only floating rice can survive. These rice varieties can elongate their stems up to 30 cm/day, keep their leaves above the surface of the water, and thus escape drowning (Cummings 1978). In our study area, farmers stated that these rice varieties can grow up to 50 cm/day in cases of flood disaster. This rice cropping system is the most extensive, requiring little labor and capital. As soon as the first rains fall, farmers start plowing, doing this twice if needed to incorporate weeds into the soil and get them to decompose. After harrowing, they sow in April or May and wait until December to harvest. In 2010 some farmers started using the herbicide Roundup to kill all weeds before incorporating them into the soil, instead of plowing twice. This rice cropping system is the most resistant to floods but is also the riskiest, because when the water is still high at the maturity stage, rats can climb on trees and eat rice panicles. This rice cropping system is almost chemical free. It has a low yield, 1.57 t/ha on average, but still more than direct seeding of rainy-season rice, which yields 1.2–1.4 t/ha. This is because lands used for rainy-season rice receive less alluvium from the floods compared to floating rice fields. Some farmers growing floating rice in zone 2 tried to use some fertilizer (50 kg/ha urea), but the yield was not different because the water caused nitrogen runoff. The ES and EDS provided by this rice cropping system are shown in Table 6.1.

6.3.3 Trade-Offs and Opportunity Cost Analysis

6.3.3.1 Trade-Offs Between Provisioning Services and Other ES in Each Production System

The key interest of ESF is to focus on trade-offs. For farmers and policy makers, the main output of ES is rice provision. As we have noted, the TSL ecosystem is the main zone of rice production in Cambodia, due to the flood pulse process. The previous analysis showed the opportunities and risks associated with different zones and rice cropping systems within the TSL ecosystem. Our fieldwork led us to identify different strategies developed by farmers to manage these opportunities and risks. These strategies are combinations of different cropping systems (i) into a specific production system (j; see equations in Sect. 6.2.2). Based on several criteria (surface, labor, capital), we distinguished 21 different combinations of rice production systems and calculated the value-added for each (Table 6.2). Value added per family labor (fl) corresponds to the capacity of one family's labor to produce on their land. It is then possible to gather these different groups into six main production system models (A–F in Table 6.2), taking into account relationships with regulating and cultural services. Because in this study it is not possible to evaluate a monetary value for all these ES or EDS for each category, we adopted a qualitative valuation (+ for positive effect ES; – for negative-effect EDS) based on our expertise in the field and interviews with farmers. Only the provisioning services, which we are able to calculate in monetary terms, are represented as value added/fl per year.

Model A is a combination of different types of short-term rice systems; some of them produce only a double cycle of short-term rice. In this production system, farmers can significantly increase their yearly revenue. Farmers in model A try as much as possible to convert all their rice fields to adopt short-term rice, and they buy water from private rice companies.[1] This model provides only a low level of cultural services related to the fact that they maintain the existing spiritual places. In contrast, in terms of EDS it degrades many regulating services, as detailed in Table 6.2.

[1] A few private companies produce short-term rice in zone 3. They own 150–200 ha of rice fields and invest in irrigation systems by making high dikes around them, with reservoirs inside, to prevent floods and drought.

Table 6.2 Rice production system typologies

No. farmers	Rice production system	Surface/fl (ha)	Provisioning services: Value added/fl ($)	Provisioning services: Natural variety	Regulating services: Agro-biodiversity	Regulating services: Habitat/ biodiversity/water quality	Regulating services: Flood regulation	Cultural services: Spiritual/ scenic/identity
11	A. System intensive, providing high provisioning services and high EDS (average ES score: −9)							
3	A1. (Early-season rice + receding rice)	1.35	1004.67	–	–	–	–	+
3	A2. (Early-season rice + receding rice) + receding rice	1.43	785.54	–	–	–	–	+
5	A3. Receding rice	1.47	478.28	–	–	–	–	+
38	B. System intensive, providing medium to high provisioning services and low regulating and cultural ES (average ES score: +3)							
5	B1. (Early-season rice + receding rice) + receding rice + floating rice	2.24	1077.81	–	+	–	–	++
5	B2. Receding rice + medium-term transplanting + long-term direct seeding	2.55	994.43	+	+	+	–	++
3	B3. Early-season rice + receding rice + long-term transplanting	2.24	908.2	+	+	+	–	++
3	B4. Receding rice + medium-term direct seeding + floating rice	1.41	456.41	++	+	++	–	++
2	B5. (Early-season rice + receding rice) + medium-term direct seeding	0.59	344.61	–	+	–	–	++
2	B6. (Early-season rice + receding rice) + medium-term direct seeding + floating rice	0.88	316.19	–	+	–	–	++
7	B7. Early-season rice + floating rice	0.61	261.45	+	+	++		++
11	B8. Receding rice + floating rice	1.00	260.92	+	+	++	–	++
22	C. Organic rice system providing medium provisioning services and medium regulating and cultural ES (average ES score: +6)							

6	C1. Organic rice + long-term direct seeding	0.75	235.86	–	++	++	–	+++
11	C2. Organic rice	0.29	132.1	–	+++	++	–	+++
5	C3. Organic rice + floating rice	0.48	106.87	–	+++	++	+	+++
6	D. System intensive, providing low provisioning services and high EDS (average ES score: –9)							
3	D1. (Early-season rice + receding rice)	0.37	127.63	–	–	–	–	+
3	D2. Early-season rice	0.15	54.3	–	–	–	–	+
37	E. Traditional system. Providing low provisioning services and high regulating and cultural ES (average ES score: +11)							
14	E1. Long-term direct seeding + floating rice	0.63	129.37	+++	++	+++	+	+++
6	E2. Long-term direct seeding	0.46	117.4	+++	++	++	–	+++
11	E3. Medium-term transplanting + medium-term direct seeding	0.41	113.64	+++	++	++	–	+++
6	E4. Long-term direct seeding + floating rice	0.32	69.13	+++	++	+++	+	+++
42	F. Floating rice system, providing low provisioning services and high regulating and cultural ES (average ES score: +15)							
42	F1. Floating rice	0.78	151.08	+++	+++	+++	+++	+++

Ecosystem service (ES) scores were calculated by adding 1 for each positive-effect ES (+) and subtracting 1 for each negative-effect ecosystem disservice (–)
fl one family's labor

Model A is a clear trade-off between maintaining other ES and provisioning services corresponding to productivity of land and labor (value added/fl). To bring the value added from \$478.28/fl to \$1004.67/fl,= farmers get an ES score of −9.[2]

Model B is a combination of short-term rice systems (double or single cycle) with other rainy-season rice cropping systems. Farmers manage their system to convert their land to an agro-ecosystem of short-term rice as much as possible. On the rice fields where they cannot produce short-term rice, they continue to cultivate rainy-season rice, including floating rice. Because of its high price, medium-term rice, grown both by transplanting and direct seeding, is their best way to get high land productivity and value added/fl. Model B shows a possibility to increase land and labor productivity while still staying in harmony with the ecosystem of the TSL floodplain. In model B, provisioning services from different groups vary significantly (from \$260.92/fl to \$1077.81/fl) but with almost the same ES score, about +3 on average.

Production model C represents organic production systems, with some in combination with rainy-season rice cropping systems in order to also produce on land where flooding cannot be controlled. Organic rice production is not the most effective model in terms of ES because it degrades natural varieties and has increased risks of flooding causing contamination and loss of organic labeling. During the seasonal flood period, farmers generally drain water from their rice field by letting water flow through neighboring rice fields by gravity. Organic rice fields forbid this, thus increasing flood risks for other fields. They also disrupt alluvial deposits for nutrient renewal. In model C, a farmer's value added/fl is between \$106.87 and \$235.86, generating a +6 ES score on average.

Model D represents the short-term rice production system of farmers with small production areas. Farmers using this model do not have capital to invest in converting their rice field to short-term rice. Thus, they get water from private companies and pay them back after harvest. They cannot get high labor productivity because of their small area, less than 1 ha/farmer. This system offers small value added/fl and comes with a high cost for society and the environment, like production model A. In this model, one farmer can make only \$54.30–127.63 while generating EDS of −9 on average.

Model E, called the "traditional system" by farmers, refers to combinations of different rainy-season rice cropping systems. This model represents the way farmers try to adapt to the floodplain ecosystem by creating rice field terraces, which let them adopt rainy-season rice cropping systems in all three zones. It symbolizes a complex manmade agro-ecosystem in harmony with an ecosystem at high risk of seasonal flooding. Floating rice is a perfect component of harmony between humans and the "roneam" (flooded forest) ecosystem because, instead of changing this ecosystem, farmers cultivate rice varieties that can adapt to floods. Farmers can get from \$69.13/fl to \$129.63/fl and accumulate a high ES score of +11 on average.

Production model F is practiced by poor farmers who own around 1 ha, only in lowland zone 1, on grassland, in the flooded forest ecosystem, or along the water-

[2] All dollar amounts are in USD.

way, with high risk of flooding. These farmers do not have enough financial capital to invest in conversion to short-term rice. Therefore, they continue to produce floating rice, which provides low provisioning services but very high regulating and cultural services. Farmers in this model produce on average $151/fl while providing the highest ES score, +15 on average.

6.3.3.2 The Most Efficient Production System Model

Table 6.2 clearly shows that it would be impossible to promote a specific production system that would conserve all ES provided by the agro-ecosystem. The more we promote quantity of rice provision, the more we destroy regulating and cultural services. Moreover, for political decisions it would be useful to calculate the opportunity cost that farmers would have to face if policy makers would decide to promote pro-ES rice production systems. To pursue this argument, we calculated the opportunity cost of each production system compared to the adoption of the most productive rice production system, model B1](double cycle of short-term rice + receding rice + floating rice; see Table 6.2). This system provides high productivity per one family's labor along with a balance between ES and EDS. In other words, model B1 provides high provisioning services (rice) at zero cost to the ecosystem. Figure 6.6 compares opportunity costs versus ES scores among each of the 21 model systems described in Table 6.2. Some of the production systems provide the same (regulating and cultural) ES score but with more or less opportunity cost compared to model B1. Thus, it is possible to define an optimal frontier of ES production systems (Fig. 6.6, solid curve) composed of the different efficient production systems.

- Systems with low opportunity cost, less than $300/fl: B1, B2, B3, A1, and A2. Model B1 is a control system, with zero opportunity cost and zero balance of ES and EDS. At the same opportunity cost, systems A1 and A2 have negative ES scores, as opposed to B2 and B3, which have positive ES scores. In this group, B2 is efficient in terms of opportunity cost to preserve ES.
- Systems with medium opportunity cost, between $600/fl and $850/fl: B4, B7, B8, C1, B5, B6, and A3. At almost the same opportunity cost, A3 and B5 have negative ES scores, while B4, B7, B8, C1, and B5 have positive scores. Thus, among these production systems, only B4 is efficient in ES preservation.
- Systems with high opportunity cost, more than $900/fl: F1, E4, E3, E2, C2, D1, and D2. With equal opportunity cost, model D (D1 and D2) has negative ES scores. In contrast, others have very positive ES scores. Among them, F1 is the most efficient for ES preservation.

When comparing production models A–F, models A and D appear to be the most dangerous for ES. Both systems have ES scores of −9, indicating strong EDS. However model D has a much higher opportunity cost, $986/fl on average, than model A, with only $321.65/fl on average. All production systems in model B are reasonable for ES and can also provide high provisioning services. They are thus

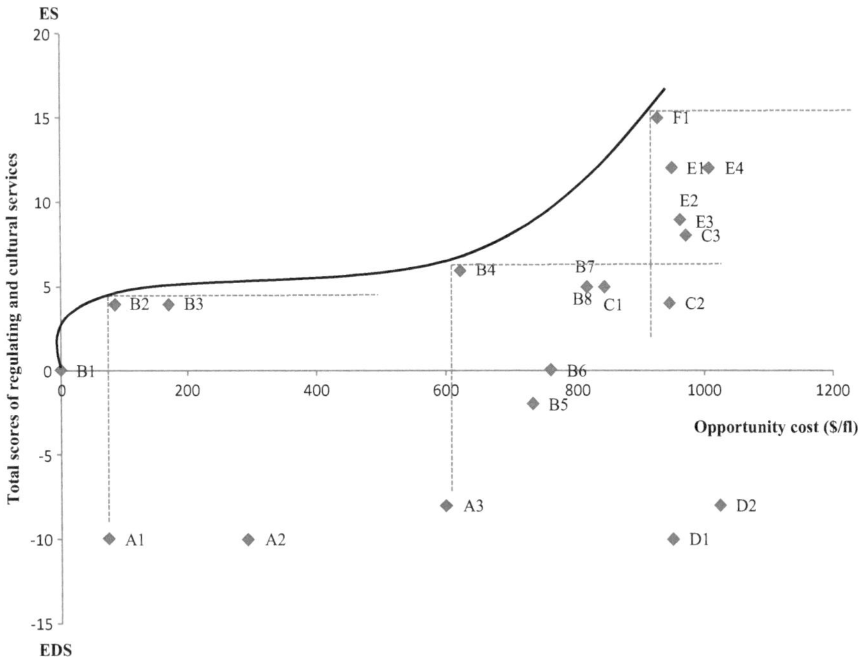

Fig. 6.6 Comparison of opportunity cost (horizontal axis) versus ecosystem service score (vertical axis) for each of the 21 model systems described in Table 6.2. Solid curve represents optimal ES production systems

able to ensure ES with low opportunity cost. Interestingly, models A and D are not productive compared to model B. In other words, producing only short-term rice with high chemical pollution and ecosystem conversion is less productive than combining short-term rice with rainy-season rice plus floating rice. The latter also helps increase positive externalities on the environment, with low chemical pollution and low ecosystem conversion, which are good for ES preservation. Models E and F are the most effective and efficient for ES preservation, but they represent the highest opportunity cost for farmers. Organic rice systems (model C) are not efficient for ES provision, while still coming with a high opportunity cost for farmers, around $919/fl, despite a price premium for organic label.

6.4 Conclusion

In most of these production systems, farmers achieve economic efficiency thanks to short-term rice, with even better performance if they can adopt a double-cycle schedule, having two harvests per year on the same land. On the other hand, rainy-season rice and floating rice enable farmers to achieve ecological performance by

ES provision. Thus, production systems that are most efficient economically for famers and also ecological in terms of ES provision are systems in which farmers combine short-term rice cropping systems with rainy-season rice cropping systems, especially floating rice.

More precisely, compared to production system model B1 (see Table 6.2 and Fig. 6.6), which is the most profitable production system, only three production systems provide an efficient trade-off between provisioning services, on one hand, and regulating or cultural services on the other:

- B2—receding rice + medium-term transplanting + long-term direct seeding
- B4—receding rice + medium-term direct seeding + floating rice
- F1—floating rice

In terms of number of farmers in these production systems, F1 includes the majority of farmers (F1 = 42 farmers, vs. B2 = 5 farmers, B4 = 3 farmers, and B1 = 5 farmers). In spite of this, floating rice production systems are practiced by the poorest farmers in the region and are on a path toward disappearance due to their low productivity, together with the high risk of yield loss caused by climate uncertainty and rats. The trend in the region is to convert floating rice fields into short-term rice fields, which require irrigation and drainage, as well as chemical inputs to provide high yield.

Consequently, for public policies aiming at promoting pro-ES production models, we recommend encouraging the readoption of rainy-season rice, especially floating rice, or increasing its production surface area in order to be both economically efficient for farmers and operationally effective for the agro-ecosystem. Producing only floating rice generates the highest ES but with very high opportunity cost for farmers. However, this cost could be reduced by diversifying toward high value-added rice cropping (short-term rice, medium-term rice, or organic rice). Conversely, producing only short-term rice generates high value added for farmers but with the highest EDS. However, EDS could be reduced by diversifying toward rice cropping systems providing high ES, such as floating rice. Hence, promoting this production will contribute to poverty reduction in Cambodia. Despite high ES provision, this production system also generates high opportunity cost, which will make it expensive for public policies to maintain.

Based on our results, organic rice production systems are not economically and ecologically efficient in ES provision. Thus, we propose three different choices:

- To promote production systems with medium efficiency for ES but low opportunity cost, promote adoption of rainy-season rice, excluding floating rice, in combination with short-term rice.
- To promote production systems with medium performance for ES and medium opportunity cost, promote adoption of rainy-season rice, including floating rice, in combination with short-term rice.
- To promote production systems with high performance for ES and high opportunity cost, promote adoption of floating rice alone in a production system.

References

Arias, M. E., Cochrane, T. A., Piman, T., Kummu, M., Caruso, B. S., & Killeen, T. J. (2012). Quantifying changes in flooding and habitats in the Tonle Sap Lake (Cambodia) caused by water infrastructure development and climate change in the mekong basin. *Journal of Environmental Management, 112*, 53–66. https://doi.org/10.1016/j.jenvman.2012.07.003.

Brooks, S. E., Allison, E. H., & Reynolds, J. D. (2007). Vulnerability of Cambodian water snakes: Initial assessment of the impact of hunting at Tonle Sap Lake. *Biological Conservation, 139*(3–4), 401–414.

CARDI. (2007). *Rice crop in Cambodia*. Phnom Penh: Cambodian Agricultural Research and Development Institute.

Cochet, H. (2012). The systeme agraire concept in francophone peasant studies. *Geoforum, 43*(1), 128–136.

Cochet, H., & Devienne, S. (2006). Fonctionnement et performances économiques des systèmes de production agricole: Une démarche à l'échelle régionale. *Cahiers Agricultures, 15*(6., novdem 2006), 5.

Cochet, H., Devienne, S., & Dufumier, M. (2007). L'agriculture comparée, une discipline de synthèse? *Economie Rural, 15*(janvier-avril), 297–298.

Cummings, R. C. (1978). Agricultural change in vietnam floating rice region. *CNRS, INIST, 37*(3), 10.

Dale, V. H., & Polasky, S. (2007). Measures of the effects of agricultural practices on ecosystem services. *Ecological Economics, 64*(2), 286–296. https://doi.org/10.1016/j.ecolecon.2007.05.009.

Dan, P., Gordon, C., & Sok, S. I. (2005). Long-term rates of sediment accumulation in the Tonle Sap, Cambodia: A threat to ecosystem health? *Journal of Paleolimnology, 33*(1), 95–103.

Dufumier, M. (2006). Etude des systèmes agraires et typologie des systèmes de production agricole dans la région cotonnière du mali. In INAPG (Ed.), *Programme d'amélioration des systèmes d'exploitation en zone cotonnière (PASE) Projet "caractérisation des systèmes agraires"*. Paris: INAPG.

Fujisaka, S. (1991). A set of farmer-based diagnostic methods for setting post 'green revolution' rice research priorities. *Agricultural Systems, 36*(2), 191–206.

Gray, T. N. E., Chamnan, H., Borey, R., Collar, N. J., & Dolman, P. M. (2007). Habitat preferences of a globally threatened bustard provide support for community-based conservation in Cambodia. *Biological Conservation, 138*(3–4), 341–350. https://doi.org/10.1016/j.biocon.2007.04.030.

Keske, M., & Huon, R. (2002). Data bases and topographic zoning, using gis for zoning of the villages. In MRC & WUP-FIN (Ed.), *WUP-FIN socio-economic studies on Tonle Sap 7*. Phnom Penh: WUP-FIN.

Kummu, M., & Sarkkula, J. (2008). Impact of the mekong river flow alteration on the Tonle Sap flood pulse. *Ambio, 37*(3), 185–192.

Kummu, M., Sarkkula, J., Koponen, J., & Nikula, J. (2006). Ecosystem management of the Tonle Sap Lake: An integrated modelling approach. *International Journal of Water Resources Development, 22*(3), 497–519.

Kummu, M., Dan, P., Sarkkula, J., & Koponen, J. (2008). Sediment: Surse or blessing for Tonle Sap Lake. *Ambio, 31*(3), 6.

Lamberts, D. (2006). The Tonle Sap Lake as a productive ecosystem. *International Journal of Water Resources Development, 22*(3), 481–495.

Ly, P., Jensen, L. S., Bruun, T. B., Rutz, D., & de Neergaard, A. (2012). The system of rice intensification: Adapted practices, reported outcomes and their relevance in Cambodia. *Agricultural Systems, 113*(0), 16–27. https://doi.org/10.1016/j.agsy.2012.07.005.

MacAlister, C., & Mahaxay, M. (2006). *Mapping wetlands in the lower mekong basin using landsat etm images and field survey data*. Paper presented at the European Space Agency, (Special Publication) ESA SP, Frascati, Rome.

Masumoto, T., Hai, P. T., & Shimizu, K. (2008). Impact of paddy irrigation levels on floods and water use in the mekong river basin. *Hydrological Processes, 22*(9), 1321–1328.
Millennium Ecosystem Assessment. (2005). In I. Press (Ed.), *Ecosystems and human well-being: Synthesis*. Washington, DC: Island Press.
Millennium Ecosystem Assessment. (2007). Millennium ecosystem assessment. In Millennium Ecosystem Assessment (Ed.), *A toolkit for understanding and action*. Millinium Ecosystem Assessment.
Mitchell, J., Cheth, K., Seng, V., Lor, B., Ouk, M., & Fukai, S. (2013). Wet cultivation in lowland rice causing excess water problems for the subsequent non-rice crops in the mekong region. *Field Crops Research, 152*, 57–64.
Mund, J-P. (2010). The agricultural sector in Cambodia: Trends, processes and disparities. *Pacific News*.
Neang, M., Meral, P., Aznar, O., & Déprés, C. (2017). Diversity of rice cropping system and organic rice adoption in agro-ecosystem with a high risk of flood in Cambodia. *The International Journal of Agricultural Resources, Governance and Ecology, 13*(4), 351.
Nguyen, Y. T. B., Kamoshita, A., Araki, Y., & Ouk, M. (2011). Farmers' management practices and grain yield of rice in response to different water environments in kamping puoy irrigation rehabilitation area in northwest Cambodia. *Plant Production Science, 14*(4), 377–390.
Pham, T. H., Takao, M., & Katsuyuki, S. (2008). Development of a two-dimensional finite element model for inundation processes in the Tonle Sap and its environs. *Hydrological Processes, 22*(9), 1329–1336.
Power, A. G. (2010). Ecosystem services and agriculture: Tradeoffs and synergies. *Philosophical Transactions of the Royal Society, 365*(1554), 2959–2971. https://doi.org/10.1098/rstb.2010.0143.
Rapidel, B., DeClerck, F. A. J., Le Coq, J. F., & Beer, J. (2011). *Ecosystem services from agriculture and agroforestry: Measurement and payment*. London: Earthscan.
Shimizu, K., Masumoto, T., & Pham, T. H. (2006). Factors impacting yields in rain-fed paddies of the lower mekong river basin. *Paddy and Water Environment, 4*(3), 145–151.
Someth, P., Kubo, N., Tanji, H., & Ly, S. (2009). Ring dike system to harness floodwater from the mekong river for paddy rice cultivation in the Tonle Sap Lake floodplain in Cambodia. *Agricultural Water Management, 96*(1), 100–110.
Swinton, S. M., Lupi, F., Robertson, G. P., & Landis, D. A. (2006). Ecosystem services from agriculture: Looking beyond the usual suspects. *American Journal of Agricultural Economics, 88*(5), 1160–1166.
Swinton, S. M., Lupi, F., Robertson, G. P., & Hamilton, S. K. (2007). Ecosystem services and agriculture: Cultivating agricultural ecosystems for diverse benefits. *Ecological Economics, 64*(2), 245 252. https://doi.org/10.1016/j.ecolecon.2007.09.020.
TKK, Baran, E., & Myschowoda, C.. (2008). Have fish cathes been declining in the mekong river basin? In M. Kummu, M. Keskinen, & O. Varis (Eds.), *Modern myths of the mekong* (pp. 55–64). Helsinki: Water & Development Publication-Helsinki University of Technology.
Tsubo, M., Fukai, S., Tuong, T. P., & Ouk, M. (2007). A water balance model for rainfed lowland rice fields emphasising lateral water movement within a toposequence. *Ecological Modelling, 204*(3–4), 503–515.
UNESCO. (2012). Tonle sap biosphere reserve. *Biosphere Reserves*. Retrieved 5 Aug 2013, from http://www.unesco.org/new/en/phnompenh/natural-sciences/biosphere-reserves/tonle-sap-biosphere-reserve/
Van Zalinge, N., Nao, T., Touch, S. T., Deap, L.. (2000). Where there is water, there is fish? Cambodian fisheries issues in a mekong river basin perspective. In I. S. A. Reviews (Ed.), In M. Ahmed & P. Hirsch (Eds.), *Common property in the Mekong: issues of sustainability and subsistence*. Penang: ICLARM--The World Fish Center.
Varis, O., & Keskinen, M. (2006). Policy analysis for the Tonle Sap Lake, Cambodia: A bayesian network model approach. *International Journal of Water Resources Development, 22*(3), 417–431.

Varis, O., Kummu, M., Keskinen, M., Sarkkula, J., Koponen, J., Heinonen, U., Makkonen, K. (2006). Integrated water resources management on the Tonle Sap Lake, Cambodia. In G. Amy, A. Annachhatre, E. Arvin, J. Chen, J. Cho (Eds.), *Water science and technology: Water supply* (Vol. 6, pp. 51–58). London : IWA Publishing

White, P. F., Oberthür, T., & Sovuthy, P. (Eds.). (1997). *The soil used for rice production in Cambodia. A manual for their identification and management.* Phnom Penh: MAFF, IRRI, AusAID.

White, P., Dobermann, A., OberthÏr, T., & Ros, C. (2000). The rice soils of Cambodia. I. Soil classification for agronomists using the Cambodian agronomic soil classification system. *Soil Use and Management, 16*(1), 12–19. https://doi.org/10.1111/j.1475-2743.2000.tb00164.x.

World Bank. (2013). Where have all the poor gone? Cambodia poverty assessment 2013. In T. W. Bank (Ed.), *A world bank country study* (p. 144). Washington, DC: World Bank.

Yen, N. T. H., Sunda, K., Oishi, S., Ikejima, K. (2007). *Tonle sap ecosystem water quality index development and fish production.* Paper presented at the WIT Transactions on Ecology and the Environment, Algarve.

Zhang, W., Ricketts, T. H., Kremen, C., Carney, K., & Swinton, S. M. (2007). Ecosystem services and disservices to agriculture. *Ecological Economics, 64*(2), 253–260.

Chapter 7
Payment for Ecosystem Services in Cambodia: Challenges and Potential

Soriya Yin and Seng Vang

Abstract Payment for ecosystem services PES has been identified as a sustainable financing strategy for conservation and local livelihood improvement in Cambodia. It is based on a market approach in which ecosystem beneficiaries or users pay money or other nonmonetary incentives to ecosystem service providers or park managers to protect their natural resources. The PES schemes in Cambodia are currently being implemented by conservation organizations and are small scale, operating with a very limited legal basis and unclear property rights. Despite these challenges, they have shown significant impacts on both conservation and local livelihoods. If well designed with legal and political supports, PES could potentially fund the conservation of shrinking natural resources, for which the Cambodian government usually lacks a budget, and improve the well-being of both local communities and business owners.

Keywords Ecosystem services · PES · Biodiversity conservation · Thmat Boey · Community-based ecotourism · Oddar Meanchey · REDD+

7.1 Biodiversity and Conservation in Cambodia

Cambodia is considered one of Asia's hotspots for biodiversity conservation. It has an abundant diversity of ecosystems, including coastal systems, wetlands, and forests, and agro-biodiversity as well. The coastal zone is considered among the richest areas in biodiversity resources, including significant aquatic resources and marine endangered species, such as coral reefs, sea grass, mangrove, green turtles, dolphins, sharks, groupers, shrimps, tortoises, and dugongs (UNEP 2005). Cambodia

Soriya Yin (✉)
Massachusetts Institute of Technology, Cambridge, MA, USA
e-mail: soriya@mit.edu

Seng Vang
Massachusetts Institute of Technology, Cambridge, MA, USA

ASEAN-US S & T Advisor, Phnom Penh, Cambodia

M. A. Stewart, P. A. Coclanis (eds.), *Water and Power*, Advances in Global Change Research 64, https://doi.org/10.1007/978-3-319-90400-9_7

is a flat country; much of its land is covered by water and wetlands that are linked to the Mekong River, the Tonle Sap Greater Lake, Se Kong, Se San, and coastal tidewater. The Cambodian Forestry Administration (2011) calculated that 57% of the country's land area is covered with forest. The largest remaining forest areas are the northern plain, the northeast, and the southwest parts of the country. Agro-ecosystems, which are perhaps most significant for the people of Cambodia, are primarily rice-based agriculture along with shifting cultivation.

Cambodia's economy is resource based, and its prosperity largely depends on how the country effectively and wisely uses and conserves its ecosystem resources, which play an important role in food security for local communities. Ecosystems services provide a variety of benefits, such as food, clean water, nutrient cycling, and climate regulation services, to millions of Cambodians. For example, fish is an essential part of the diet in Cambodia, providing more than 80% of total animal protein (Fisheries Administration of Cambodia 2010). Fisheries generate more than 1.6 million full-time jobs, and production is estimated to be worth US$200–300 million annually. According to the United Nations Environment Programme, mangrove forests and sea grass are estimated to generate US$109 million annually for local communities (UNEP 2007).

However, with recent economic development and growth, the country's ecosystems are in decline and are facing many threats, including logging, deforestation, increasing human population, lack of policies, global trading, climate change, loss of habitat, overexploitation of biological resources, and pollution (MoE 2002). The loss of ecosystems seriously affects the production of food, both today and in the future, since much of the country's economy is based on biological resources. The threats need to be addressed if ecosystems are going to continue to provide food security and livelihoods for local communities and the country.

Recognizing the important role of ecosystems in conservation, as well as in securing food security for local people, the Royal Government of Cambodia has made great efforts to conserve the country's natural resources. It has made a commitment under the Convention on Biological Diversity to protect 10% of its marine and coastal ecoregions by 2020 (MoE 2002). The government has designated the country's first large marine protected area around Koh Rong Archipelago, about 20 km off the coastal city of Sihanouk Province. The marine protected area is known in Cambodia as a Marine Fisheries Management Area, which aims to ensure that fishery resources are managed sustainably while also encouraging tourism, reducing poverty, and maintaining biodiversity. The marine protected area, encompassing approximately 350 km^2, is designed to protect a significant diversity of corals, fish, invertebrates, sponges, and algae.

Thousands of hectares of economic land concessions (ELCs) have been canceled, returning them to public land and protected forest. Five forest areas with nearly 1 million hectares of land have recently been designated as protected forest. One of the five forest protected areas is Prey Lang Forest, the largest remaining lowland deciduous forest in Southeast Asia and one of the world's top 10 hotspots for biodiversity conservation. If well protected and conserved, such biodiversity conservation will not only protect and conserve valuable habitats and threatened species but also help provide food security for local communities.

However, conservation of ecosystems is not free—it comes with costs, and countries are constantly searching for ways to increase the resources available for the conservation and protection of the environment. The Cambodian government's National Biodiversity Conservation Plan has identified payment for ecosystem services (PES) as a sustainable financing strategy for conservation and sustainable livelihoods. In line with the Aichi Targets 2011–2020 adopted by the United Nations Convention on Biological Diversity COP-10 in Nagoya, Japan, the government has recently set 20 biodiversity indicators, one of which is to promote PES adoption throughout the country (MoE 2014). If well planned and supported with a legal framework, PES could be an effective tool for biodiversity conservation and poverty reduction in Cambodia.

The basic PES concept is that people in local communities are paid to protect their natural resources and/or to change behaviors and activities that are harmful to the environment, including sometimes "doing nothing" to allow an intact ecosystem to continue to provide services. In more theoretical terms, PES is defined as a voluntary transaction for a well-defined ecological service, with at least one buyer and at least one ecosystem service provider, based on the condition that the buyer(s) continues to pay only if the provider(s) continues to deliver the defined ecosystem service over time (Wunder 2005). The payments recognize the importance of specific ecosystem services and give a direct monetary reward to communities or individuals that protect them.

PES was introduced into Cambodia by conservation organizations in 2002, and it was well accepted by both local communities and the government. A number of PES projects and PES-like activities have been piloted and implemented in various fields, including REDD+,[1] ecotourism, and bird habitat protection (Milne and Chervier 2014). PES is understood as a strategy for conservation and poverty reduction wherein community members are paid to protect their environment. In 2009, a PES law was proposed in Cambodia for its large hydro dam projects but was rejected by the government due to the concern that the law would increase electricity prices (Milne and Chervier 2014). At that time Cambodia desperately needed to be developed, and natural resources were required for its development goals.

Now the country has some stability and a growing economy, with reduced poverty. Ecosystems are shrinking and becoming scarce resources, and people are becoming more aware of the consequences of not managing them. In this context, it is now appropriate to reintroduce and adopt PES policies in Cambodia.

7.2 Case Studies of Payment for Ecosystem Services

The following two case studies of PES, implemented by conservation organizations on small scales, demonstrate the potential for PES to have significant impacts on both conservation and local livelihoods.

[1] REDD+ refers to the UN Framework Convention on Climate Change program to reduce emissions from deforestation and forest degradation, recognizing the role of conservation, sustainable management of forests, and enhancement of forest carbon stocks in developing countries.

Fig. 7.1 Giant Ibis in Thmat Boey community, Preah Vihear

7.2.1 Thmat Boey Community-Based Ecotourism Project

The Thmat Boey community-based ecotourism project in Preah Vihear Province presents an excellent case of how PES could be done at the local level. Tourists were asked to pay more if they saw the white shouldered ibis or giant ibis (Fig. 7.1), two critically endangered birds. The giant ibis is the key species that has attracted many bird watchers from around the world to visit the Thmat Boey community.

Income from tourists has significantly contributed to bird conservation and improved local livelihoods. Community members are employed as local tourist guides, cooks, and transportation and other service providers. Some amount of income from tourism is spent for bird nest protection. Community members are paid when they discover giant ibis nests and look after them. The payment has motivated the community to conserve the birds and love the birds much more.

The Thmat Boey community is located in the north of Cambodia, about a 4-h drive from Siem Reap. It is in the Kulen Promtemp Wildlife Sanctuary, the largest remaining area of dry deciduous forest in Southeast Asia. The Thmat Boey community-based ecotourism project was supported by the Sam Veasna Center and the Wildlife Conservation Society. The project tries to stop hunting and poaching of endangered bird species and to improve local livelihoods. Since its establishment in 2005, the project has significantly contributed to growth in the bird population and to poverty reduction in the community.

7.2.2 Oddar Meanchey Community Forest REDD+ Project

To tackle deforestation and improve their livelihoods, 13 communities in the northwest province of Oddar Meanchey have organized themselves as community forestry organizations to manage a total forest area of over 68,000 ha, including evergreen, semi-evergreen, and deciduous forests. In 2008, with support from

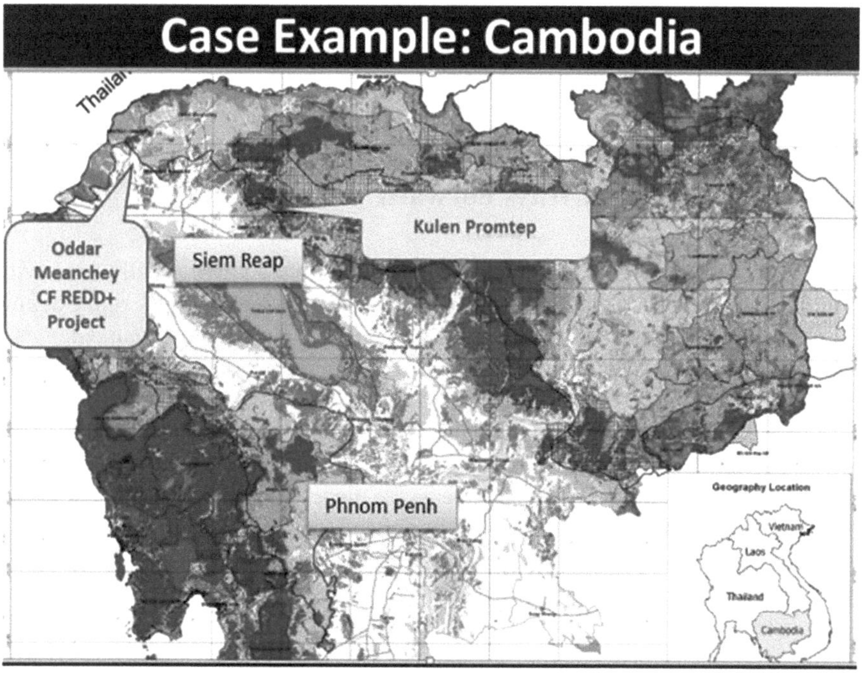

Fig. 7.2 Map of Oddar Meanchey community forest REDD+ project

Community Forestry International, Pact Cambodia, and the Cambodian government, the community started its first REDD+ pilot project (Fig. 7.2). The project aims to reduce poverty, mitigate climate change, and protect biodiversity through the sale of carbon credits. The project expects to generate US$40 million over its 30-year life-span. Based on an agreement with the Cambodian Forestry Administration, 50% of the net income generated from carbon credit sales will be given to the community for local development projects.

The success of the project largely depends on how well community forest land rights are respected. The government has granted the community the right to manage its forest for 15 years, which can be renewed upon satisfactory review. However, community forest land tenure rights are not always fully respected. For example, economic land concession (ELC) companies have expanded their forest land into community forest land. Eight ELCs have been granted, with some overlapping and unclear boundaries between ELCs and community forests. ELC is mainly designed for industrial agro-projects, while the REDD+ project aims to protect forests. Many times ELC projects have been implemented without proper consultations with local communities. Illegal logging and hunting occur in the project area. The issues have worsened with the increasing presence of the military to increase security along the border with Thailand. The military has built bunkers and roads in and through the community forest.

Efforts have been made to empower the community to better manage its forest, including legal consultation, raising awareness, and improving management structures and community forestry networks.

7.3 Challenges and Ways Forward

The current PES schemes in Cambodia are small scale and are usually done with limited legal support, lack of political will, and have unclear property rights definitions and protections and limited science-based research to support them (Clements et al. 2009; Milne and Chervier 2014; Development Alternatives and World Resources Institute 2015; WWF 2016).

Legal frameworks are important tools to implement PES successfully. PES involves contracting and making terms and conditions between sellers and buyers before services are delivered and payments are made. Thus, the support of laws and regulations from legal institutions are important for successful PES implementation. A specific legal framework to support PES does not exist in Cambodia, making implementation difficult.

It is highly recommended that PES be reintroduced and adopted at the highest policy level in Cambodia, and PES should be encouraged where applicable. If well planned and developed, PES is an alternative policy instrument to generate revenue to pay for conservation, promote social and environmental businesses and improving local livelihoods. The Cambodian government often does not have sufficient funds to protect biological resources, which are important for sustainable economic growth. PES is a sustainable financing strategy for conservation, providing financial incentives to those who protect ecosystem resources. In its policy formation process, the Cambodian government should set national PES targets—how many PES schemes or sites are planned. This would attract resources to materialize PES projects or programs.

In addition, PES requires strong political will and commitment, which the current government is lacking, especially when it involves big projects such as hydro dams that profoundly impact the protection of ecosystem services. The government has put PES schemes on hold since 2009; since then there have been no PES projects or programs through the government. Small PES schemes such as nest bird conservation, agri-environmental services, and ecotourism implemented by non-governmental organizations (NGOs) are allowed, however, because they are less of a political concern for the government.

The Cambodian government has perceived PES as a form of tax, with potential to increase energy prices and create barriers to investment and development (Milne and Chervier 2014). However, some senior government officers believe PES schemes could lead to win-win solutions for businesses, local community, and conservation if well planned and developed. The Ministry of Environment has recently made recent remarkable conservation efforts to cancel several ELC areas and designate almost 1 million hectares of forest land for conservation. This gives some hope for conservation and opportunities for PES policy adoption. The current PES schemes, which are implemented by conservation NGOs, are small scale but provide significant lessons and good practices for greater PES program design. The government should collaborate with the NGOs to develop and expand the current PES projects into a national PES program.

Successful PES implementation depends on clear property tittles and property rights. Most PES-designated places in Cambodia are public land. Many rural communities do not have land titles and the rights to manage public properties. The current PES communities with support from NGOs have formed themselves to claim their rights to manage public properties in the forms of community forest, community fisheries, community-based ecotourism, and community-protected areas. However, the process of community-based organization registration is complicated, and even with registered community-based organizations, these collective rights are not always respected.

Successful PES schemes depend on clear and secured land titles and property rights. Clearly defined property rights and land titles—either individual or collective ones—enable implementation of PES schemes. With clear land ownership, it is easy to determine who to pay and PES contracts can be legally enforced. Thus, it is important that property rights and land tittles in specific areas should be secured and enforced before PES schemes are introduced.

Last but not least, research is important. Currently, there is a lack of independent research on ecosystem service mapping, economic valuation, and best practices for sustainable PES schemes in Cambodia. Much PES research has been conducted and/or commissioned by NGOs and donors that implement and fund PES schemes. Because of this, the government often questions the credibility of the research.

If PES is going to become a politically acceptable strategy for conservation, more independent and science-based research is needed to inform policy makers and enable them to make better decisions. The research should identify and map out available ecosystem services and current status of ecosystems at specific sites, assess economic valuation of ecosystem services, perform cost-benefit analyses, and provide policy recommendations to planners and policy makers. The research findings would help land planners and policy makers make better decisions on land use planning and policy adoption.

References

Clements, T., John, A., Nielsen, K., An, D., Tan, S., Milner-Gulland, E. J. (2009) *Case study: Payments for biodiversity conservation in the context of weak institutions: Comparison of three programs from Cambodia*. New York: Wildlife Conservation Society.

Development Alternatives and World Resource Institute. (2015). Valuing ecosystem services in the lower Mekong Basin: Country report for Cambodia, USAID, Cambodia, DAI, WRI, USAID Mekong ARCC.

Fisheries Administration of Cambodia. (2010). *The Strategic planning framework for fisheries: 2010–2019*. Phnom Penh: FiA.

Forestry Administration of Cambodia. (2011). *Forest resources and cover in Cambodia in 2010.* Phnom Penh: FiA.

Milne, S., & Chervier, C. (2014). *A review of payments for environmental services (PES) experiences in Cambodia.* Cambodia: USAID.

Ministry of Environment. (2002). *National biodiversity strategy and action plan.* Phnom Penh.

Ministry of Environment. (2014). *The fifth national report to the Convention on Biological Diversity.* Phnom Penh.

UNEP. (2005). Sixth meeting of the regional working group for the wetlands subcomponent of the UNEP/GEF project: "Reversing Environmental Degradation Trends in the South China Sea and Gulf of Thailand" Sihanoukville, Cambodia, 12th–15th September 2005.

UNEP. (2007). Procedure for determination of national and regional economic values for ecotone goods and services, and total economic values of coastal habitats in the context of the UNEP/GEF project entitled: "Reversing Environmental Degradation Trends in the South China Sea and Gulf of Thailand". South China Sea Knowledge Document No. 3. UNEP/GEF/SCS/Inf.3.

Wunder, S. (2005). *Payments for environmental services: Some nuts and bolts* (Occasional Paper No. 42). Bogor: Center for International Forestry Research.

WWF. (2016). *Executive summary and policy recommendations for valuation of ecosystem services.* Cambodia: WWF.

Chapter 8
Ecosystem Services from Tonle Sap Flood Pulse: Spatial and Economic Analysis in Aek Phnom and Sangkae Districts of Battambang Province, Cambodia

Kimchhin Sok, Philippe Méral, Didier Pillot, and Stéphanie Defossez

Abstract Livelihoods of people in the Tonle Sap Lake (TSL) area of Cambodia are complex combinations of rice-based cropping, fishery systems, mixed cash crops/ home gardens, natural pond culture/aquaculture, cattle and livestock, collection of flooded forest products, and nonfarm and off-farm activities. The productivity of these activities is intimately linked to a number of ecosystem services and natural resources derived from the TSL. However, in recent decades the fish stock and the flooded forest have been degraded. This situation is made even worse as the TSL faces dramatic hydrological changes in the flood pulse regime and water levels. These changes have seriously impacted local livelihoods. In this situation, the environment becomes a significant source of vulnerability. This chapter provides an overview of this situation through economic and mapping analysis of two districts of Battambang Province. The implementation uses agrarian system diagnosis based on geographic information system mapping and qualitative interviews with informants during 2 years with low and normal flood pulses to identify types of household activities and their economic performance, changes in farming systems, agro-ecological zones, levels of poverty and resilience, and the country's related gain and loss in gross domestic production. We discuss the usefulness of such analyses in the field of ecosystem services mapping, which may contribute solutions to the Royal Government of Cambodia and to development partners to recognise this impact of climate and flood changes on rice production, natural resources, household livelihoods, and the country's economy as a whole.

Kimchhin Sok (✉)
Ecosystem Services and Land Use Research Centre, Royal University of Agriculture, Phnom Penh, Cambodia
e-mail: sokkimchhin@rua.edu.kh

P. Méral · S. Defossez
French Research Institute for Sustainable Development, UMR GRED, Montpellier, France

D. Pillot
SupAgro, Montpellier, France

M. A. Stewart, P. A. Coclanis (eds.), *Water and Power*, Advances in Global Change Research 64, https://doi.org/10.1007/978-3-319-90400-9_8

Keywords Livelihoods · Agrarian system analysis · GIS mapping · Household strategy · Resilience

8.1 Introduction

The importance of Tonle Sap Lake (TSL) for Cambodian people is well described in the literature (e.g., Bonheur et al. 2002; Hap et al. 2016). Over 1.7 million people live in 1555 villages in the area and rely on TSL ecosystems to a large extent for their food production and income (NIS 2008; Pillot 2008). More than for a typical lake, TSL's special features are based on a flood pulse connected to the level of the Mekong River during the dry and rainy seasons (Lamberts 2006).

Using typology provided by Millennium Ecosystem Assessment (2005), one can identify several ecosystem services (ES) provided by the TSL. First, *supporting services* are linked to all the nonmaterial benefits that play important roles in supporting soil formation, nutrient cycling, and photosynthesis (Ahmed et al. 1998; Hap et al. 2016). Second, TSL provides *provisioning services* related to all products and resources local people extract from their environment, including fish, nontimber forest products, and fuelwood. Save Cambodia's Wildlife (WCS 2010) estimated that between 289,000 and 431,000 tonnes of fish are caught annually in TSL. It is the fourth most productive captive fishery in the world, providing some 70% of the protein intake for the entire Cambodian population. In addition, the TSL floodplain contributes to one third of agriculture gross domestic production (GDP) in Cambodia (Van Zalinge et al. 2000). Third, *regulating services* are the ways in which living organisms mediate or moderate the ambient environment that affects human performance, such as flood regulation, climate regulation, and water purification. The World Bank (2006) also reported that the TSL provides habitats for more than a hundred species, such as water birds, 89 of which are abundant and 14 of which are considered to be of international significance, as well as a large variety of fish that migrate in large numbers throughout the lower Mekong system to spawn during high season, returning with the receding waters to the open lake and river system beyond. Other regulating services include water regulation and siltation services that impact soil fertility for rice production, and transport by inland waterways. Fourth, *cultural services* are all the nonmaterial, and normally nonconsumptive, outputs of ecosystems that affect physical and mental states of people: spiritual, recreational, educational, and so forth. For instance, TSL and the river play roles in Cambodian cultural ecotourism activities, such as birds, floating villages, and TSL river trips (World Bank 2006).

These benefits from the TSL are essential for sustaining local livelihoods. Livelihoods are organized around complex aquatic agricultural systems characterized by occupational pluralism and diversified income-generating activities. In turn, these aquatic agricultural systems are essentially based on complex combinations of a variety of rice-based cropping systems and fisheries systems, but also include cash crops/home gardens, natural pond culture and aquaculture, livestock, and the collection of nontimber forest products. Yet despite diversified livelihood strategies, many local households are entrenched in poverty, and this is particularly the case for those with a high reliance on provisioning ES who live in villages in the ecological zone most affected by seasonal water level changes. On the World Food Programme's poverty and vulnerability map, communes located near the TSL have high rates of poverty (above 50% poor), especially in the northern area. Various reports also point out that poverty and vulnerability have multiple dimensions in these particular landscapes, going beyond low income to include low welfare indicators (education, health, and nutrition) and limited access to financial capital and productive resources (flooded forests, fisheries, land).

The relationship between the TSL ecosystem and livelihoods is closely linked to a regular seasonal variation in the lake's water level (Nuorteva et al. 2010). If the flood during the rainy season is too high, risks of impacts on infrastructure, such as destruction of roads and houses, are high. At this opposite end of the spectrum, a long period of drought leads to decreased fish catch or a lack of water for rice crops (Sithirith 2011). With hydrological changes, maintaining a regular water level is becoming more challenging. In recent decades the TSL has faced dramatic hydrological changes due to climatic events, hydropower dams in the Mekong, overfishing, deforestation of flooded forest, and other changes (Nuorteva et al. 2010; Serrat et al. 2005; Sothorn et al. 2011; Kummu et al. 2008). These changes have seriously impacted local livelihoods that rely on natural resources.

Few studies on socioeconomic impacts of these changes are available. Bonheur et al. (2002) and Keskinen and Varis (2012) brought new knowledge concerning the TSL's governance and policies. Others like Keskinen (2006) focused on TSL zones related to village locations. Marschke and Berkes (2011) and Nuorteva et al. (2010) provide useful analyses of livelihood strategies based on qualitative data. Neang (2015) investigated strategies in quantitative analysis but only for rice cropping.

In this chapter, we add to these studies by using quantitative analysis based on agrarian system analysis to study how changes in TSL ecosystem functioning impact household livelihoods and the country's economy. Our hypothesis is that the local livelihoods are characterized by a diversity of economic activities, which are highly dependent on the ecosystem. The flow of ES used by local farmers, fishermen, and forestry resource suppliers contributes to capital accumulation of rural households in the TSL area and increases resilience to economic difficulties. This study also defines the following specific hypotheses: *First*, the TSL area is cultivated with systems that rely heavily on ES linked to the annual flood, but with uneven access to ES among the population. *Second*, economic changes from intensification of agriculture and fishery are important, but this highly increased use of inputs (use of financial capital) has uneven effects on local livelihoods. Land concentration tends to generate landless farmers side by side with large-scale farm production; those with fewer means of production are more dependent on ES than those who have intensified, leading to uneven fragility and risk exposure.

8.2 Concept, Methodology, and Data Sources

8.2.1 Selection of Study Area and Household Sampling

This research used the framework of adapted agrarian system analysis and diagnosis (Barral et al. 2012; Cochet 2012, 2015), while considering nonfarm and off-farm activities,[1] to identify different drivers of change in the institutional and ecological environment, types of household asset activities, main agro-ecological zones, economic performance of farming systems, migration, and key poverty/development-related issues. Agrarian system analysis is a systematic approach to study agricultural activities through a nested scale analysis (from the regional to the plot level) to account for both ecological and socioeconomic dimensions. It allows us to determine how individual farmers organize their own farms and how they manage the combination of cropping and livestock systems by using their inputs—labour, capital, and land—in an effecting manner (Neang 2015).

This work has been organized into six steps: (1) landscape analysis; (2) historical analysis; (3) pre-typology; (4) questionnaire interviews with farmer households; (5) agrarian system diagnosis with in-depth household interviews; and (6) Global Positioning System (GPS) mapping of land use changes.

Step 1: Landscape Analysis In this step the different natural conditions of the area and the way they are used are described in terms of climate, geology and soils, land cover, animals, and topography. The analysis of the links among natural conditions, human settlement, and land use enabled us to identify homogeneous zones (i.e., the agro-ecological zones) through field observations, discussions with local farmers, and the use of maps. This qualitative mapping was realized for each district, commune, and village, based on our personal observations as well as community mapping with eight focus group discussions.

Step 2: Historical Analysis The objective of this step is to gather information on the evolution and dynamics of the agrarian area since the beginning of its use. Understanding these dynamics will help us identify factors have impacted the area. This step was conducted through individual historical interviews with key informants, such as farmers of different types, representatives of government agencies and nongovernment organizations, and elders from the communities.

Step 3: Pre-typology A provisional typology was drafted at this point, based on the field interviews and the qualitative information collected in steps 1 and 2. The sample of farmers for the following individual interviews on the farming systems

[1] *Off-farm activities* generate a portion of household income obtained off the farm, such as selling labour and short-term businesses, either in local areas and by migration, but the household still engages in its own farming activities. *Nonfarm activities* comprise all activities associated with wage work or self-employment in income-generating activities (including income in kind and remittances) that are not agricultural.

and off-farm and nonfarm activities was proposed on this basis. The pre-typology is also a key driver for the individual household questionnaire survey.

Step 4: Individual Questionnaire Interviews with Farmer Households The different farmer types were identified for the interviews in order to better understand the various activity systems. A total sample of 290 farmers were selected for interview, beginning in 2014. The researchers decided which particular groups to interview (reflecting the different categories of the pre-typology) while attempting to minimize bias and to select a sample that best represented the population under study (Glenn 2009).

Step 5: Agrarian System Diagnosis with In-Depth Household Interviews After grouping the activity systems into types, the study conducted an agrarian system survey to confirm and validate the preliminary outcomes of the typology. We performed in-depth interviews members of 56 households. In addition, in 2016–2017 we conducted follow-up economic surveys with all 346 households/correspondents (from steps 4 and 5 combined) to compare their economic situation between a "bad" year (2013–2014) and a "good" year (2016–2017).

Step 6: GPS Mapping of Land Use Changes and Localization of Households During and after the agrarian and household interviews, we also conducted GPS mapping on land use and fishing area changes in two districts in the TSL vicinity. We collected the geographic information system (GIS) data point of every correspondent and used Google Earth images to create production zone representations of household accessibility to resources for stakeholders and communities and to aggregate with GIS data sources provided by the Battambang Department of Land Management, Land Administration, and Construction.

8.2.2 *Estimation of Economic Outputs of the Activity Systems*

Since it was assumed that the main reasons behind household practices were their economic impacts,[2] we used a production-function-based valuation approach to specify the feasible output of goods and services that could be produced with each cropping system, with a given set of inputs (labour, machinery, natural resources, etc.). Production-function-based valuation approaches are based on the contribution of a given ES to the production of a commodity that is traded in an existing market (Tallis and Polasky 2009). Some important elements of economic performance of

[2] We do not ignore that a number of practices and strategies are also related to social or cultural values, especially when these practices require collaboration among the stakeholders (collective actions). However, we clearly limit our analysis to the economic justifications of the choices made by the stakeholders. We also consider that the choices were made while they were linked to social or cultural values and often corresponded to economic foundations.

cropping systems refer to land productivity, defined as the gross value added (GVA) per unit of land, or the labour productivity (Barral et al. 2012). Simultaneously, to determine a household's system of activities, our study simply considered the GVA of each activity (rice, fishing, animals, aquaculture, nonfarm and off-farm activities, etc.) up to the total income of a household. Thus, in this case study, GVA can be considered consistently in interpreting the results from all households' systems of activities. We used the following formulas to assess economic performance at the farm and plot scale:

- Gross output (GO) = production × unit price
- Intermediate cost (IC) or intermediate input = all variable costs on necessary inputs, such as seed/plant, fertilizers, pesticides, petrol, and eventual service delivery (hired labour costs and family labour costs excluded)
- Gross value added (GVA) = GO – IC
- Total income = GVA – labour hired – land rent

8.3 Study Area

Battambang is located in the northwest of Cambodia. It borders Banteay Meanchey Province to the north, the TSL to the east, Pursat Province to the south, and Pailin Province and Thailand to the west. Battambang Province, with an area of 11,622 km^2, has a variable topography from the TSL floodplains in the east through a broad belt of lowland paddy terraces in the centre to lowland/upland mosaic and upland forested areas to the west. Battambang is classified as a rural province with a population of 1.15 million inhabitants. With a tropical climate, the province's fertile rice fields have led to a mostly agricultural economy, giving rise to the moniker "the Rice Bowl of Cambodia." The TSL is under two water regimes: during the dry season, from November until June, the lake flows into the Mekong; during the rainy season, from mid-June to late October, the Mekong River flows into the TSL area—the river swells from rainwater and snowmelt from the Himalayas, and its lower delta does not allow enough flow into the sea to eliminate all the excess water quickly, so the delta, up to the TSL, becomes flooded. At the peak of the flood in August, the TSL normally triples its surface, and then it recedes.

Administratively, the province is normally divided into five zones starting from the TSL:

Zone I, permanently flooded zone
Zone II, low-potential agricultural zone
Zone III, high-potential agricultural zone
Zone IV, agro-industry and forestry
Zone V, mosaic agro-industry

The study was conducted in Aek Phnom and Sangkae Districts located along the Sangkae River and next to the TSL floodplain. The study zone therefore covers

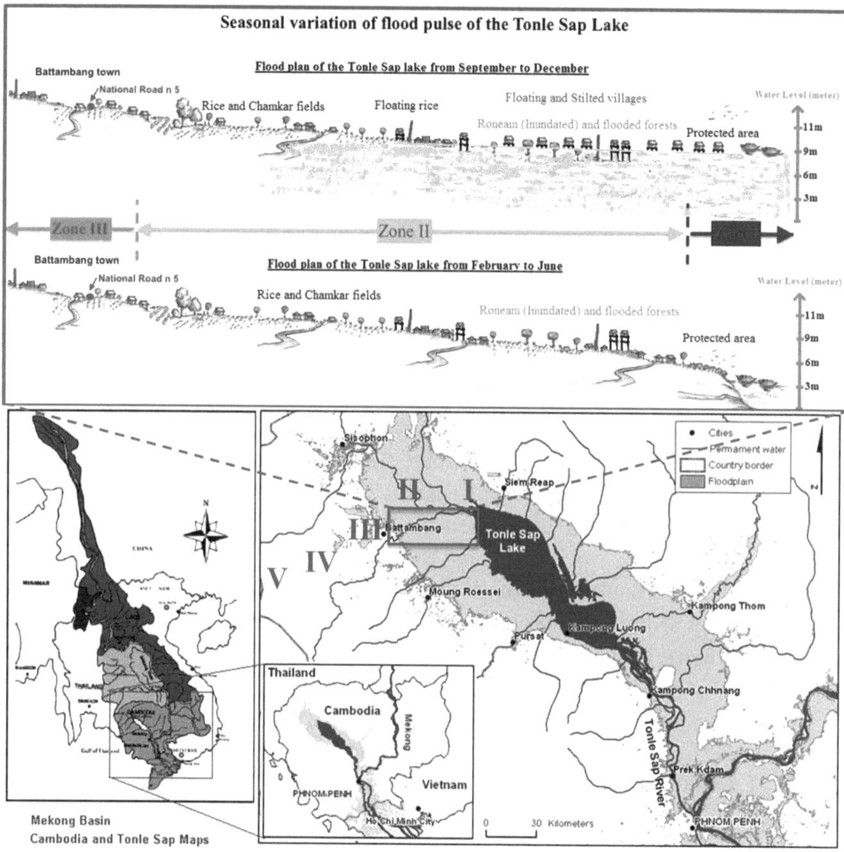

Fig. 8.1 Tonle Sap Lake zones (top) and study area in Battambang Province (bottom)

zones I–III. Fig. 8.1 shows the study area in Battambang Province, and a virtual transect image of the different agro-ecology and cropping and fishery systems in zones I–III during rainy and dry seasons.

8.4 Results

8.4.1 Household Strategies: A Diversified Situation Based on the Distant Location from the TSL

Our first analysis captures the location of households and production zones in different ecosystems of the floodplain. Economic activities and general livelihoods of households clearly depend on the location of the villages. Figure 8.1 shows that the villages in the agricultural area are situated relatively close to one another and at some distance from the lake. This leads to good living conditions for people who extract benefits from agricultural activities, typically rice-based cropping systems, cash crops, fishing, animal husbandry and aquaculture, and nonfarm and off-farm activities. Therefore, the agricultural land covers all of zone II and a small part of zone III (Fig. 8.1), at a distance of 5–15 km from National Road no. 5 (NR5) to downstream of TSL, where rice can be cultivated in both rainy and dry seasons. The villages in this agricultural area have higher living styles and livelihood diversification compared to fishing areas. Access to social and public services such as roads, hospitals, credit, and agricultural markets is also easier there.

In contrast, the fishing villages are located between 10 and 32 km from NR5. Their area covers all of zone II's water-land shown in Fig. 8.1, where generally only fishermen live, unevenly extracting benefits from fishery products (see Fig. 8.2). The fishermen irregularly move their residences between rainy and dry seasons. In fact, large fishermen (defined below) mostly live in the villages of the eastern part (Aek Phnom), whereas smaller fishermen are more widespread in the territory: some live in the fishing villages of the eastern zone already mentioned, while others may live in the villages of the agricultural area. Social and public services are limited.

A more detailed analysis of the households' economic strategies identifies six main categories of households: landless households, rice monocroppers, diversifiers, larger farmers, small fishermen, and large fishermen (see Sok forthcoming, for more details).

8.4.1.1 Landless Households

These households are represented by 29 households in our sample of 346 stakeholders. Some had been farmers in the past who inherited from the state land distribution during 1980–1989 or from their parents after the Khmer Rouge[3] but did not have enough land and capital to secure their household's viability since around 1990 to present; when a difficulty occurred, such as a health problem in the family, they had no other solution but to sell their land. Others live by selling short-term or long-term labour, migrating to other towns in Cambodia and Thailand.

These families are very poor. They often share their habitation land by building an additional wooden house beside the one of their host, planning to join their efforts to improve their situation. They live mainly by selling agricultural labour, especially for harvesting in the first-cycle crops and for performing rice and intensified crop tasks (harvest and sowing) in the second cycle. According to the survey, they can sell 30–40 days' labour per year (20 days during each of the first and second cycles), which may be an underestimate. In most cases, these people are young adults and adults and have many young children. Some of the younger children look

[3] The civil war known as Khmer Rouge in Cambodia lasted 3 years, 8 months, and 20 days from 1975 to 1979.

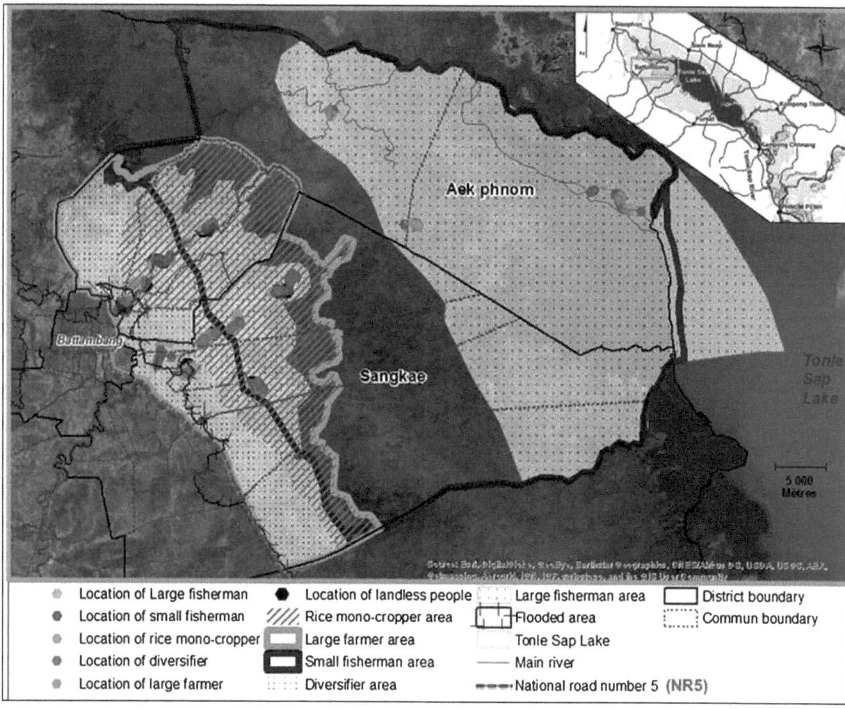

Fig. 8.2 Location of households and production zones in Aek Phnom and Sangkae Districts, Battambang Province, Cambodia. (Source: Google Earth map in 2007 in dry season from Google Image *Landsat/Copernicus* (Google map elevation: location attitude, 17 m/56 ft; latitude, 13.09573 N; longitude, 103.20221 E))

after cattle through a "sharing system",[4] while older children provide additional external labour. These families are especially fragile when faced with disease or accident. Moreover, one to three adults in the family have migrated to other areas, mostly Thailand, to find work and provide financial support, sending money home to support their elders and younger children.

These landless households may manage land that belongs to someone else. They may "occupy" the land for a transitional period on behalf of absent owners who would lose the land if it were not cultivated within 3 years of its official receipt. Depending on the individual arrangement, the transitional farmer knows that the agreement can end at any time by the owner's choice. Their goal is therefore to save enough money to purchase their own land elsewhere and to be able to pass lands down to their children. No long-term investments are undertaken on these transitional

[4] The sharing system for tending cattle can be defined in two ways: a landless family may sell the labour of two or three of their children to look after another household's cattle, with generally 10–30 heads, and can earn 500 Khmer Riel per head of cattle per day (0.12 USD/cattle head/day); or families will take turns looking after their combined cattle, with no money exchanged.

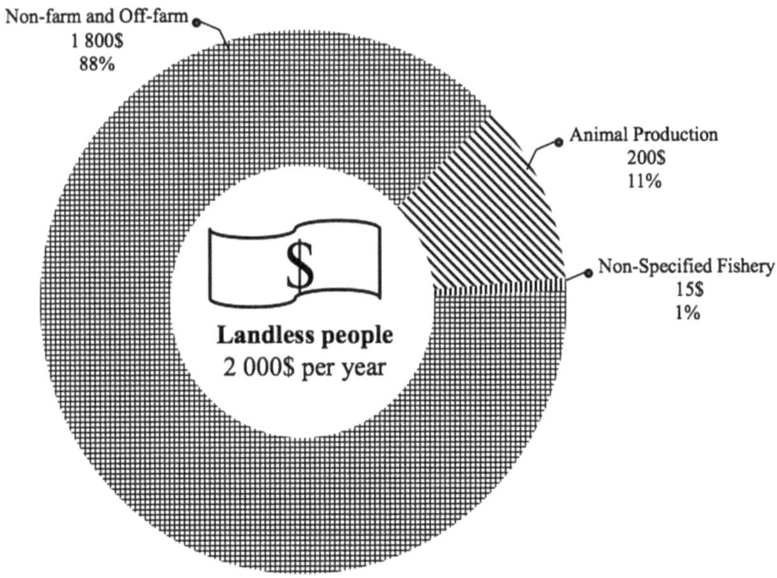

Fig. 8.3 Economic performance of landless households

farms. Therefore, the maximum annual income of these families is lower than for other households (Fig. 8.3). They earn only around US$2000/year, 88% of which is from nonfarm and off-farm activities (including income from migration), about 11% from animal rearing, and 1% from nonspecified fishing, which can be done during very short periods compared to other households since most household members are either too old or too young to fish for long hours.

8.4.1.2 Rice Monocroppers

This group of stakeholders, with 76 households in the sample, are small farmers who mostly specialize on floating and long-term rice. Historically, they have invested in floating rice and long-term rice (also called heavy rice) cultivated in the wet season. They do not invest much in intensified paddy cultivation, such as early-season rice and receding rice, or in fishery, nonfarm, or off-farm activities. The condition of their cropping pattern either does not provide or provides just enough income for the basic expenditures of the household. This group typically has two to four plots comprising about 3.5 ha, located in the lowland and water-land areas, farther from residential lands and close to the flooded forest. They mostly rely on rainfall and water flow from the Sangkae River and TSL, since they cannot access irrigation. Their cropping pattern provides either not enough or just enough income for the basic expenditures of the household (food, clothes, health, etc.). This group of households has an average five members and has a low capacity to invest.

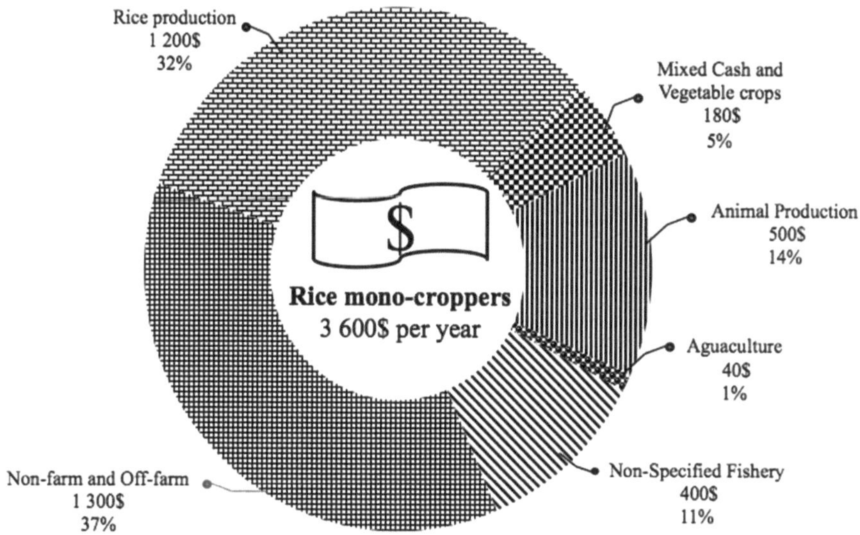

Fig. 8.4 Economic performance of rice monocroppers

Small cash crops and vegetable gardens around their residential home may be undertaken, as well as small animal production and nonspecified fishery during the rice fallow periods. They may also attempt to invest in perennial crops, but it provides delayed income and seems to be limited. There is also high rate of migration in these households; in interviews they confirmed that one or two members of their family have migrated, mostly to Thailand, Phnom Penh, and Siem Reap, to earn money to send back to their family, who use the remittance to support agricultural production and daily expenses.

This type of household as an average annual income of around US$3600 (see Fig. 8.4), from 37% nonfarm and off-farm activities,[5] 32% rice production, 14% animal production such as cattle and buffalo and poultry for home consumption, 11% nonspecified fishing, and 5% from mixed cash and vegetable crops, or double *chamkar.*[6]

[5] Nonfarm and off-farm activities undertaken by landless, rice monocropper, diversifier, and small fishermen households include small fish trading, carpenter, construction worker, agricultural wage labourer, construction wage labourer, local petty trader (including petroleum street seller), food processing (including small-scale food processing), moto taxi, factory worker, tailor, palm collector, port worker, small local cake producing and selling, brick worker, junk collector, etc.

Nonfarm and off-farm activities undertaken by large farmer and large fishermen households include shop and grocery, rice mill and businessman, boat and car driver, mechanic, urban services, teacher, civil servant, nurse/doctor/midwife, small industries, etc.

[6] Mixed cash crop and vegetable farms, or double *chamkar* farms, are defined differently by Cambodians based on various locations; in this study, informants confirmed that mixed cash crops and vegetables are those crops which are cultivated with several cycles per year on the same plot. Generally, farmers practiced it on the plot(s) close to residences and as a home garden.

8.4.1.3 Diversifiers

This household group includes farmers who diversify their agricultural production as much as possible, especially by going for a second cycle of cultivation in the dry season after the flood recedes. With 109 households in the sample, they are the group most represented in the area. Most of them got their rice lands and started cultivation during the distribution time. These households started by growing heavy and floating rice like the rice monocroppers did after the Khmer Rouge and in the 1980s, but since then they widely shifted to two cycles per year. These households have around 3 ha of cultivated paddy and intensified *chamkar*.[7]

They seem to generate enough income to secure all expenditures of the household and make them ready for investments. They also make extensive use of their family labour force to run all the activities of the household, but they still require more labour, which they hire from landless households and rice monocroppers, for land preparation, weeding, and harvesting. This is costly in their production.

Generally, the agricultural plots of diversifiers are located near permanent sources of water where they can easily manage irrigation for the dry-season crops. Their farms have a positive balance due to their nonspecified fishery and animal production like cattle, when they can tend to their cattle while fishing at the fields during the rainy season. Income is also generated from the combination of high yields at both the first and second cycles of rice and other intensified crops. Early season-rice and intensified *chamkar* crop production provides high GVA, but the household faces cash shortages after paying for land preparation, seeds, fertilizers, and herbicides for the both crop cycles, until after harvest. Selling some of the early rice can help farmers delay the intensified crop sale to get higher prices on the market. However, this requires manual weeding throughout the cropping cycle, and at least two to three active family members are necessary for the cropping. Nonfarm and off-farm activities are also an important income source for the family, but they invest more in the local area than in migration.

Again, the most profitable diversification is the early-season rice, intensified crops, and mixed cash and vegetable crops. Rice production can be significantly profitable if the yield is high (2.1 MT/ha from short-term rice and water-receding rice). Almost all of them have invested in a hand tractor (power tiller) and apply fertilizers in their rice and crop production. The maximum total income of this type of household is around US$4000 (see Fig. 8.5): 46% from nonspecified fishery, 23% from nonfarm and off-farm activities; 11% from long cycle rice, 2% from early-season rice and intensified *chamkar* crops, 8% from mixed cash and vegetable crops, 8% from animal production, and 2% from aquaculture.

[7] Various authors define intensified *chamkar* differently; in this study, we take it as the combination of mixed cash crops (such as corn, soybean, groundnut, watermelon, sesame, or jute) and vegetables in the second cycle of cultivation following early-season rice.

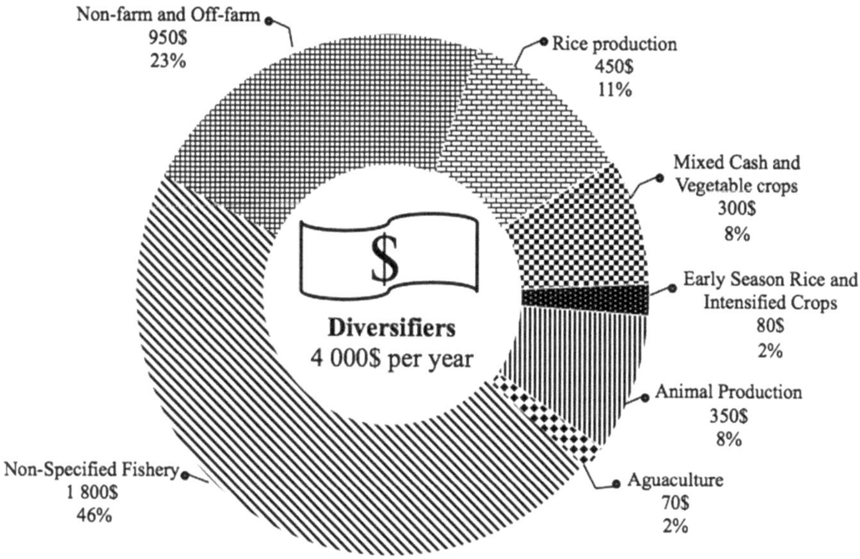

Fig. 8.5 Economic performance of diversifiers

8.4.1.4 Large Farmers

This group, with 52 households in the sample, has a high income, based on their land size: 5–20 ha, with 3–8 paddy plots located in lowland to upland water areas. All of them have invested in various types of agricultural activities, and they also run medium and large businesses. Some of them may have been former members of Khmer Rouge military and solidarity groups, and they may be government officials such as village chiefs, commune chiefs, and councillors.

They typically have had the capacity to invest in high diversification from their initial long-term rice and floating rice toward early-season rice, receding rice, and dry-season rice and then to expand with multicrop production systems. Their yields, however, are lower than those of diversifiers, mostly because of the dimensions of the areas where they grow *chamkar* crops and dry-season rice. Compared with monocroppers and diversifiers, large farmers have very high annual income from rice production and nonfarm and off-farm activities. They also have large agricultural equipment, hand tractors, and tractors (80% of them already invested in a tractor; sometimes they have two hand tractors and one tractor per family).

Diversification has the same importance for these farmers as for diversifiers. Their investment capacities are high and are suitable to take risks, but these farmers also often share their land among their relations and share labour and big machinery. All use fertilizers, and they have all the same rainy-season rice, receding rice, dry-season rice, and intensified crops in the dry season. They have enough money to

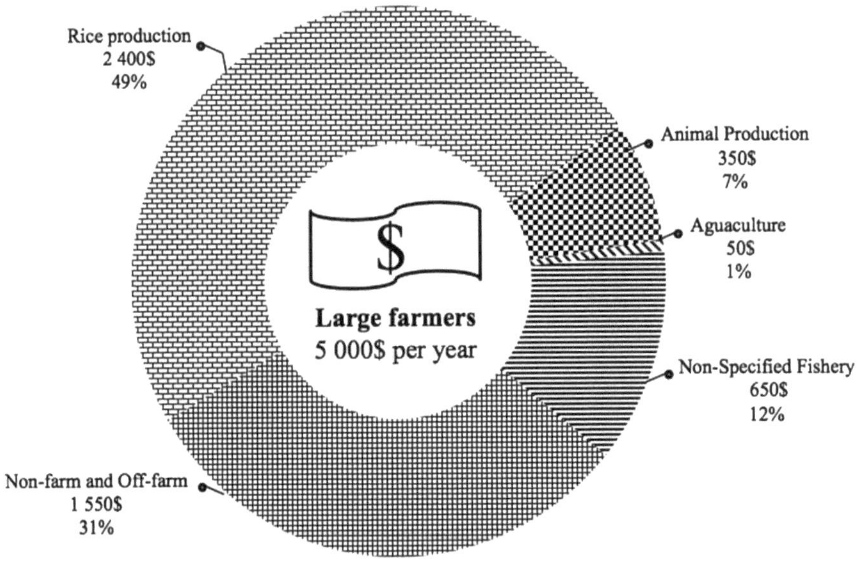

Fig. 8.6 Economic performance of large farmers

hire labour from landless households and rice monocroppers, but most of them are reluctant to use it as they are afraid of soil depletion and becoming more dependent on fertilizers.

The maximum total net income is around US$5000 for this group (see Fig. 8.6), with 49% from rice production, 31% from nonfarm and off-farm activities, 12% from nonspecified fishing, and the rest from animal production and aquaculture. With good income, around 60% of them have the capacity to send their children to school and have access to other social services, such as health care.

8.4.1.5 Small Fishermen

This type of household (with 55 households in the sample) relies mainly on income from fishing activities, from selling labour to large fishermen and to the farmers on agricultural land area, and from migration. They are former workers for fishing lot owners, and in some cases they subleased fishing lots before the lots were terminated in 2010.

They fish with small boats, catching 1–5 kg per day from the TSL in the rainy season and around 1–2 kg during the dry season. They also rely on different fishery resources, such as eel, frog, snake, turtle, rat, and other wildlife, as well as honey collecting, during both seasons. They live permanently on floating and stilt villages, located close or even inside the inundated forest. They raise a few pigs, chickens, and ducks also on floating houses during the rainy season and on high land during

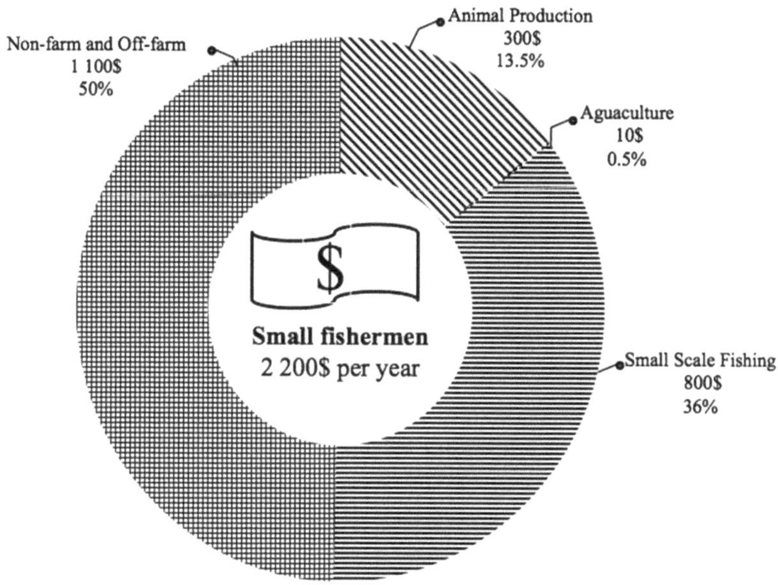

Fig. 8.7 Economic performance of small fishermen

the dry season. Most of them have a good fishing sharing system by engaging with their relatives and neighbours so they don't need to hire labour.

In addition to the income they can generate from the fishery, they also undertake nonfarm and off-farm activities, including migration, which helps the rest of the family survive. They are extremely fragile when faced with disease or accident, since their uneven residence is farther from the central villages, and they cannot access public services. Moreover, they face difficulty in accessing loans from official microfinance institutes; they can only access private lenders such as fishery traders and large fishermen, with high interest rates—60–180% per year.

Figure 8.7 shows that the main sources of income for their households are from fishery products. They can earn a maximum total net income of $US2200 from only three main sources: 50% nonfarm and off-farm activities, 36% small-scale fishing, and 14% animal production, with a small amount of small aquaculture.

8.4.1.6 Large Fishermen

This household type (with 25 households in the sample) relies mainly on large-scale fishing. They can fish 50–550 kg or more per day in the rainy season and up to 50 kg in the dry season. They also have high income from nonfarm and off-farm activities and have high capacity in intensified fishing systems and aquaculture, such as crocodile and fish rearing. An estimated 83% of their fish catch is sold at market, and the rest is kept for home consumption and aquaculture input.

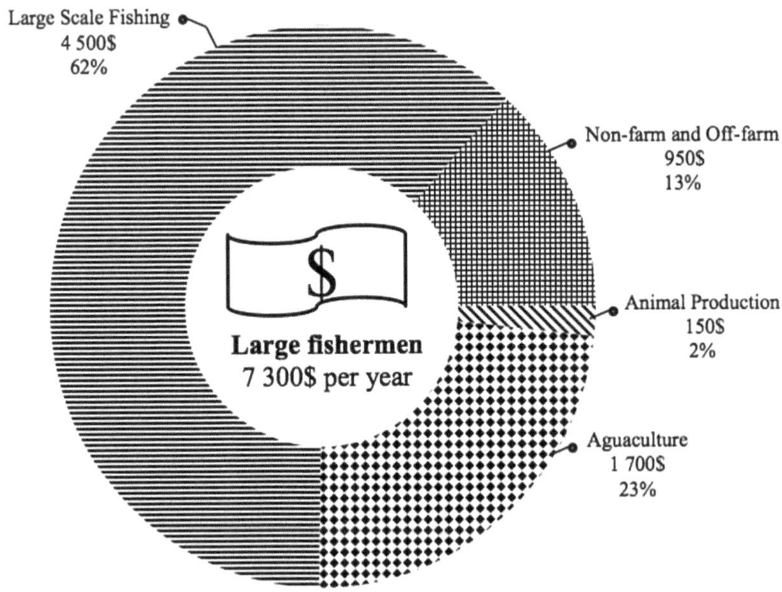

Fig. 8.8 Economic performance of large fishermen

They are former Khmer and Vietnamese fishing lot owners and subleasers, before fishing lot termination in 2010, and are now fishermen with a high capacity to invest. They also live permanently on floating and stilt villages, located in the inundated forest with the small fishermen, but this group of households more often performs activities in the deep waters of the Sangkae River and the TSL basin, where they can also extend their business, access health care, and send children to school in neighbouring Siem Reap Province. Their fishing materials and engines are much larger than those of small fishermen.

The correspondents confirmed that, with their previous strong profiles as former fishing lot owners and subleasers, most large fishermen have very good connections and relationships with such authorities as fishery administrators and river police officers, as well as journalists, which allows them to expand their fishing illegally with fish traps with pens (Nor-Rav),[8] while small fishermen can do the same only rarely or not at all.

Figure 8.8 shows that the main sources of income for these households are fishery products (62%), aquaculture (23%), and nonfarm and off-farm activities (13%), with the rest from animal production, for a total annual income of about US$7400. All of them have already invested in a large boat with an engine and large fishing tools. They increase their annual income with aquaculture, such as crocodile or fish

[8] Arrow-shaped traps with pens (Nor-Rav) are illegal large-scale fishing traps used in the middle width of TSL, usually with a small holding pen attached, placed in deep water at depths of 500–2000 m. These are used by people with powerful connections or are hidden from government authorities.

farming, which they feed with a part of their fish catch during both seasons. These households mostly came from Vietnam initially.

8.4.2 Mapping the Impacts of the TSL Water Regime on Households

Flood pulse changes can be caused by different factors, such as climate change and the construction of hydropower dams along the Mekong River upstream from Battambang Province. It is believed to impact local livelihoods and to lead to changes in the environment. We investigated the historical changes in livelihoods in the TSL area of this province at the landscape and economic levels. Key informants were interviewed about unusual environmental events in the past, most importantly years with significantly higher (2016–2017) or lower (2013–2014, 2015) water levels, as well as their impacts on the six types of households described above.

The research was conducted with different stakeholders: Battambang provincial sectoral departments, especially the Departments of Agriculture, Forestry, and Fisheries, Department of Water Management and Meteorology, and Department of Land Management, Land Administration, and Construction; nongovernmental organizations; fishery protected communities; authorities in both districts; and correspondents from the six household types. The study also conducted GIS village mapping on land use changes and localization of the households, and GPS mapping on land use of the three types of rice farmers (monocroppers, diversifiers, large farmers). The study collected the GIS data points of every correspondent and used Google Earth to represent households' access to production zones and resources.

Figure 8.9 shows the differentiation of flood pulse varieties in rainy and dry seasons of a good year and a bad year, illustrating their impact on the expansion of the flood and the households' production zones in the study area, and shows example locations of rice monocroppers and large fishermen in the case study. "Good" or "bad" characterize the availability of water during the dry season: a good year is when the water remains available in the lowest canals and rivers up to NR5, facilitating irrigation.

It is important to note that, although the flood pulse is closely linked with local livelihoods, the flood is generally considered to be a both positive and negative phenomenon. For this reason, the term "bad year" is characterised by significantly less water/flood than an average year, and "good year" refers to a high, but not too high, flood level, with clearly more water and/or higher water level than normal. In this study, the informants were interviewed individually and in groups to describe good and particularly bad years and floods. However, the stakeholders reported similarly that bad years happened in 2005 and in a few years again from 2012 to 2015. And the good years happened in 2010 and then in 2016–2017, when the water level in the TSL was remarkably high.

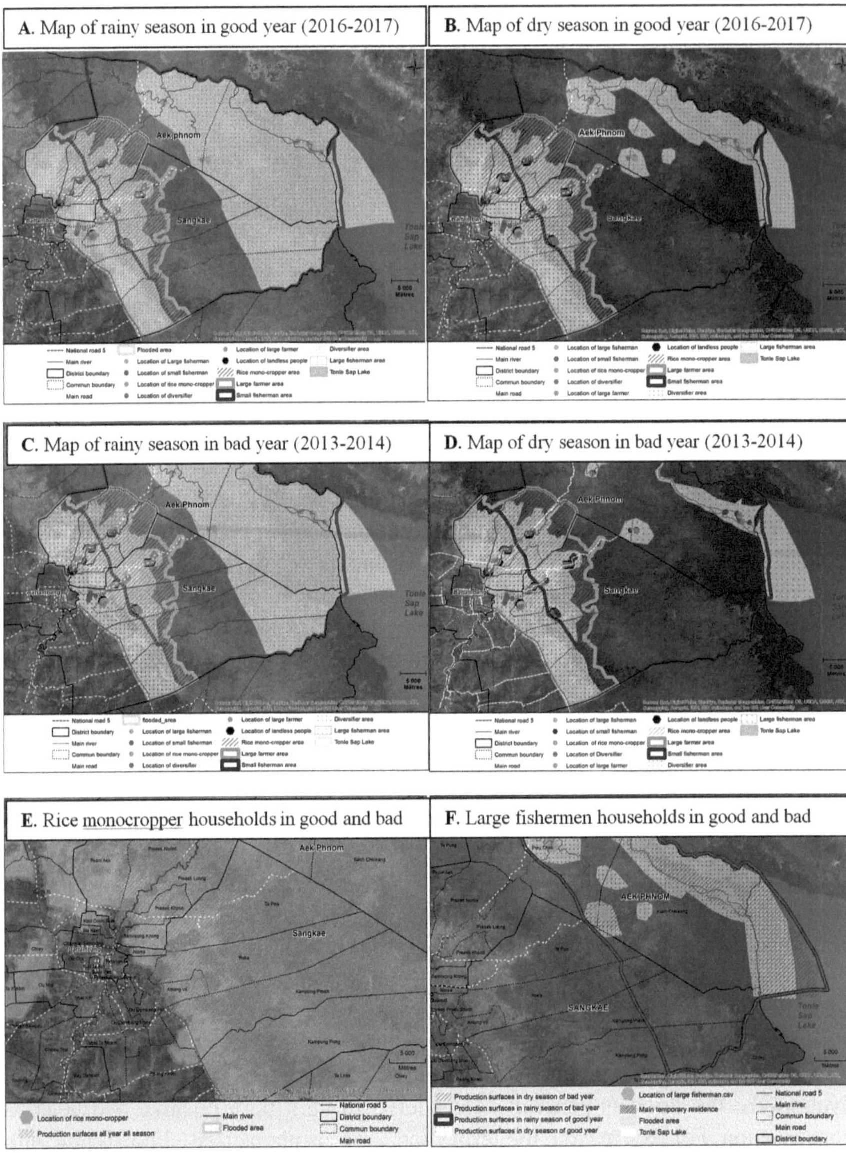

Fig. 8.9 (**a–d**) Maps of rainy season (**a**, **c**) and dry season (**b**, **d**) for good years (**a**, **b**) and bad years (**c**, **d**) in Aek Phnom and Sangkae Districts. (**e**, **f**) Locations of rice monocroppers (**e**) and large fishermen (**f**), households that stay in the same place when flooding is either low or high

The overall difference in water level is not very great between good and bad years, but due to the very flat slope of the area, small differences can have a very large impact on the territory benefitting from the flood—and on access to water in the following dry season. In good years the water depth at a reference point of the

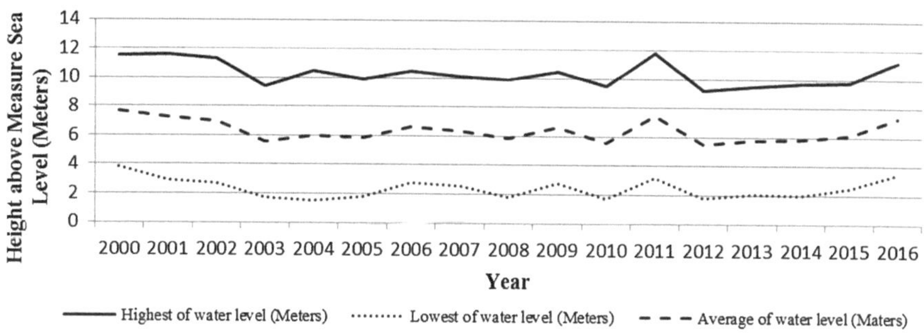

Fig. 8.10 Annual highest and lowest water level of Tonle Sap Lake in Battambang Province from 2000 to 2016. (Source: Battambang Provincial Department of Water Resources and Meteorology)

flooded area was estimated to average 2.6 m, and the flooding could reach around 1 km past NR5 (see Fig. 8.9a), whereas in bad years it averaged only 1.3 m, with the flooding never reaching NR5, stopping around 2.5–4 km away from it (see Fig. 8.9c). In bad years, the area covered by the flood is more limited, and thus the total quantity of siltation is reduced, impacting the fertility of the soils in the following crop season. In Fig. 8.9e, f, show the locations of households that stay in the same place when flooding is either low or high. In the fishing area, the depth was estimated to average 4–5 m in the good years and only 2.2 m in the bad years, with an average difference of around 1–1.5 m at the peak of the flood. This is 50 cm less than shown in the records of the Battambang Provincial Department of Water Resources and Meteorology in the Ministry of Water Resources and Meteorology.

In Fig. 8.10, data from the Department of Water Resources show that, over the last 16 years, the annual average water level of the TSL at Battambang exceeded the overall average level (7.5 m above sea level) in 2000, 2001, and 2002, and then in 2011 and 2016. By contrast, the average level decreased to around 6 m or less in 2003–2010. These years were considered low flooding by the farmers, fishermen, and relevant stakeholders. Västilä et al. (2010) also noted the rising water level in the TSL in Pursat and Kampong Chhnang Provinces in 2011 and the succession of low floods from 2003 to 2010. The flood pulse exceeded 7.5 m in 2016 and 2017, as indicated by our study surveys. We can assume that an average annual water level lower than 6.5 m is a disaster for farmers and fishermen in both fishing and agricultural areas. A review of rainfall data from the Department of Water Resources also indicated that rainfall over the last 15 years has been variable but increased in 2011 and again in 2016 and 2017. The heaviest rainfall was seen in 2011, which coincided with flooding in the entire country of Cambodia.

In terms of extension of the flood, the differences between the villages are clearly visible (Fig. 8.9). In the agricultural villages where the rice monocroppers, diversifiers, and large farmer households depend on agriculture, there is actually not much difference between good and bad years in where farmers raise their crops—whether the flood is a good or a bad year, they still stay and work in the same agricultural plots. This is because these cropping systems start in the early season at the latest, when the nurseries for the early rice are prepared and land preparation for sowing floating rice occurs. These operations need to be done several weeks before people can assess the importance of the flood that year. Nevertheless, the flood impacts their production, which in turn impacts income. For instance, higher application of chemical inputs may be required to help compensate for the lack of natural siltation, whereas the yields of the traditional and high-yield rice varieties tend to decrease.

We also found effects on the livelihoods of the small and large fishermen households. The small fishermen especially have low capacity to adapt to bad floods, mainly due to the external limitations they face, such as lack of agricultural land, challenges with fisheries management and related power inequalities, poor transport capacities, lack of fishing equipment, and low income from other sources. Figure 8.9f shows how the production zone of the large fishermen is highly modified by the level of water. With higher water, they can expand their fishing area and benefit from new opportunities to diversify their fishery production. Low floods, when they occur, can damage stilt houses and other infrastructure. On the other hand, the floating houses in the villages closest to the lake remain highly unaffected by high floods and actually benefit from them in terms of increased fish productivity in the good years.

People in the TSL area have adapted to past changes, which offers an opportunity to assess their vulnerability and their adaptive capacity toward future changes. The current level of resilience in the study villages was considered in two ways: adaptive capacity of the villages in different parts of the floodplain, and adaptive capacity of the social groups and households within the villages. In both cases, considerable differences between the levels of resilience were found. In terms of the household strategy in different parts of the floodplain, agricultural areas have higher levels of resilience to risks and access to social services due to their diversification in agricultural productions and the importance of nonfarm and off-farm activities. For instance, when agricultural crops are destroyed, farmers are able to find more sources of income by selling labour and increasing other nonfarm and off-farm activities, including migration to Battambang and Siem Reap Cities, Phnom Penh, Poi Pet, and even Thailand. Moreover, they benefit from easier access to social services from the public administration than the fishermen do when facing risks.

In the long term, beyond the possible impact of climate change, structural changes might affect Mekong River levels. The extension of irrigation schemes upstream of the TSL, in China, Myanmar, and Laos, will increase infiltration of part of the river flow or even possibly drain some flow to other catchments. Dams and reservoirs currently under construction will certainly contribute to more regulation between rainy and dry seasons and will alter total and seasonal sediment loads of the river. Since the rising water brings alluvial fertility for crops and favours fishery production in the lake, it provides benefits for and impacts on habitats and biodiversity, typically affecting household economics. Higher flooding allows people to diversify production, for example, rice and fish. Lower flooding reduces opportunities for this kind of diversification.

8.4.3 Economic Impacts of the Water Regime

According to the differences in household strategies between the agricultural and fishing areas, internal differences between the households and social groups within the villages were observed. The poorest households were the most vulnerable in all study areas. The poor households among rice monocroppers, small fishermen, and landless households often rely on a single livelihood source only, while the better-off households among diversifiers, large farmers, and large fishermen typically had more possibilities for supplementary livelihood strategies. The better-off households also had more savings and other assets and access to public services, while poorer households depended on external assistance in cases of emergency. Many of the low-income households among rice monocroppers, small fishermen, and landless households also suffer chronically from food shortages, as well as unclean drinking water, health problems, and lack of secondary education opportunities, making them particularly vulnerable to additional shocks and stresses.

The six types of households that depend on environmental resources for livelihoods expressed another important strategy for coping and adapting to the flooding: relying more strongly on secondary livelihood sources not affected by environmental changes, as well as extending to altogether new livelihoods. In agricultural areas, rice can provide the main source of income and food security; fishing can provide additional income and food sources for households in times when agriculture is affected, for example, by floods, and fishery and other aquatic wildlife and creatures were considered important for additional income for people in fishing areas. Changes in the flood pulse in bad years can negatively impact yields of crops and fisheries, rotation of crops, insect and diseases, and market prices of crops and fishery products, as well as the period of employment for nonfarm and off-farm activities. In addition, different forms of short-term paid employment in agriculture and fishing were mentioned in all study areas as important additional livelihood sources. The source of employment varies according to the livelihood background and the production location; in agricultural areas employment is linked to all farming and nonfarm and off-farm activities, while in fishing areas the employment is generally only fishing related and includes working for large-scale fishing operations.

Based on the economic impact analysis in Table 8.1, we found that the large farmer households were highly impacted by the bad flood in 2013–2014 compared to the good year in 2016–2017, suffering a 35% decrease in average annual income, from US$4197 in the good year to US$1985 in the bad year. These households rely much on hired labour from outsiders for farm production, with high costs of inputs and labour and high capital investment. During the bad year, there were major concerns of rat/rodent infestations, insects, disease, and weeds due to low flooding and drying of the plots, leading to loss of yields of rice and other crops. For instance, total rice yields, which average 3.3 tonnes per hectare, decreased by 600–1500 kg per hectare. Despite this negative economic impact, large farmers still have a higher annual income than the other groups of households except for large fishermen.

Table 8.1 Economic impacts of good and bad years on the six types of households

Type of household	GVA in good year (2016–2017; $/household/year)			GVA in bad year (2013–2014)					
				Agriculture		Non- and off-farm		Household total	
	Agriculture	Non- and off-farm	Household total	GVA ($/year)	Percent loss	GVA ($/year)	Percent loss	GVA ($/year)	Percent loss
Landless households	248.14	1879.18	2127.32	230.74	7	1766.81	6	1997.55	6
Rice monocroppers	2773.25	1423.90	4197.15	1985.52	28	1333.93	6	3319.45	21
Diversifiers	3585.25	954.24	4539.49	2585.53	28	868.85	9	3454.38	24
Large farmers	4335.21	1574.93	5910.14	2407.60	45	1415.45	10	3823.05	35
Small fishermen	1244.11	1163.35	2407.46	979.48	21	1028.62	12	2008.10	17
Large fishermen	6711.16	1029.70	7740.85	6191.18	78	909.17	12	7100.35	8

Diversifier and rice monocropper households are also impacted by the low flood in the bad year compared to the good year, losing around 20% of their annual income, from US$4200 in a good year to around US$3400 in a bad year. Many of the these households also suffer chronically from floating and heavy rice and fishery catch shortages, as well as from other aspects of poverty, such as health problems and lack of income from nonfarm and off-farm activities. However, these households are more adapted to such situations, relying on family resources to operate the farm, such as labour, work animals and hand tractors, and small capital to invest on production, including fishing.

Small fishermen were more impacted by the low flood in the bad year; their average annual income decreased US$399 (16%) compared to the good year. This is mostly due to the lower availability of fishery products, but also lower income from nonfarm and off-farm activities. In contrast, large fishermen, while also suffering from decreased fish catches in bad years, still have income from aquaculture, such as crocodile and fish rearing, and nonfarm and off-farm activities.

Landless households are in the worst absolute position, with a decrease of their income from only US$2127 in the good year to around US$1997 in the bad year. These households are generally elders and grandchildren, while their children have migrated to Phnom Penh or Thailand. They can only sell their labour to agricultural and nonfarm activities and receive remittance money from their migrated children.

Most of these income strategies were related to the informants' current livelihoods, as well as other livelihoods practiced in the area. In the fishing villages next to the TSL, many hoped to broaden their livelihoods into fish raising, fish processing, and utilisation of wetland products, while in the agricultural villages raising livestock and broadening to other crops such as vegetables were considered as possible diversification strategies. Starting small businesses such as shopkeeping and involvement in different forms of paid labour were also mentioned by informants in all study villages. One approach to diversifying the livelihood base is through migration, which was recognised as a potential future adaptation strategy in all study villages (Rigg 2006).

To gain a broader view on loss for our respondents and the critical level of resilience to disasters, we approximated their loss of total income between good and bad years. Table 8.2 shows the economic gain and loss of the two different years and critical level of resilience of the six types of households. For each of the six types characterized, we have correlated the total income from the livelihoods to access to all farm, off-farm, and nonfarm activities. Where the land and fishery production factors are now getting rarer and where land and fish have to carry environmental services other than strict provisioning services, this allows us to identify the household types now more sensitive to land access. We also compared total income coming from livelihoods, include nonfarm and off-farm activities, with a theoretical overall minimum poverty threshold (MPT)[9] (MOP 2013, 2014; Gibson et al. 2006;

[9] In 2013, the overall poverty line (food poverty line + nonfood allowance) per capita per day was 6347 Khmer Riel per capita per day ($1.54 US/capita/day) in Phnom Penh, 4352 Riel per capita per day ($1.06 US/capita/day) in other urban areas, and 3,503 Riel per capita per day (0.85 US/

Table 8.2 Critical level of resilience by household type

Type of household	Household MPT (US$)	Annual income (US$) Good year	Bad year	Bad year income – MPT (US$)	Critical level of resilience (%)
Landless households	1559 (5 members)	2127.3	1997.6	438.3	+22
Rice monocroppers	1715 (5.5 members)	4099.5	3221.9	1506.7	+47
Diversifiers	1715 (5.5 members)	4414.5	3329.4	1614.3	+48
Large farmers	1871 (6 members)	5727.8	3640.7	1769.6	+49
Small fishermen	1559 (5 members)	2339.0	1939.6	380.3	+20
Large fishermen	1559 (5 members)	7523.9	6883.4	5324.1	+77

Critical level of resilience is the difference between the bad-year income and the MPT, expressed as percentage of the bad-year income. For example, for large fishermen, even in a bad year 77% of their income exceeds the MPT and thus may be available for investment
MPT minimum poverty threshold

Kimsun and Bopharath 2011). When families have incomes below the MPT, they cannot cover their basic needs unless they sell land or other assets or increase the cultivation cycle, threatening fertility. In contrast, when they have incomes higher than the MPT, they have a surplus for possible capitalization. The annual MPT was calculated to be US$1559–1871, depending on the number of household members, derived by multiplying the average size of families (five to six members; see Table 8.2) by the cost of consumption per capita for rural people (US$311.85/person/year).

We also compared total income in the bad year with the MPT to identify which household types have the highest resilience in bad years. The income in bad years is still higher than the MPT for all six types of households. The level of resilience is calculated by subtracting the income in the bad year from the MPT and expressing the difference as a percentage of the bad-year income. Landless households and small fishermen have the lowest resilience: around 20–22% of their bad-year income exceeds their MPT, due to lower intensification and diversification of activities—they rely more on only one traditional rice cropping or fishery system and have lower income from nonfarm and off-farm activity. For monocroppers, diversifiers, and large farmers, more than +47% of their bad-year income exceeds their MPT. Thus, the land and fisheries play essential roles in contributing income for these three types of households. The more means of production they have, the more they can intensify their income from those systems. Large fishermen have the highest resilience, with 77% of their bad-year income exceeding the MPT. This indicates that the low and high level of flooding has little impact on their annual total income:

capita/day) in rural areas. The study considers 2013 data for the overall poverty line per capita per day because it is close to the current MPT of rural people at 3503 Riel per person per day (US$0.85/capita/day) per person per day.

their income is already high, and they might be able to shift to other activities, such as off-farm and nonfarm activities.

Consequently, the variability of the water regime of the TSL provides both positive and negative impacts on household livelihoods, with higher impacts for households that depend significantly on the provisioning services of the TSL. The situation is worst for people who have low capacity to shift their income-generating activities—landless and small fishermen households—and they will become more and more fragile as the bad years occur more often; indeed, it drives them to migrate to other places.

Cambodia's economy grew by an estimated 6.9% in 2016 (US$20 billion), the same pace as in the previous years. A mild slowdown in industry and services was mitigated by a slight pickup in agriculture. Gross domestic production (GDP) is foreseen to grow by an additional 7.1% per year in 2017 and 2018 (MOEF 2016).[10]

Agricultural production accounts for 35% of Cambodia's GDP but employs 56% of the labour force. Reviving agriculture is critical to sustaining rapid growth and reducing poverty (MOEF 2016). Main products from the sector are rice, rubber, corn, vegetables, cashews, and cassava. Agricultural gross production grew by 8.7% between 2004 and 2012, driven by crop production, mainly paddy rice (annual growth of 9%), maize (20%), cassava (51%), sugarcane (22%), and vegetables (10%).

At the country level, rice is clearly the most important source for food and income security for the people of Cambodia. We analysed the impact of the bad flood years on the Cambodia's GDP (Table 8.3). The contribution of rice to the Cambodian national GDP in good and bad years was calculated by multiplying the GVA per year that we found from each rice production system at the plot level by the total amount of the country's paddy land surface (MAFF 2016), and then calculating the difference in GDP between good and bad years.

We found that the contribution of rice to the national GDP is US$1,128,653,027 in good years but US$557,654,881 in bad years, resulting in a GDP loss of US$570,998,146 in bad years, almost 3% of the national GDP. Thus, bad years can affect national economy as much as 40% of annual growth. What is worse, all agricultural production has decreased with bad climate conditions during 2005–2015, with only 2 years of good flooding, which also reduces the country's GDP.

8.5 Discussion and Conclusion

Based on this agrarian system analysis of rice-based aquatic agricultural systems, including both agriculture and fishing, the Tonle Sap Lake (TSL) area is highly vulnerable to negative changes in the water environment. This vulnerability is further intensified by the country's low capacity to adapt to environmental shocks and

[10]The Asian Development Bank also reported the national GDP of Cambodia; see https://www.adb.org/countries/cambodia/economy.

Table 8.3 Contribution of rice to Cambodian national gross domestic production (GDP) between good and bad years

Rice production system	GVA per hectare (US$/ha)			Land surface (ha)	Contribution to Cambodia's GDP (US$/ha)		
	Good year	Bad year	Loss in bad year		Good year	Bad year	Loss in bad year
Floating rice	370.64	236.11	134.53	46,759	17,330,635	11,040,270	6,290,365
Wet-season rice	468.27	232.77	235.49	580,227	271,701,466	135,061,294	136,640,172
Early-season rice	333.17	152.73	180.43	1,898,188	632,411,244	289,918,942	342,492,302
Dry-season rice	423.35	248.51	174.84	489,455	207,209,682	121,634,375	85,575,307
Total				3,014,629	1,128,653,027	557,654,881	570,998,146

stresses (RGC 2001; Sithirith 2015). In the TSL area, people are generally well adapted to the seasonal changes caused by the flood pulse, and both livelihood sources and income levels have a strong seasonal component (Keskinen et al. 2011).

Our research findings indicate, however, that this adaptive capacity has limits, and the people and their livelihoods are actually relatively vulnerable to significant changes in their environment, including changes to the flood pulse system. Nuorteva et al. (2010) argued that the vulnerability of households is highly driven by climate changes and declining resources. Our findings also show that livelihood diversity and a sufficient living situation provide the foundation for the people's household strategies and capacity to adapt to these kinds of environmental changes. The level of livelihood diversity in the study area is already relatively high, as individual households commonly complement their main livelihood sources with supplementary strategies, especially nonfarm and off-farm activities, including migration to Battambang and Siem Reap Cities, Phnom Penh, and Thailand to send remittances to their families back home. However, strong dependence on just one main livelihood source, habitually either fishing or rice cultivation, within each village increases overall vulnerability to sudden environmental changes, and this situation is growing worse in seasons of low floods in bad years. Varis and Keskinen (2006) also noted that diversifying the livelihood base both within the households and more generally within village locations and production zones increases resilience to risks of shock and stress from the environment, as has been noted by other studies in Cambodia and elsewhere.

In this study, the production zones were described by informants and supported by spatial GIS mapping analysis to generate images of the locations of the households' activities in the study area, as well as reading the landscape and correlating various cropping and fishery production systems in the territorial units. The changes of these systems when a bad or good flood occurs show the supplementary livelihood strategies which enabled the stakeholders to increase or protect their asset base and their living situation overall. The importance of livelihood diversity has been highlighted by this study. Among the six household types (landless households, rice monocroppers, diversifiers, larger farmers, small fishermen, and large fishermen), our findings demonstrate that most vulnerable people in terms of income and access to social support services are the households with low capacity to invest in diversification of their income sources: landless households, rice monocroppers, and small fishermen. Low levels of income from farming production caused by low siltation from low flooding significantly decrease their annual income and nutrition and force them to risk migration.

This research suggests further studies on how supplementary livelihood strategies that diversify the existing livelihood base should build on existing livelihoods and on specific characteristics of each village. They should also consider the initiatives of the villagers themselves. This is particularly important among the poorest groups, who already have the weakest level of resilience and whose living conditions are expected to deteriorate further. An improved standard of living brings several benefits that were also visible in the interviews: better housing, additional assets to support and diversify existing livelihood sources, and savings that can be used during difficult times. It also has more indirect consequences, leading, for example, to better health conditions and improved school attendance by their children. All of these factors can be seen to build, directly or indirectly, the household's resilience to environmental shocks and stress.

Our calculations of the critical level of resilience (Table 8.2) show that all annual incomes were higher than the minimum poverty threshold (MPT), but only slightly positive for landless households, rice monocroppers, and small fishermen. In consequence, their total income is not enough for daily expenses, maintaining their farms, plus investing in the coming year. The situation will continue to worsen if natural resources continue to decline, creating more difficulties for those households.

Our research also examined the impact of low flooding on rice production, from the plot level to the nation as a whole, by calculating GDP gain and loss between the "good" and "bad" years in our study. The result shows that the low flood of the bad year reduced the country's GDP (Table 8.3). This clearly shows that the impacts on flood pulse patterns by different key drivers such as climate change and hydropower dam construction are likely to bring new challenges as well as opportunities and to magnify the challenges that people in developing countries are already facing. The majority of the impacts to people and their livelihoods are mediated through alterations in the hydrological cycle and, consequently, in the spatial and temporal availability of water caused by changing climate. This is likely to be the case also in Cambodia's TSL area that forms a unique lake floodplain system with remarkable seasonal changes in its water level.

In conclusion, we point out that one of the most capable strategies for enhancing people's adaptive capacity is to improve their ability to maintain a productive livelihood and thus to raise their general living conditions. This study found a low level of existing adaptive capacity especially of poor households. Providing short-term support is not likely to increase their long-term adaptive capacity unless other actions are taken in parallel to increase resilience. Since the households' livelihoods are significantly linked with the environmental services provided by the TSL, their household strategies to adapt to the flood pulse changes must thus be considered as a driving force complementary to already existing actions aimed at water management and suitable livelihood development in the TSL and Cambodia.

Acknowledgments The authors are very much indebted to the villagers and relevant sectoral department governors/authorities for their warm hospitality and great cooperation during the field research. Intellectual and institutional support is acknowledged from the Ecosystem Service and Land Use Research Center of the Royal University of Agriculture in Phnom Penh, Cambodia, and the MOISA laboratory of École Doctorale Économie et Gestion in Montpellier, France. Data collection, analysis, and writing were funded by the UMR-GRED research unit of the French Research Institute for Sustainable Development (IRD)-2015 and Institut Agronomique, Vétérinaire et Forestier de France—Agreenium-2016, respectively.

References

Ahmed, et al. (1998). *Socio-economic assessment of freshwater capture fisheries of Cambodia; report on a household survey, 1998*. Phnom Penh: Mekong River Commission.

Barral, S., Touzard, I., Ferraton, N., Rasse-Mercat, E., & Pillot, D. (2012). 'Assessing smallholder farming. Diagnostic analysis of family-based agricultural systems in a small region', Illustrated with the Case Study of the Giham Pioneer Front, Sumatra, Indonesia. Southeast Asian Regional Center for Graduate Study and Research in Agriculture, Philippines.

Bonheur, et al. (2002). Natural resources management for human security in Cambodia's Tonle Sap biosphere reserve. *Environmental Science and Policy, 5*(1), 33–41.

Cochet, H. (2012). The systeme agraire concept in francophone peasant studies. *Geoforum, 43*(1), 128–136.

Cochet, H. (2015). *Comparative agriculture*. Dordrecht: Springer.

Gibson, E., Fedorenko, E., & Rohde, D. (2006). The nature of working memory capacity in sentence comprehension: Evidence against domain-specific working memory resources. *Journal of Memory and Language, 54*(4), 541–553.

Glenn, D. I. (2009, November). *Determining sample size 1. Fact sheet* (pp 1–5).

Hap et al. (2016). *A review of socioeconomic studies in the fisheries sector in Cambodia* (p 14).

Keskinen, M. (2006). The Lake with Floating Villages: A socio-economic analysis of the villages around the Tonle Sap Lake. *International Journal of Water Resources Development, 22*(3), 463–480.

Keskinen, M., & Varis, O. (2012). Institutional cooperation at a basin level: For what, by whom? Lessons learned from Cambodia's Tonle Sap Lake. *Natural Resources Forum, 36*(1), 50–60.

Keskinen, M. et al. (2011, December). *Exploring Tonle Sap Futures study: Baseline results from hydrological and livelihood analyses*.

Kimsun, T., & Bopharath, S. (2011). *Poverty and environment links: The case of rural Cambodia*. CDRI Pub.

Kummu, M., et al. (2008). Sediment: curse or blessing for Tonle Sap Lake? *Ambio, 37*(3), 158–163.

Lamberts, D. (2006). The Tonle Sap Lake as a productive ecosystem. *International Journal of Water Resources Development, 22*(3), 481–495.

MAFF. (2014). *Annual report of Agriculture Forestry and Fisheries 2013–2014 and Direction 2014–2015*. Ministry of Agriculture Forestry and Fishery, Phnom Penh.

MAFF. (2016). *Annual report of Agriculture Forestry and Fisheries 2015–2016 and Direction 2016–2017*. Ministry of Agriculture Forestry and Fishery, Phnom Penh.

Marschke, M., & Berkes, F. (2011). Exploring strategies that build livelihood resiliance: A case from Cambodia. *Addiction*.

Millennium Ecosystem Assessment (2005). *Ecosystems and human well-being: Synthesis*. Available at: http://www.who.int/entity/globalchange/ecosystems/ecosys.pdf%5Cn, http://www.loc.gov/catdir/toc/ecip0512/2005013229.html

MOEF. (2016). *Cambodia macroeconomic monitor mid-year assessment 2016*. Ministry of Economy and Finance, Phnom Penh.

MOP. (2013). *Poverty in Cambodia – Redefining the poverty line* (pp. 1–12).

MOP. (2014). *Poverty alleviation – An approach to an action plan*.

Neang, M. (2015). *Maintaining ecosystem services provided by rice production systems in Cambodia: Identifying costs for farmers and consumers' preferences*. Clermont-Ferrand: University of Auvergne.

NIS. (2008). *General population census of Cambodia 2008* (pp. 1–37). Ministry of Planning.

Nuorteva, P., Keskinen, M., & Varis, O. (2010). Water, livelihoods and climate change adaptation in the tonle sap lake area, Cambodia: Learning from the past to understand the future. *Journal of Water and Climate Change, 1*(1), 87–101.

Pillot, D. (2008). *Jardins et rizières du Cambodge: les enjeux du développement agricole (Gardens and Rice lands of the Cambodia – The agricultural development issues)*. Paris: GRET Association Ed. Karthala.

RGC. (2001). *Royal degree on the establishment and management of the Tonle Sap biosphere reserve*. Available at: http://www.tsbred.org/docs/lawandregulation/RoyalDecreeoncreationandmanagemetoftsbrEng.Pdf

Rigg, J. (2006). Land, farming, livelihoods, and poverty: Rethinking the links in the Rural South. *World Development, 34*(1), 180–202.

Serrat, O., Gallego-Lizon, T., & Moffatt, D. (2005). *The Tonle Sap basin strategy*. Asian Development Bank. Available at: http://www.adb.org/sites/default/files/institutional-document/33050/files/tonle-sap-basin-strategy.pdf

Sithirith, M. (2011). *Political geographies of the Tonle Sap: Power, space and resources political geographies of the Tonle Sap: Power, space and resources*.

Sithirith, M. (2015). The governance of wetlands in the Tonle Sap Lake, Cambodia. *Journal of Environmental Science and Engineering, 4*(B), 331–346.

Sothorn, K. et al. (2011). *Policy coherence in agricultural and rural development: Cambodia* (55).

Tallis, H., & Polasky, S. (2009). Mapping and valuing ecosystem services as an approach for conservation and natural-resource management. *Annals of the New York Academy of Sciences, 1162*, 265–283.

Van Zalinge et al. (2000). *Where there is water, there is fish? Cambodian fisheries issues in a Mekong River Basin perspective*.

Varis, O., & Keskinen, M. (2006). Policy analysis for the Tonle Sap Lake, Cambodia: A Bayesian network model approach. *International Journal of Water Resources Development, 22*(3), 417–431.

Västilä, K., et al. (2010). Modelling climate change impacts on the flood pulse in the lower mekong floodplains. *Journal of Water and Climate Change, 1*(1), 67–86.

WCS. (2010). Conservation of threatened grassland biodiversity in the Tonle Sap basin, Cambodia. Briefing document, April 2011. Phnom Penh.

World Bank. (2006). Cambodia, halving poverty by 2015? *Poverty Assessment 2006 4*(35213), 1–250.

Chapter 9
Rice Straw: An Alternative for Energy Generation by Anaerobic Co-Digestion to Pig Manure

Nguyen Vo Chau Ngan, Nguyen Huu Chiem, Tran Sy Nam, Le Hoang Viet, and Kjeld Ingvorsen

Abstract Anaerobic digestion technology (biogas) has been used to treat husbandry wastes in the Mekong Delta of Vietnam for more than 30 years. However, biogas applications are limited. One major barrier is lack of input material for biogas digesters, because husbandry is normally conducted at the household scale. In contrast, rice straw (RS) is available in huge quantities in the Mekong Delta, but it is seldom put to use. This study tested the suitability and efficiency of anaerobic co-digestion of RS and pig manure (PM) to produce biogas. The study used 21-L lab-scale batches of anaerobic co-digestion of PM and RS, testing different sizes of RS. The mixing ratio for co-digestion was 50:50 based on organic dry matter values of input materials. The results showed that temperature, pH, alkalinity, and redox potential of almost all RS sizes were in the suitable range for activities of methanogenic microorganisms. The biogas yields per kilogram of fermented organic dry matter for the treatments were, for 1.0 cm RS, 691.05 L; 10.0 cm, 687.79 L; 20.0 cm, 685.08 L; and original RS size (uncut), 680.44 L. These values did not significantly differ from that of the 100% PM treatment. The methane content tended to increase with time, starting with the second week, and was well qualified for energy use. Our results strongly confirm that RS can be added to PM to produce biogas and that chopping RS is unnecessary.

Keywords Anaerobic co-digestion · Biogas · Rice straw · Mekong delta · Vietnam

Nguyen Vo Chau Ngan (✉) · Nguyen Huu Chiem · Tran Sy Nam · Le Hoang Viet
College of Environment & Natural Resources, Can Tho University, Can Tho, Vietnam
e-mail: nvcngan@ctu.edu.vn

K. Ingvorsen
Department of Bioscience, Aarhus University, Aarhus, Denmark

M. A. Stewart, P. A. Coclanis (eds.), *Water and Power*, Advances in Global Change Research 64, https://doi.org/10.1007/978-3-319-90400-9_9

9.1 Introduction

The Mekong Delta (MD)—the "rice bowl" of Vietnam—produces about 30 million tons of paddy rice straw (RS) annually, but the RS is normally burned in place due to high intensive farming activities. Open burning of paddy RS causes loss of precious nutrients and pollution to the environment. According to Ngo and Duong (2009), burning one ton of RS causes a loss of 91.3% carbon (C), which is equivalent to 291.2 kg C and releases 1067.6 kg CO_2 and 12.6 kg NO. Thus, it is necessary to explore ways to make good use of paddy RS to limit gas emissions from such on-field burning practices.

In addition to growing rice, livestock is raised on the MD, with around 3.6 million pig herds (GSO 2016). A large quantity of livestock waste is not managed and treated correctly, polluting the surrounding area and negatively affecting local residents' health. In recent years, biogas technology has been introduced to local farmers as an appropriate way to eliminate pollution from livestock. Nevertheless, small-scale pig farming, combined with sudden diseases and/or unstable markets, causes pig herds with unsteady numbers, which can lead to a lack of pig manure (PM) to feed digesters, so the operation efficiency of biogas digesters is rather low. This has a negative impact on local people's attitude toward economic efficiency of biogas applications.

Research on using local biomass to produce biogas in the MD has been published. One study suggested that farmers in the MD can use water hyacinth as a potential supplement to PM in their anaerobic digesters in cases of shortages of PM (Nguyen et al. 2011). Another study found that spent mushroom compost may be an acceptable material for energy recovery in an anaerobic fermentation process (Nguyen and Fricke 2012). Research also found that some garden weeds could be applied as extra input material for biogas production (Nguyen et al. 2014). In searching for biomass resources available in the MD, we found that paddy RS could be a promising supplemental input to generate biogas (Mallik et al. 1990; Zhang and Zhang 1999), but no prior research has studied this possibility in the MD.

Anaerobic technology was introduced in the MD in the 1980s, but it is not popular for several reasons, such as lack of input material (Vo et al. 2012). RS, which used to be a highly popular input material source in the MD, has been suggested for use in anaerobic co-digestion with PM to generate biogas. Co-digestion of various solid biowastes that helps balance the C:N ratio could optimize the digestion process and improve biofermentation efficiency (Wang et al. 2012). A previous study indicated that anaerobic co-digestion between PM and cornstalks at a C:N ratio of 20 increased cumulative biogas production up to 11-fold and increased cumulative net methane volume up to 16-fold compared to digesting PM alone (Wu et al. 2010). A similar observation on both batch and semicontinuous anaerobic co-digestion of PM and spent mushroom compost also demonstrated that optimum methane production was achieved with an adjusted C:N ratio of 22 and 28 (Nguyen et al. 2012). By optimizing the substrate's C:N ratio, anaerobic co-digestion of wastes of different C:N ratios could greatly enhance the efficiency of biogas digestion. Furthermore, regarding pretreatment of RS, one study found that large particles of RS could

Fig. 9.1 Preparing the materials for testing. (**a**) Cutting rice straw to various sizes. (**b**) Weighting rice straw for loading. (**c**) Drying pig manure in a shadowed area. (**d**) Grinding pig manure for experiments

decrease gas production compared to small particles (Sharma et al. 1988). Therefore, this study aimed to apply RS as additional material to PM through anaerobic digesters for biogas production, using different sizes of RS to test the efficiency of co-digestion.

9.2 Materials and Methods

9.2.1 Material Preparation

The input materials loaded for the anaerobic process were prepared as follows (Fig. 9.1):

- Rice straw (RS) of the 50,404 rice variety was collected from a rice field at Dong Hung 2 Village, Dong Thanh Ward, Binh Minh District, Vinh Long Province. The RS was air-dried at ambient temperature for 5 days up to unchanged weight

Fig. 9.2 Different sizes of rice straw cut for the testing: 1 cm, 10 cm, 20 cm, and uncut straw

and cut as needed into different sizes: 1.0 ± 0.3 cm), 10.0 ± 0.8 cm, 20.0 ± 1.3 cm, and original, uncut size (72 ± 13 cm) (Fig. 9.2).

- Pig manure (PM) was collected from a livestock household at Phu Loi Village, Phu Thanh Ward, Chau Thanh District, Hau Giang province and was air-dried up to unchanged weight. The dried PM was mashed and mixed until it became a homogeneous mixture.
- Inoculum was effluent taken from an existing active biogas plant, used as the seeding liquid to shorten the time of gas production. The source was an 8 m^3 polyethylene biogas tube currently used to treat the piggery wastewater at a farm household at Long Hoa Ward, Phong Dien District, Can Tho City.

9.2.2 Experimental Setup

Five anaerobic digestion apparatuses were randomly installed, with testing times of 45 continuous days. Each apparatus consisted of a 21-L plastic bottle connected to a gas collection pipe (Fig. 9.3). To avoid the gas produced pressing the substrate against the gas collector, only 17 L of each digester was filled with substrate; the remaining 4 L was left empty for gas production.

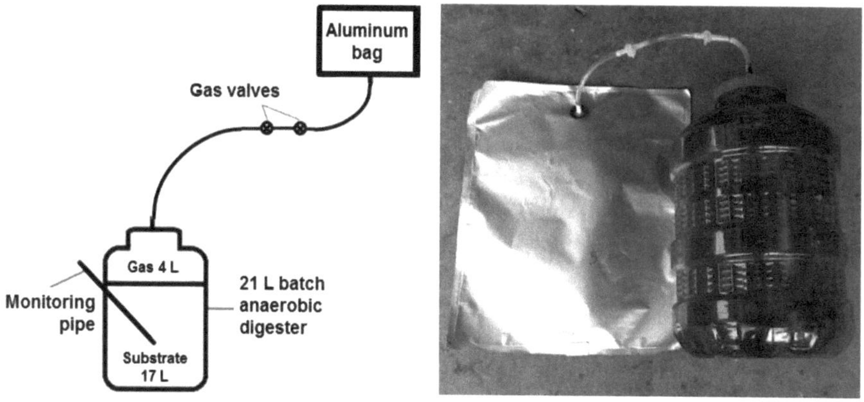

Fig. 9.3 Experimental apparatus

Table 9.1 Experimental treatments: co-digestion of rice straw (RS) and pig manure (PM)

Treatment	Material	Straw size (cm)	Mixing ratio	C:N ratio
NT1	RS+PM	1	50: 50	25.0:1
NT2	RS+PM	10	50: 50	24.9:1
NT3	RS+PM	20	50: 50	24.9:1
NT4	RS+PM	Uncut (72 ± 13)	50: 50	24.8:1
NT5 (control treatment)	PM	–	100	20.8:1

All bonds for the reactor (the inlet, outlet, and gas pipe) were connected by soft joints to be airtight and watertight. To minimize the growth of algae that could create oxygen inside the reactors, all reactors were covered by black nylon bags throughout the testing period. A 15-L aluminum gas bag was connected directly to each of the reactors through its gas collection pipe to collect the gas generated. Two air valves were installed on the pipe connecting the reactor and the gas bag, to switch the gas off when the gas was being recorded. Another polyvinyl chloride pipe was submerged into the substrate through each reactor body for monitoring purposes (Fig. 9.3).

For co-treatment sets, RS of different sizes was mixed with the PM at a ratio of 50:50 (Table 9.1). The mixing ratio was based on the materials' organic dry matter (ODM) values. A nonmixed treatment (100% PM) was also set up as a control. Each treatment was set up in triplicate.

Regarding anaerobic processes, the degradation of organic matter was measured based on volatile solids (VS). Values of the input material. For an anaerobic digester, the optimal input should be 1–4 kg VS × day^{-1} × m^{-3} (Eder and Schulz 2007). Based on this suggestion, the RS used in the experiments was 1 g VS × day^{-1} × L^{-1}. For each of the co-digestion reactors, we used 497.03 g RS (VS = 382.5 g) and 552.87 g PM (VS = 382.5 g), for a total VS of 765 g; the control reactor contained 1105 g PM (VS = 765 g) and no RS. In this study, the retention time of the material was 45 days, but in practice all treatments were monitored for 60 days to check the fermentation status of the input material. The total material input was 1 × 17 × 45 = 765 g VS per 17 L in each reactor for 45 days.

Just after RS was put into reactors, 10 L of effluent from an activated biogas digester was added to each reactor to pretreat RS in aerobic conditions. After 5 days, PM was fed to each reactor and 7 L tap water and 200 mL inoculum were added to each reactor. All treatments were mixed by manually shaking the reactors gently every day to improve operating conditions. This mixing also prevents the formation of scum within the substrate.

9.2.3 *Analysis and Statistics*

The substrate before and after the experiments was sampled and analyzed for total Kjeldahl nitrogen (TKN) and organic dry matter (ODM) according to standard methods suggested by APHA et al. (1995). All analyses were conducted at the Biogas Laboratory, College of Environment and Natural Resources, Can Tho University, Can Tho, Vietnam.

The reactors' operation parameters—pH, temperature, alkalinity, and redox potential—were recorded daily directly from each treatment by portable equipment through the monitoring pipe attached on each reactor (Fig. 9.4).

The first gas record was taken after 3 days of operation for gas production. The gas produced was recorded daily by a Ritter gas meter (with the smallest scale of 20 mL), and gas composition was monitored each day by a Biogas 5000 gas analyzer. The gas records were expressed at ambient temperature and at stable atmospheric pressure (1004–1006 mbar); gas records were taken at the same time on all days.

The Duncan test was performed to determine significant differences among treatments using SPSS software. p-Values >0.05 indicate significance.

9.3 Results and Discussion

9.3.1 *Substrate Characteristics*

9.3.1.1 C:N Ratio

A high C:N ratio indicates that nitrogen is consumed rapidly by methanogens, which results in a lower gas production. In contrast, a lower C:N ratio causes ammonia accumulation and pH values exceeding 8.5, which are toxic to methanogenic bacteria (Fabien 2003). In this study, the C:N ratio of RS was 53.2:1, which was higher than the ideal range (20:1–30:1), showing rapid nitrogen consumption leading to lower gas production. The C:N ratio of PM was around 20.8:1, in the lower range of ideal values.

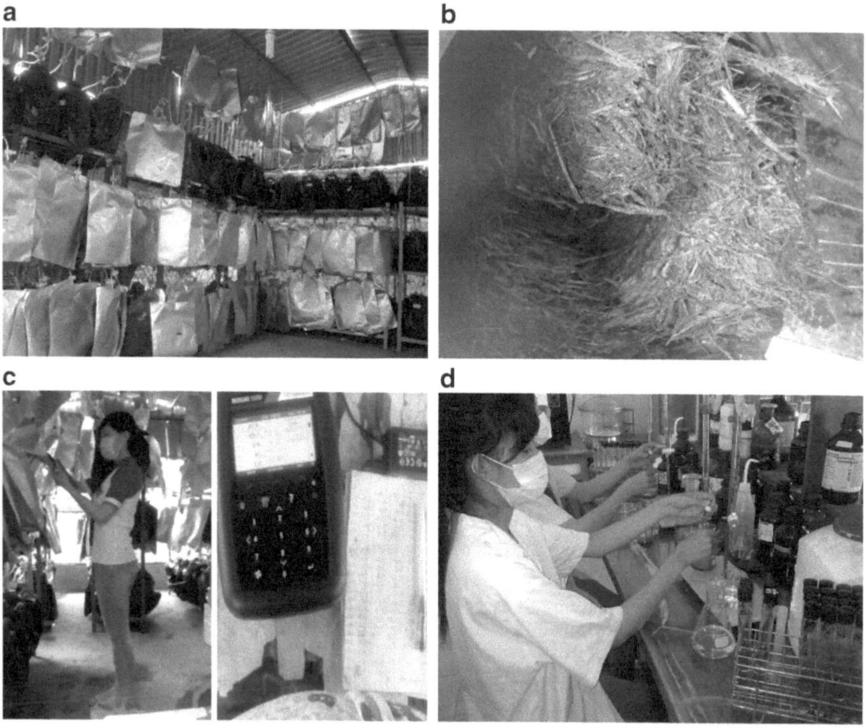

Fig. 9.4 Experimental setup and sampling activities. (**a**) Digesters with different treatments. (**b**) Rice straw after 60 fermentation days. (**c**) Recording operating units and gas components. (**d**) Sample analysis in the lab

The C:N values of each treatment with different sizes of RS are shown in Table 9.1. In the co-digestion of PM and RS, the C:N values were adjusted and reached the ideal range (from 24.8:1 to 25:1). There was a slight difference in the C:N values between treatments NT1, NT2, NT3, and NT4, which could be due to the analysis technique.

In this study, the co-digestion of PM and RS could balance the C:N value of input materials, optimizing the digestion process and improving biogas yield.

9.3.1.2 Temperature

The average temperature of co-digestion treatments recorded daily was 25–30 °C (Fig. 9.5). These temperature values are within the mesophilic temperature range for anaerobic microbes, 25–40 °C according to Fabien (2003).

According to Fabien (2003), thermophilic digesters are more efficient in terms of retention time, loading rate, and nominally gas production, but they need a higher heat input and greater control of digester operation. However, since the goal of our

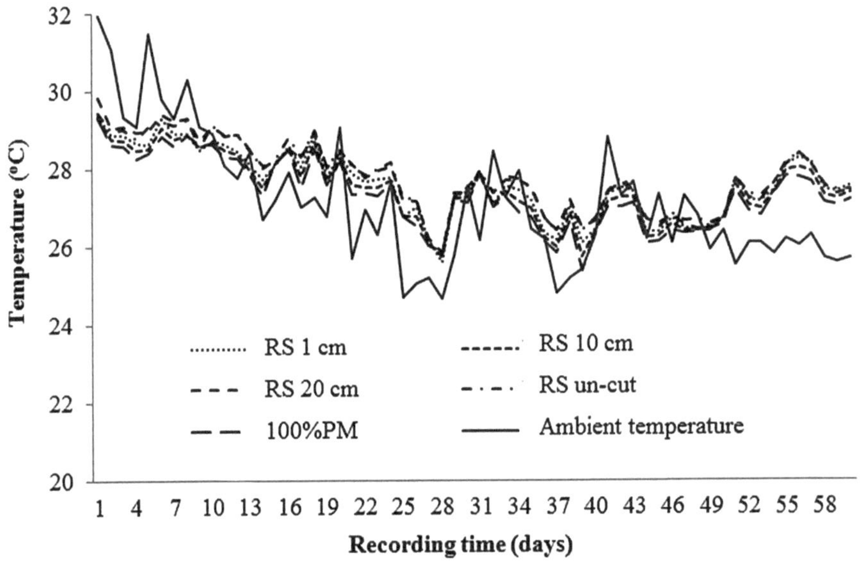

Fig. 9.5 Daily average temperature of each treatment

study is to apply this result to farmers in the Mekong Delta, the ambient temperature is preferable.

The temperature of substrates depended on biochemical fermentation reactions. In the first 3 weeks, due to more organic matter in the substrate, the bacteria operated under good conditions and the inner temperature was rather high. From the fourth week onward, with less of organic matter remaining from the substrates, biochemical processes were reduced and the temperature gradually decreased until reaching a stable temperature in the last 10 days.

9.3.1.3 pH

In the first 10 days, the pH values of all treatments were lower than 6.6 due to acidogenesis. During the acidogenesis stage, acetic, lactic, and propionic acids were formed, causing falling pH; on the fifth day the pH values of all co-digester treatments were the lowest (RS 1 cm, 6.19; RS 10 cm, 6.26; RS 20 cm, 6.22; RS uncut, 6.19), in the range that can be toxic for methane-forming bacteria (Fig. 9.6). For the control treatment of 100% PM, the pH values were in a suitable range at all times.

From the 10th day, the pH values of all treatments were within a suitable range of 6.4–7.2 (Fabien 2003) and remained stable for the later digestion stage. In this period, almost all organic matter in the substrate was fermented, and hydrolysis and acidogenic bacteria fell to an inactive state. However, the quantity of methanogenic

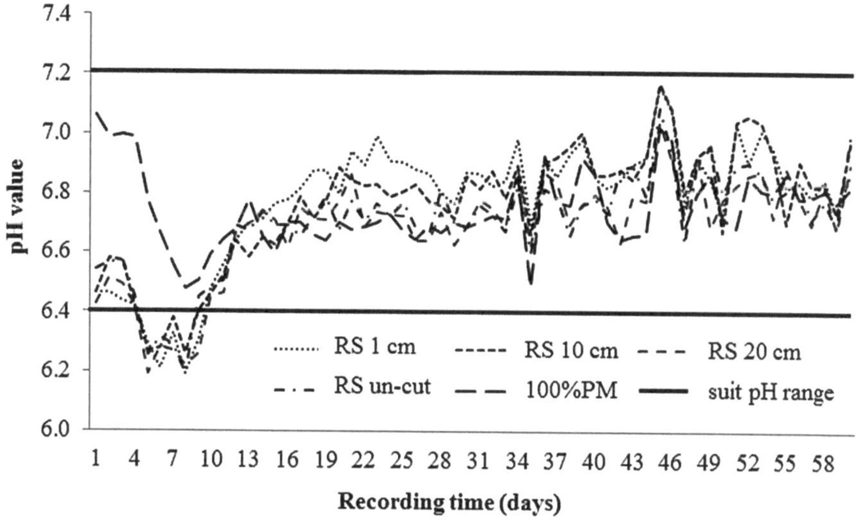

Fig. 9.6 pH variation of each treatment

bacteria increased and converted the organic acidity and simple compounds into biogas.

9.3.1.4 Alkalinity

Alkalinity is the buffering capacity of water to neutralize acid. The methanogens in anaerobic digestion are affected by small changes in pH, while the acid producers can function satisfactorily across a wide pH range. Digestion stability depends on the buffering capacity of digester contents. Higher alkalinity values indicate a greater capacity for resisting pH changes (Ken 2012).

The alkalinity values of all treatments tended to increase with fermentation time (Fig. 9.7). For the influent substrates, the alkalinity values ranged from 1230 mg $CaCO_3$/L (control 100% PM treatment) to 2427 mg $CaCO_3$/L (RS 10 cm treatment). The alkalinity value of the 100% PM treatment was lower than the ideal value of 1500 mg $CaCO_3$/L but then increased and reached the optimum value after 20 days.

The alkalinity values of the effluent from co-digestion treatments ranged from 2357 to 2798 mg $CaCO_3$/L, which was significantly different from the 100% PM treatment (1718–2107 mg $CaCO_3$/L). Alkalinity values of co-digester treatments ran between the ideal range of 1500 and 5000 mg/L (Ken 2012), creating a good buffer for pH values in the substrate.

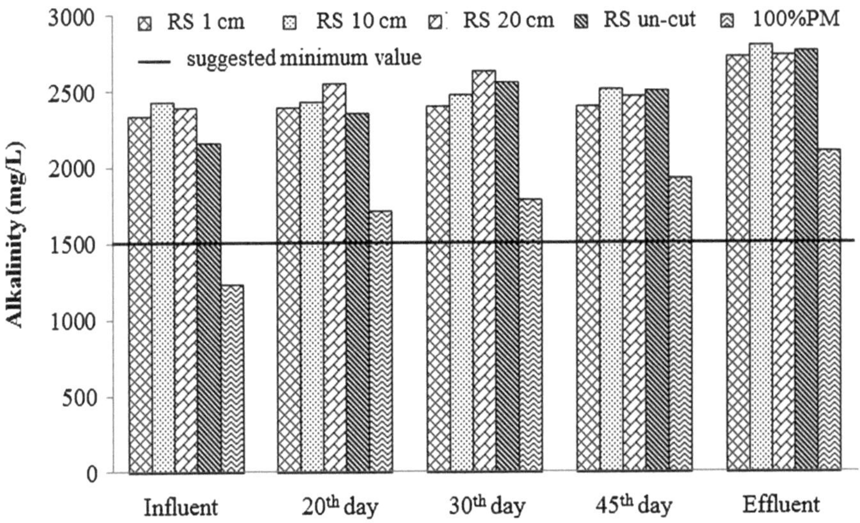

Fig. 9.7 Alkalinity values for each treatment

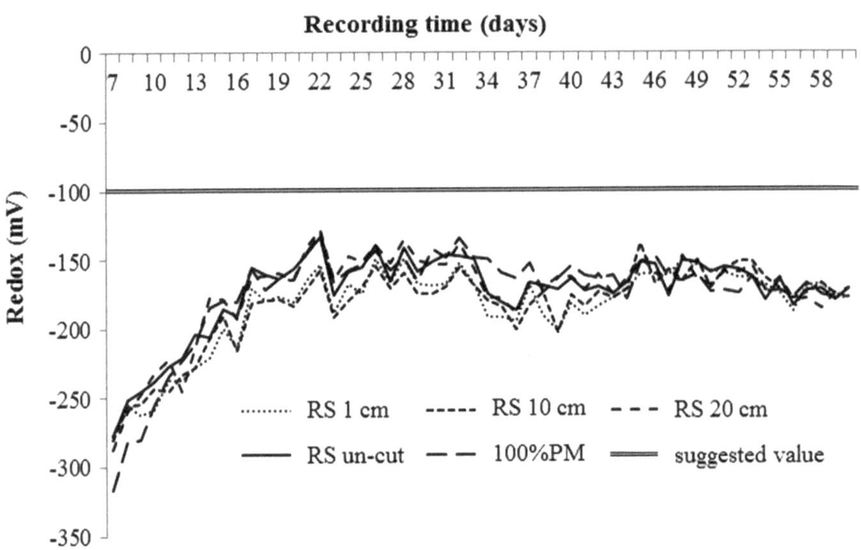

Fig. 9.8 Redox values for each treatment

9.3.1.5 Redox Potential

The redox potential of a digester is a measure of oxidizability or reducibility of its content. In this study, the redox potential of all treatments fluctuated between −316.4 and −128.2 mV (Fig. 9.8). These values show that all reactors operated efficiently under anaerobic conditions of less than −330 mV (Wiese and König 2007).

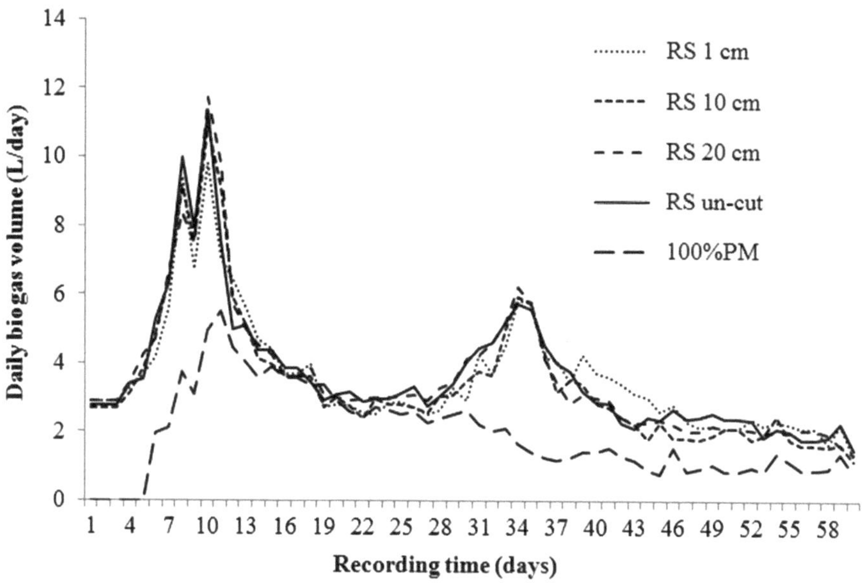

Fig. 9.9 Mean daily biogas production for each treatment

In the first 2 weeks, within the hydrolysis phase, the organic matter broke down to smaller monomers. Acetogenic bacteria converted these monomers to acetic acid, propionic acid, and butyric acid, leading to low redox values in this period. In the next phase, the methanogenic microorganisms converted these short-chain fatty acids into acetic acid, hydrogen, and CO_2, increasing the redox potential of substrates. In the final step, the redox values remained stable due to the methanogenic bacteria in the biogas.

Measurement of redox potential is an important activity to give early warning before the shift in pH occurs, which could be harmful to the fermentation process.

9.3.2 *Biogas Production*

9.3.2.1 Daily Biogas Volume

The mean daily biogas production of the treatments of different RS sizes during 60 days of digestion is shown in Fig. 9.9. All the co-digestion treatments produced gas earlier (from the first day) than the 100% PM treatment (from the sixth day). This could be because RS was pretreated for 5 days, helping shorten the time to generate biogas (Fig. 9.10).

From the 12th day, the biogas produced from all treatments decreased fewer to less organic compounds remaining in the substrates. However, from the 30th day, the biogas increased again and created a second peak for co-fermentation treatments

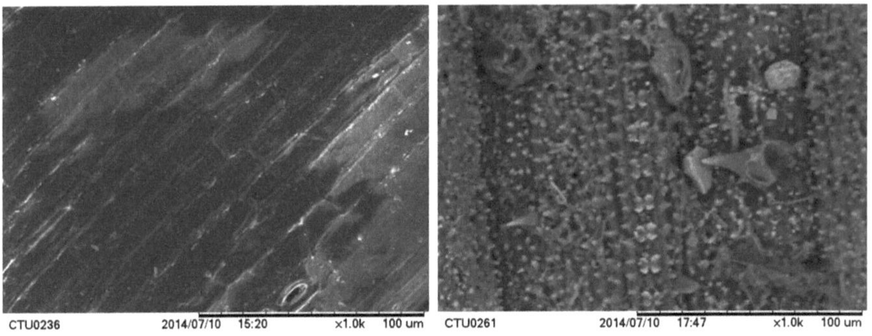

Fig. 9.10 Rice straw before pretreatment (left) and 5 days after pretreatment (right)

(Fig. 9.9). This may have resulted from the mixing actions that sank the RS into the substrate, and the second biogas peak in the co-reactors may have been produced mostly from RS. Because of its high cellulose content, RS fermented slowly, causing the second peak of biogas volume. No similar second peak occurred in the 100% PM treatment.

In the same context, due to the small size of input material, the 1-cm RS could have been easier to degrade than the other co-treatments. In fact, there was a third peak of biogas production from the 1-cm RS treatment on the 39th day (Fig. 9.9).

9.3.2.2 Cumulative Biogas Volume

The results of the cumulative biogas production by various sizes of RS treatment are shown in Fig. 9.11. The total biogas volumes after 60 days from RS treatments of 1 cm, 10 cm, 20 cm, and original size were 216.12, 214.65, 218.42, and 217.69 L, respectively. In contrast, the cumulative biogas volume of treatment of 100% PM was only 117.55 L.

The cumulative biogas volumes of the co-digestion treatments were not significantly different ($p > 0.05$) but were significantly higher ($p < 0.05$) than that of the 100% PM treatment. In this study, the smaller RS size did not greatly improve biodegradability and biogas production. This means that farmers do not need to chop RS into small particles before applying it to an anaerobic digester, and this can save labor costs.

We applied RS with thin fibers; thus, increasing the surface area of RS by chopping did not significantly increase biogas volume compared to treatment with larger-sized RS. On the contrary, the materials from smaller-sized treatments could easily float up on the surface of the substrate and could form scum, which could limit gas release from the substrate. Therefore, mixing methods should be taken into account when feeding fiber materials of small size to digesters so as to help break floating scum and release biogas.

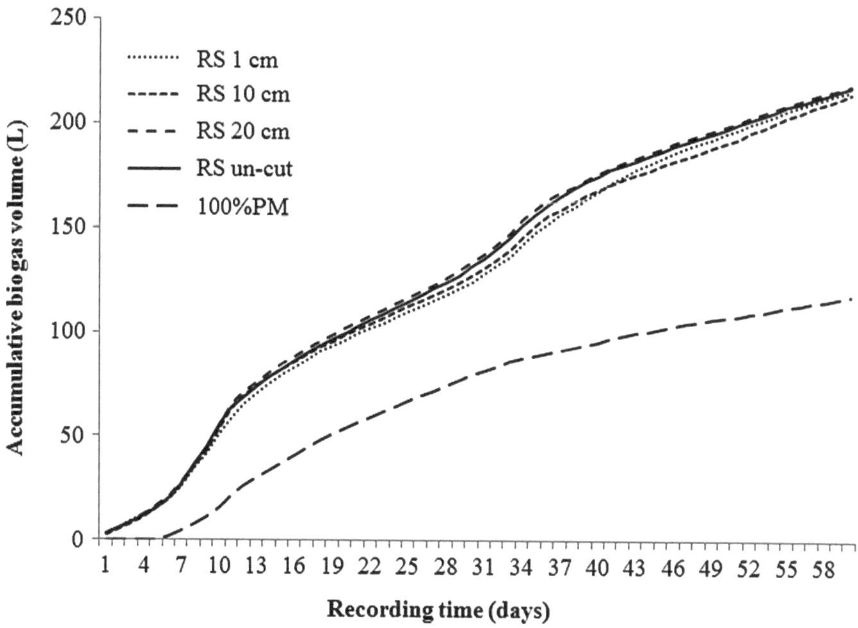

Fig. 9.11 Cumulative biogas production for each treatment

9.3.2.3 Biogas Yield

For biogas yield, the treatments of co-digestion with 1 cm, 10 cm, 20 cm, and uncut RS produced 658.90, 690.19, 695.61, and 678.16 L/kg fermented ODM and were not significantly different from one another (Fig. 9.12). However, these results showed an increase by 50.8%, 58.0%, 59.2%, and 55.2% compared to the control treatment of 100% PM (436.95 L/kg fermented ODM). Thus, biogas yield of co-digestion treatments was significantly higher ($p < 0.05$) than that of the 100% PM treatment. The results clearly indicate that co-digestion of PM and RS is effective to prolong the period of highest gas generation and improve biogas yield.

9.3.2.4 Biogas Composition

In the first week, all experiments were in the acidogenesis stage, causing lower methane content in the biogas produced. Methane content from the co-digestion treatments was rather low and fluctuated from 35.91% to 42.39%. The 100% PM treatment recorded the lowest methane value of 27.71% (Fig. 9.13).

From the second week, the percentage of methane increased in all treatments due to stable fermentation conditions and more active methanogenic microorganisms. Methane from co-digestion treatments ranged from 48.05% to 67.89% (RS 1 cm,

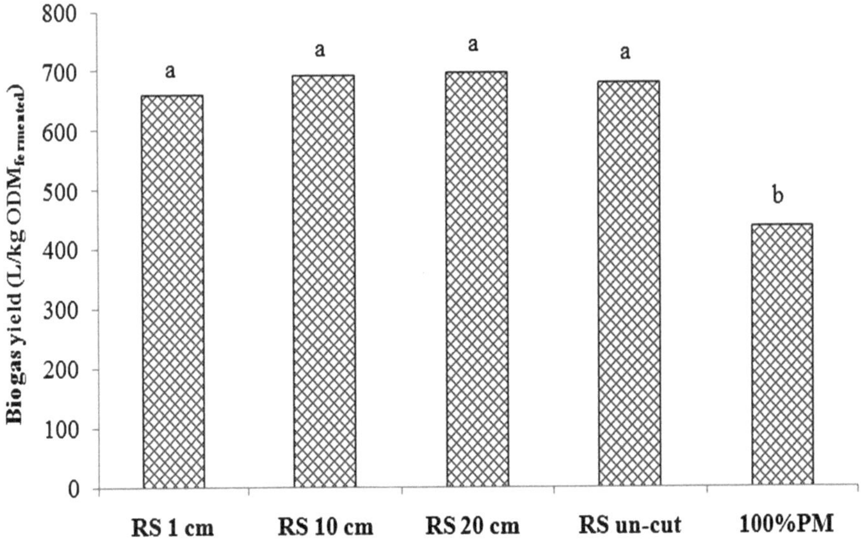

Fig. 9.12 Biogas yield for each treatment. Different letters indicate statistically significant differences

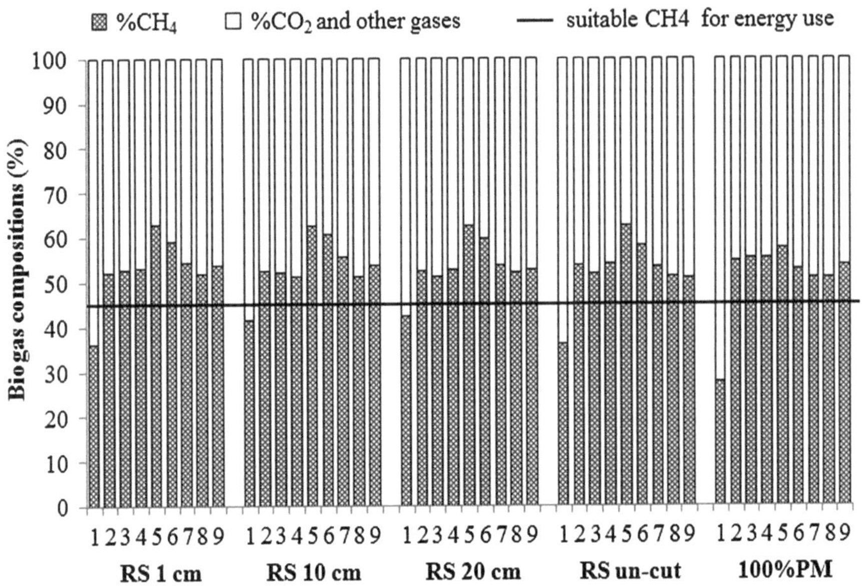

Fig. 9.13 Gas composition for each treatment

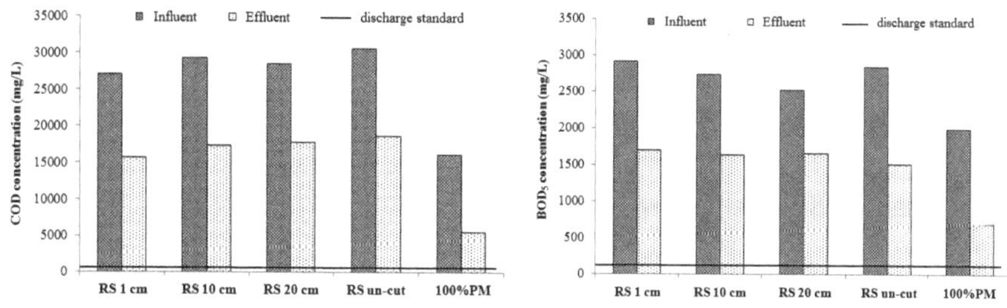

Fig. 9.14 Chemical oxygen demand (COD; left) and 5-day biochemical oxygen demand (BOD_5; right) values from influent and effluent for each treatment

48.13–66.79%; RS 10 cm, 48.05–67.89%; RS 20 cm, 48.79–67.53%; RS uncut, 48.45–64.89%), compared to 100% PM treatment, which ranged from 47.25% to 60.13% (Fig. 9.13).

The percentage of methane of all treatments was highest in the fifth week. After the fifth week, the percentage of methane decreased but remained high enough for cooking, which requires a methane content of at least 45%. The percentage of methane was not significantly different ($p > 0.05$) in all treatments.

9.3.3 *The Effluent Quality*

9.3.3.1 Organic Remaining

The chemical oxygen demand (COD) values from co-digestion treatments (27,141–30,591 mg/L) was higher than for the 100% PM treatment (16,139 mg/L) due to high organic matter content in RS (Fig. 9.14). After 60 fermentation days, the COD values from effluents of co-digestion treatments dropped to 42.02% (for RS 1 cm), 41.14% (for RS 10 cm), 38.01% (for RS 20 cm), and 39.06% (for un-cut RS), which were much lower than that of the 100% PM treatment (65.36%). Although RS has its own high organic matter, its organic matter is hard to degrade.

Five-day biochemical oxygen demand (BOD_5) values of effluents ranged from 696 to 1698 mg/L, which were lower than the values from influents of 1991–2925 mg/L (Fig. 9.14). In the same context, the treatment efficiency of 100% PM was 65.04%, which was higher than that of co-digestion treatments (34.34–46.84%).

The organic values from effluents were higher than the discharge requirement proposed in column A of Standard QCVN 62-MT:2016/BTNMT—the National Technical Regulation on the effluent of livestock (100 mg/L of BOD_5 and 300 mg/L of COD). The effluents in this case are not allowed to discharge into open sources but need additional treatment stages.

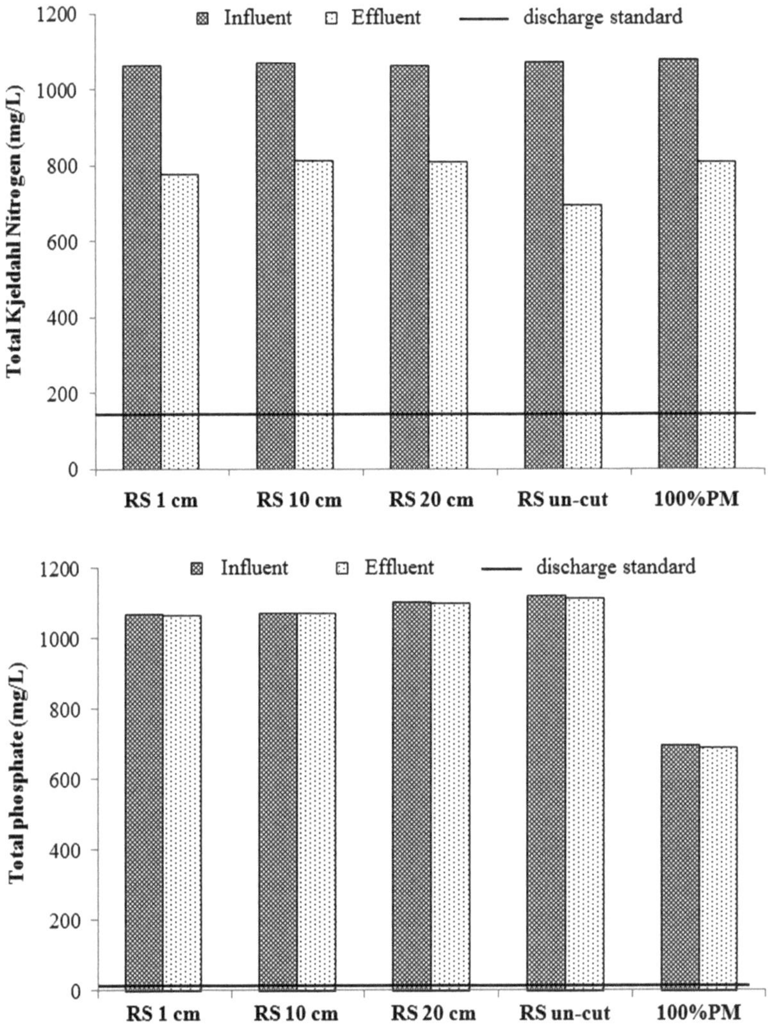

Fig. 9.15 Total Kjeldahl nitrogen and total phosphorus values from influent and effluent for each treatment

9.3.3.2 Nutrients Remaining

TKN from influent of all treatments was high, from 1060 to 1076 mg/L, and was little reduced after 60 days of fermentation (Fig. 9.15). The treatment efficiencies of treatments of 1 cm, 10 cm, 20 cm, and uncut RS and 100% PM were 26.85%, 23.87%, 23.95%, 35.18%, and 24.88%, respectively. The remained TKN from effluents was 4.6–5.4 times higher than the discharge requirements stated by Standard QCVN 62-MT:2016/BTNMT (150 mg/L of column B).

The total phosphorus influent values were also high in the co-digestion treatments (1068–1121 mg/L), but the 100% PM treatment was lower, reaching only 695 mg/L (Fig. 9.15). This shows that the RS contained more phosphorus than the PM and that chopped RS could not release more phosphorus than larger-sized RS treatments. The phosphorus treatment efficiencies were low, only 0.26–0.68% in the co-digestion treatments, and reached 0.82% in the 100% PM treatment. In the anaerobic fermentation process, microorganisms did not absorb much phosphorus, but it released more phosphorus after dying. The recorded remaining phosphorus from all treatments was 115–185 times higher than the discharge requirements stated by Standard QCVN 40:2011/BTNMT—National Technical Regulation on industrial wastewater (6 mg/L of column B).

The high values of remaining nutrients and organic matter from the effluents are not safe to be discharged into open sources. However, farmers could apply the effluents as organic fertilizers to their cultivation. To ensure application profits, calculation for a balance ratio of N-P-K composition is needed.

9.3.3.3 Total Coliform

The total coliform from influent of all treatments was high, with values of 1.3×10^7–5×10^8 MPN/100 mL (Fig. 9.16). After 60 fermentation days, total coliform decreased to 1.2×10^5–1.7×10^5 MPN/100 mL in the co-digestion treatments, and to 2.3×10^3 MPN/100 mL for the 100% PM treatment. The highest value of total coliform in the influent but lowest value in the effluent was recorded in the 100% PM treatment, which could be due to the medium in the substrate. For the co-digestion treatments, the RS itself creates a medium in which bacteria could attach and grow.

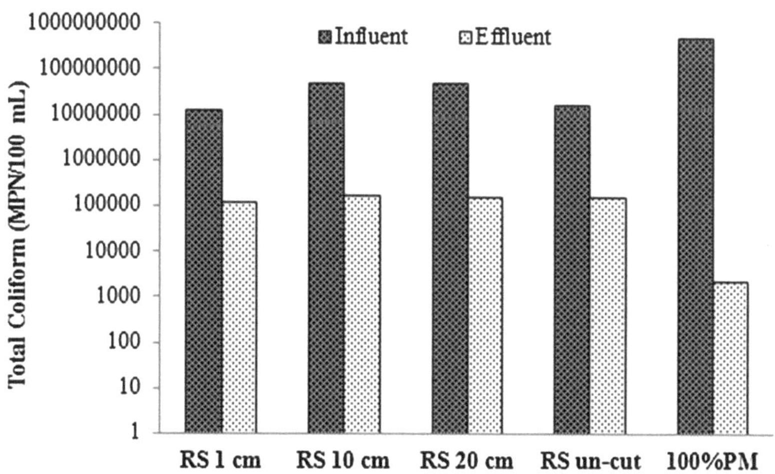

Fig. 9.16 Total coliform values from influent and effluent for each treatment

The treatment efficiencies were 99.07%, 99.66%, 99.68%, 99.05%, and 99.99% for RS 1 cm, RS 10 cm, RS 20 cm, uncut RS, and 100% PM, respectively. Even though all reactors had very high treatment efficiencies, total coliform in effluents of the co-digestion treatments was 24–34 times higher than required by Standard QCVN 62-MT:2016/BTNMT (5000 mg/L of column B).

9.4 Conclusion

In this study, anaerobic co-digestion of rice straw (RS) and pig manure (PM) was efficient and produced more biogas than did anaerobic digestion using solely PM. Gas production time was prolonged with co-digestion, and the biogas produced was highly qualified for energy use. All operation parameters of the digestion process were in suitable ranges for methanogenic bacteria growth and for biogas production. There was no significant difference in biogas yield among treatments of co-fermented PM with different sizes of RS, which implies that RS can be used for digesters without spending time chopping it into small particles. Although there was a limit in the floating phenomenon of RS within the digested substrate, the study results strongly confirmed that RS can probably be used as additional input material in an anaerobic digester to produce biogas. In addition, effluents with high nutrients and high organics remaining in digesters could be applied as an organic fertilizer source for aquaculture and agriculture.

The findings from this study are important for the promotion of biogas applications in small farm households in the Mekong Delta, as they show the economic and environmental benefits of biogas application. Since small household farmers are usually poor, they are unlikely to consider investment in biogas digesters when digesters using solely PM bring them little benefit. By using RS as an additional input to digesters, farmers will be able to maintain operation of their digesters to make use of biogas even in the face of shortages of PM. More important, RS can replace PM and be used as a sole input to digesters. As such, biogas application can be developed also for those rural areas where farmers primarily grow rice only. In addition, applying the effluents of digesters to farming saves farmers cultivation costs. This optimizes the economic benefits of biogas application for small household farms and encourages them to invest in biogas application, as well as to maintain operation of biogas digesters.

Moreover, keeping biogas digesters active by using RS as additional input helps reduce environmental pollution. It is a local common practice to dispose of livestock waste directly into open watercourses such as rivers and canals, and to burn RS after harvest in the Mekong Delta. Such practices cause pollution to the environment. By applying RS as additional material to digesters, small farmers can keep digesters in operation and this helps dissolve livestock waste and RS in a manner that is ecologically friendly to the environment, reducing carbon dioxide emissions.

For huge quantities of RS left on the field, it is necessary to have an environmental strategy from local or national constituencies to use the RS in an appropriate way instead of burning it, which causes pollution to the environment. The result of this study suggests an opportunity to scale up small household digesters to larger digesters for electricity generation for local communities. To do this, further studies are needed on RS collection and storage, RS transportation, relevant electricity grid networks, and so forth. RS is not waste—it is a source for energy generation.

However, the benefits of biogas applications for small farm households and availability of alternative inputs to digesters can be materialized only if small farmers are well aware of such information and can afford biogas digesters. To do this, campaigns are needed to raise awareness of local people on biogas applications and alternative inputs. At the same time, because small farmers in Vietnam, as well as in other developing countries, are generally poor, financial support by local governments or community-based organizations could increase the number of small farm households using biogas.

Acknowledgment The authors acknowledge financial support of Danish International Development Agency project "Sustainable Production of Biogas from Waste Rice Straw" (11-016AU).

References

APHA, AWWA, WPCF. (1995). *Standard methods for the examination of water and wastewater* (19th ed.). Washington, DC: American Public Health Association, American Water Works Association, and Water Pollution Control Federation.

Eder, B., & Schulz, H. (2007). Biogas-Praxis: Grundlagen, Planung, Anlagenbau, Beispiele, Wirtschaft-lichkeit: Ökobuch Magnum.

Fabien, M. (2003). An introduction to anaerobic digestion of organic wastes. Accessed from http://www.biogasmax.co.uk/media/introanaerobicdigestion_073323000101124042007.pdf. Last accessed on 9 May 2015.

GSO—General Statistics Office of Vietnam. (2016). *Statistical yearbook of Vietnam 2015*. Vietnam: Statistical Publishing House.

Ken, S. (2012). What every operator should know about anaerobic digestion. Accessed from http://www.wef.org/WorkArea/Downloadasset.aspx?id=12884902472. Last accessed on 9 June 2015.

Mallik, M. K., Singh, U. K., & Ahmad, N. (1990). Batch digester studies on biogas production from *Cannabis sativa*, water hyacinth and crop wastes mixed with dung and poultry litter. *Biological Wastes, 31*(4), 315–319. https://doi.org/10.1016/0269-7483(90)90089-B.

Ngo Thi Thanh Truc, & Duong Van Ni. (2009, November 16–8). Mitigation of carbon dioxide emission: An environmental assessment of rice straw burning practice in the Mekong Delta. In *Workshop proceeding MEKARN—Livestock, climate change and the environment*. Vietnam: An Giang University.

Nguyen Cong Thuan, Bui Thi Nga, Truong Van Qui, & Taro Izumi. (2014). Biogas productions of garden weed in biogas digesters. *Science and Technology Journal of Agriculture and Rural Development, 14*, 27–32 (in Vietnamese).

Nguyen Vo Chau Ngan, & Klaus, F. (2012). Energy recovery from anaerobic co-digestion with pig manure and spent mushroom compost in the Mekong Delta. *Journal of Vietnamese Environment, 3*(1), 4–9.

Nguyen Vo Chau Ngan, Le Hoang Viet, Nguyen Dac Cu, Nguyen Huu Phong. (2011). Biogas production of pig manure with water hyacinth juice from batch anaerobic digestion. In M. A. Stewart, P. A. Coclanis (Eds.), *Environmental change and agricultural sustainability in the mekong delta* (pp. 355–371). Springer.

Nguyen Vo Chau Ngan, Nguyen Truong Thanh, Nguyen Huu Loc, Nguyen Tri Nguon, Le Ngoc Phuc, & Nguyen Truong Nhat Tan. (2012). Potential use of water hyacinth and rice straw as additional loading materials for biogas digester. *Can Tho University Journal of Science, 22a*, 213–221 (in Vietnamese).

Sharma, S. K., Mishra, I. M., Sharma, M. P., & Saini, J. S. (1988). Effect of particle size on biogas generation from biomass residues. *Biomass, 17*(4), 251–263. https://doi.org/10.1016/0144-4565(88)90107-2.

Vo Chau Ngan Nguyen, Trung Hieu Phan, & Hoang Nam Vo. (2012). Review on the most popular anaerobic digester models in the Mekong Delta. *Journal of Vietnamese Environment, 1*(1), 8–19.

Wang, X. J., Yang, G. H., Feng, Y. Z., Ren, G. X., & Han, X. H. (2012). Optimizing feeding composition and carbon-nitrogen ratios for improved methane yield during anaerobic co-digestion of dairy chicken manure and wheat straw. *Bioresource Technology, 120*, 78–83. https://doi.org/10.1016/j.biortech.2012.06.058.

Wiese, J., & König, R. (2007). *Monitoring of digesters in biogas plants*. Düsseldorf: Hach Lange GmbH.

Wu, X., Yao, W. Y., Zhu, J., & Miller, C. (2010). Biogas and CH_4 productivity by co-digesting swine manure with three crop residues as an external carbon source. *Bioresource Technology, 101*(11), 4042–4047. https://doi.org/10.1016/j.biortech.2010.01.052.

Zhang, R., & Zhang, Z. (1999). Biogasification of rice straw with an anaerobic-phased solids digester system. *Bioresource Technology, 68*(3), 235–245. https://doi.org/10.1016/S0960-8524(98)00154-0.

Chapter 10
Assessment of Groundwater Quality and Its Suitability for Domestic and Irrigation Use in the Coastal Zone of the Mekong Delta, Vietnam

Nguyen Dinh Giang Nam, Goto Akira, Osawa Kazutoshi, Nguyen Hieu Trung, and Nguyen Vo Chau Ngan

Abstract This study evaluated the suitability of groundwater quality for domestic use and irrigation in Vinh Chau District, Soc Trang Province, Vietnam. Water samples from 30 domestic-use and 10 irrigation groundwater wells were collected and their physico-chemical parameters were analyzed. Results indicated that, based on WHO and Vietnamese standards, the groundwater quality had good suitability, generally to the medium level, but for drinking purposes some areas were found to require water treatment. Six water quality parameters—sodium adsorption ratio, soluble sodium percentage, permeability index, sodium percentage, residual sodium carbonate, and magnesium adsorption ratio—were calculated and employed for evaluating suitability for irrigation, in addition to electric conductivity. The results showed that the six parameters stayed at the permissible level for irrigation, whereas high sodium and salinity were a concern for irrigation. Use of the Piper trilinear diagram to analyze hydrochemical facies and groundwater types showed that the groundwater samples were in the sodium-potassium-chloride-sulfate category and that sodium water type was dominant.

Keywords Groundwater quality · Domestic water use · Irrigation · Suitability

---

Nguyen Dinh Giang Nam (✉)
United Graduate School of Agriculture Science, Tokyo University of Agriculture and Technology, Tokyo, Japan

College of Environment and Natural Resources, Can Tho University, Can Tho, Vietnam
e-mail: ndgnam@ctu.edu.vn

Goto Akira · Osawa Kazutoshi
Department of Hydrological and Environmental Engineering, Utsunomiya University, Utsunomiya, Japan
e-mail: goto@cc.utsunomiya-u.ac.jp; osawa@cc.utsunomiya-u.ac.jp

Nguyen Hieu Trung · Nguyen Vo Chau Ngan
College of the Environment & Natural Resources, Can Tho University, Can Tho, Vietnam

M. A. Stewart, P. A. Coclanis (eds.), *Water and Power*, Advances in Global Change Research 64, https://doi.org/10.1007/978-3-319-90400-9_10

10.1 Introduction

The Mekong Delta is one of the most vulnerable regions in the world, especially under the impacts of climate change. Until now, surface water has been the primary water supply source for domestic use and agriculture in the area. However, the changing hydrological regime upstream of the Mekong Delta has caused drought in recent years, which has impacted quality and quantity of surface water (Tuan et al. 2007). Therefore, groundwater is being considered as an option for domestic use and irrigation in the Mekong Delta, especially in the coastal zone, which is the most vulnerable area due to water shortages; here the trend of groundwater use has increased strongly (Nam et al. 2017). The quality of groundwater is as important as its quantity in evaluating the suitability of water for various purposes (Schiavo et al. 2006). However, the aquifer is heavily polluted with microbial and inorganic pollutants and is considered unfit for drinking water (Danh 2008). In addition, concentrations of natural contaminants and saltwater intrusion have caused a decline in groundwater quality, and it may not be suitable either for domestic water supply or for irrigation (IUCN- International Union for Conservation of Nature 2011). The major factors driving a decline in the quality of groundwater in the Mekong Delta are poor environmental practices in the delta contributing to surface and aquifer pollution; overexploitation, inducing seawater intrusion and mixing and concentration of contaminants; and poor well construction that creates a direct pathway for inferior-quality aquifer water and surface pollutants to mix with an otherwise good-quality aquifer (IUCN- International Union for Conservation of Nature 2011). Groundwater used for domestic and irrigation purposes can vary greatly in quality, depending on the type and quantity of dissolved salts (Sarath Prasanth et al. 2012). The chemical quality of groundwater is related to the geological history of the aquifers and can reveal important information on the suitability for domestic and agricultural use (Povinec et al. 2006). Therefore, understanding hydrochemical characteristics is crucial for groundwater planning and management.

The study area, Vinh Chau District (Fig. 10.1), is located in the coastal zone of Soc Trang Province, with a mean land surface elevation of about 1.0 m above mean sea level, and land use is mainly for agricultural production. This is a former salinity-controlled area where rice farming systems dominated. From the year 2000 onward, farmers protested against the protection of the salinity control measures, leading to a diversification of local land use (Kakonen 2008). At the present, serious impacts of salinity intrusion on surface water resources and freshwater shortages are increasing (DONREs – Department of Natural Resources and Environment of SocTrang 2012). Therefore, groundwater has been accessed as the water supply resource for domestic use and agricultural production (Nam et al. 2017). In addition, the study area belongs to a vulnerable zone of climate change impacts, and sea level rise can increase salt intrusion (Nhan et al. 2008). Those issues have been recognized by the local government and people through environmental education programs (DONREs – Department of Natural Resources and Environment of SocTrang 2012). However, due to a lack of experiment and practical studies on the environment in

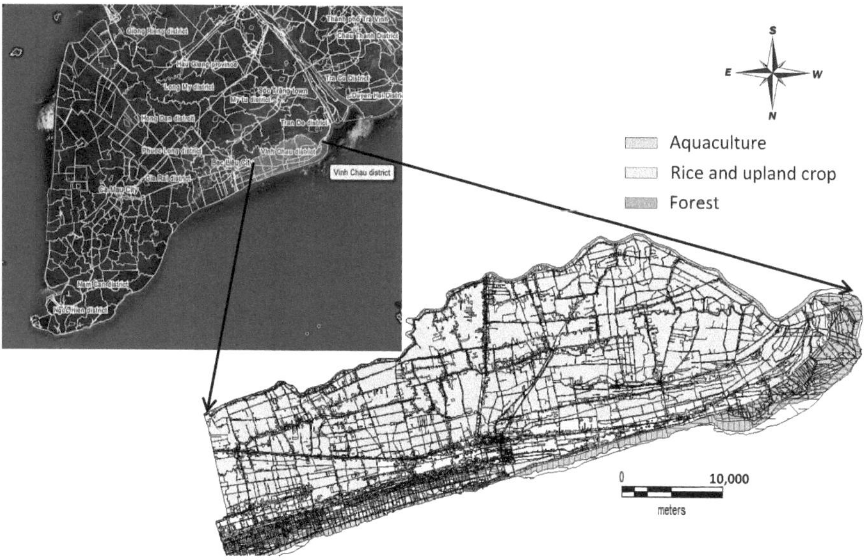

Fig. 10.1 The coastal zone study area, Vinh Chau, Soc Trang Province, Vietnam

general and on groundwater in particular, local governments are unable to make effective strategic plans for natural resource protection and management (Department of Environment in Vinh Chau 2011).

10.2 Materials and Methods

10.2.1 Sample Collection

Groundwater samples were collected from 10 wells for irrigation (samples VC1–VC10) and 30 wells for domestic use (samples D1–D30), ranging in depth between 90 and 120 m below ground level, during the post-monsoon season. The location of sampling points is shown in Fig. 10.2. High-density polyethylene bottles were used for sample collection. Bottles were completely filled with water, taking care that no air bubbles were trapped within the water samples. Bottles were sealed with double plastic caps, and precautions were taken to prevent evaporation and avoid sample agitation during transfer to the laboratory. Samples were then immediately transferred to the laboratory. During sample collection, standard procedures recommended by the American Public Health Association (APHA) were followed to ensure data quality and consistency.

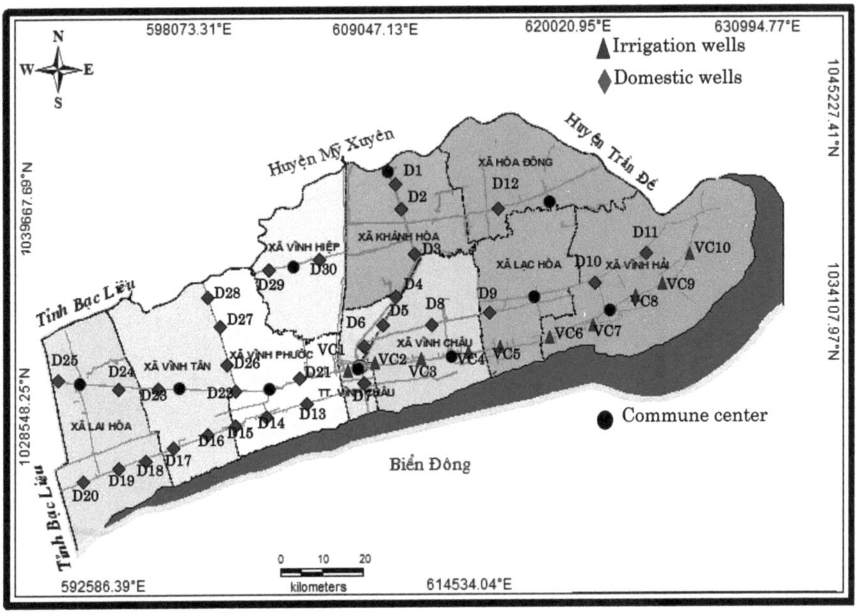

Fig. 10.2 Administrative map of study area and groundwater sample locations. D1-D30, domestic-use well samples; VC1-VC10, irrigation well samples

10.2.2 *Laboratory Measurements*

Temperature, electrical conductivity (EC), pH, and dissolved oxygen were measured in situ with the use of appropriate multiparameter instruments at the same points of water sample collection. The labeled samples were analyzed in the laboratory for concentrations of major ions (Ca, Mg, Na, K, HCO_3, SO_4, Cl, F) using APHA standard methods (APHA – American Public Health Association 1995). Sodium (Na) and potassium (K) were determined by flame photometer. Total hardness as $CaCO_3$, calcium (Ca^{2+}), magnesium (Mg^{2+}), bicarbonate (HCO_3^-), and chloride (Cl) were analyzed by volumetric methods, and sulfates (SO_4^{2-}) were estimated by using the colorimetric method.

10.2.3 *Classification Methods*

The suitability of groundwater for agricultural purposes was evaluated using the following parameters: sodium adsorption ratio (SAR), soluble sodium percentage (SSP), permeability index (PI), sodium percentage (%Na), residual sodium carbonate (RSC), and magnesium adsorption ratio (MAR). The results of the analyses were interpreted using graphical representations from the U.S. Salinity Laboratory and the Wilcox diagram. To determine the suitability of domestic water supply, we compared the values of different water quality parameters with those of the World Health Organization (WHO) (2004) and the Vietnamese standard (QCVN09) (MONRes – Ministry of Natural Resources and Environment 2015). The results of the analyses are represented using a Piper diagram of the relationships between the different cations and anions.

10.3 Results and Discussion

10.3.1 Domestic Water Supply

The following analytical results were considered to determine the suitability of groundwater in the study area for domestic water supply. The chemical analysis results of groundwater samples are presented in Table 10.1.

pH The pH of a solution is the negative logarithm of the hydrogen ion concentration, and the pH scale commonly ranges from 0 to 14. Acids have lower pH values, from 0 to 7, and alkaline solutions have higher pH values, from 7 to 14; 7 is neutral. Drinking water pH mainly ranges from 4.4 to 8.5 (Devendra et al. 2014). The pH of water provides vital information for many types of geochemical equilibrium or solubility calculations (Hem 1985). The pH values in the study area met the standard value for drinking water, specified as 6.5–8.5 (WHO) (WHO – World Health Organization 2004) and 5.5–8.5 (QCVN09) (MONRes – Ministry of Natural Resources and Environment 2015) (see Table 10.1). Four sample points had pH <7 (6.6–6.8); most of the groundwater samples had pH >7 (7.1–8.4); thus, the groundwater in the study area is slightly alkaline. Alkalinity of groundwater may be due to the presence of one or more of a number of ions, including hydroxides, carbonates, and bicarbonates. The phenomenon may be attributed to salt intrusion, which affects pH. Alkalinity results when the salts of a strong base and a weak acid are dissolved in the water.

Sulfate Natural water contains sulfate ions, and most of these are soluble in water. Sulfate is a combination of sulfur and oxygen and is part of naturally occurring minerals in some soil and rock formations that contain groundwater. The mineral dissolves over time and is released into groundwater. The maximum contaminant level is 400 mg/L (WHO and QCVN09) (WHO – World Health Organization 2004; MONRes – Ministry of Natural Resources and Environment 2015). The sulfate concentration in the study area ranged between 53 and 250 mg/L (Table 10.1), with an average of 146.5 mg/L, indicating that all samples fell within the desirable limit.

Nitrate Nitrogen is present in raw water, mainly in the form of HNO_3 (in its oxidized state). Levels of nitrates in groundwater in the study area ranged between 0.0 and 5.4 mg/L (Table 10.1), implying that all samples were within safe levels proposed

Table 10.1 Chemical analysis results for evaluating groundwater suitability for domestic use in the study area

Sample	pH mg/L	COD mg/L	SO_4^{2-} mg/L	Fe mg/L	Cl^- mg/L	NO_3^- mg/L	Total Coliform MPN/mL
D1	8.1	14	161.6	10.75	113	0	–
D2	7.1	5	163	2.02	204	0.0	–
D3	6.8	3	180	8.00	158	5.4	–
D4	7.1	13	163	0.71	161	0	–
D5	7.4	7	195	0.29	164	2.5	–
D7	7.6	5	190	0.95	220	1.4	–
D8	7.0	4	190	0.86	136	0	7
D9	7.3	3	183	0.07	132	1.5	–
D10	7.5	4	173	0.17	160	1.8	–
D11	6.7	3	173	1.16	137	0	–
D12	6.8	8	189	2.04	145	1.1	7
D13	8.1	4	163	0.30	124	1.4	–
D14	7.4	2	172	0.09	148	0.8	–
D15	7.3	3	234	0.20	123	0.1	–
D16	7.2	7	54	0.08	124	0.0	–
D17	6.6	4	57	29.60	170	0.0	–
D18	8.4	0	67	0.09	181	0.3	–
D19	7.5	2	78	0.17	338	0.8	4
D20	7.5	2	155	0.26	320	1.1	–
D21	7.7	9	75	0.01	210	0.4	–
D22	8.2	5	121	0.44	149	0.3	–
D23	8.3	5	189	0.34	480	3.2	–
D24	7.0	7	140	2.39	176	2.7	–
D25	7.6	8	110	0.61	210	0.0	–
D26	7.2	3	250	0.29	276	0.0	–
D27	7.6	0	168	0.49	363	0	–
D28	7.5	11	127	0.86	321	0.7	–
D29	8.0	3	75.6	0.01	340	1.7	–
D30	7.9	2	53	0.50	162	0.7	7

COD chemical oxygen demand, *MPN* most probable number

by WHO (50 mg/L) and QCVN09 (15 mg/L) (WHO – World Health Organization 2004; MONRes – Ministry of Natural Resources and Environment 2015).

Chloride Chloride is a negative ion of the element chlorine (Cl) and is widely distributed in the environment. Chloride is found naturally in groundwater through the weathering and leaching of sedimentary rocks and soils and the dissolution of salt deposits. Chloride concentrations in some groundwater wells of the study area were slightly higher than the WHO and QCVN09 standard of 250 mg/L (WHO – World Health Organization 2004; MONRes – Ministry of Natural Resources and Environment 2015) (Table 10.1), which may be caused by saltwater intrusion. The average chloride concentration was about the 205 mg/L, with a maximum of 480 mg/L and minimum of 113 mg/L.

Iron Iron (Fe) is a metal that occurs naturally in soils, rocks, and minerals. In the aquifer, groundwater comes in contact with these solid materials, dissolving them, releasing their constituents, including Fe, to the water. Samples at D1, D3, and D17 had very high Fe values. At concentrations approaching 0.2 mg/L Fe (WHO – World Health Organization 2004), water use efficiency may become seriously affected. However, most of sample points had safe values based on WHO and QCVN09 standards (WHO – World Health Organization 2004; MONRcs – Ministry of Natural Resources and Environment 2015).

Chemical Oxygen Demand (COD) COD is a measure of the oxygen required for the chemical oxidation of organic matter with the help of a strong chemical oxidant. In the study area, the COD of most groundwater samples indicated organic contamination (Table 10.1). This clearly demonstrates that the contamination generated from the surface is affecting groundwater quality in the adjacent areas through percolation in the subsoil.

Total Coliform Normally, groundwater does not contain this contaminant, and most samples in the study area were negative. However, coliform was found in wells D8, D12, D19, and D30 at very low concentrations. The presence of COD may indicate recent contamination of the groundwater by human sewage in the study area.

10.3.2 *Irrigation Suitability*

To assess overall water quality for irrigation, we assessed the water quality parameters SAR, SSP, PI, %Na, RSC, and MAR. In addition, EC is a good measure of salinity hazard to crops, and its relationship with water quality parameters also reflects the groundwater classification. Their corresponding values are presented in Table 10.2.

10.3.2.1 Sodium Adsorption Ratio (SAR)

SAR is a measure of the sodicity of the soil determined through quantitative chemical analysis of water in contact with it. Excess HCO_3^- and CO_3^{2-} ions in water react with Na^+ in soil, resulting in a sodium hazard (Subramani et al. 2005). SAR values are plotted against EC values (in μmhos/cm) over the U.S. Salinity diagram to categorize analyzed water samples according to their irrigational suitability quotient. SAR was calculated as

$$\mathrm{SAR} = \mathrm{Na}^{+} / \left(\left(\mathrm{Ca}^{2+} + \mathrm{Mg}^{2+}\right) / 2\right)^{0.5},$$

where concentrations of all ions are expressed in meq/L.

Table 10.2 Values of calculated groundwater quality parameters for irrigation well samples in the study area

Sample	SAR	SSP	RSC	PI	MAR	%Na
VC1	24.41	63.97	−74.10	66.94	47.38	63.97
VC2	20.90	58.94	−85.50	61.74	52.74	58.94
VC3	25.91	59.80	−122.43	62.49	57.49	59.80
VC4	19.26	53.73	−109.85	56.72	50.14	53.73
VC5	22.47	61.98	−68.56	66.00	53.77	61.98
VC6	15.90	55.34	−53.08	61.09	35.07	55.34
VC7	29.67	66.90	−84.12	69.77	41.73	66.90
VC8	31.73	76.57	−35.75	79.25	43.92	76.57
VC9	27.77	62.74	−112.80	65.07	48.81	62.74
VC10	21.94	58.28	−100.71	60.92	51.24	58.28

SAR sodium adsorption ratio, *SSP* soluble sodium percentage, *RSC* residual sodium carbonate, *PI* permeability index, *MAR* magnesium adsorption ratio, *%Na* sodium percentage

In the study area, the SAR values ranged from 15.90 to 31.73. Based on the SAR values, all samples have high sodium hazard. Plotting the results on the U.S. Salinity diagram (Fig. 10.3) shows that 80% of groundwater samples fall in the C3-S4 category, and 20% fall in the C4-S4 category, which means very high sodium content and high salinity in the sampled groundwater. For irrigation purposes, salinity should be controlled and plants with good salt tolerance should be proposed in the study area.

10.3.2.2 Permeability Index (PI)

Based on the PI, a water suitability classification for irrigation water was developed (Doneen 1964). The PI was calculated as

$$PI = \left(Na^+ + \left(HCO_3^-\right)^{0.5}\right) \times 100 / \left(Na^+ + Ca^{2+} + Mg^{2+}\right),$$

where all the ions are expressed in meq/l.

The PI values vary from 56.72 to 79.25, with the average value of about 65.00. A classification based on PI as recommended by the WHO for assessing suitability of irrigation (WHO – World Health Organization 2008) reveals that all samples belong to class 2, with PI ranging between 25% and 75%.

10.3.2.3 Sodium Percentage (%Na)

The %Na in the groundwater samples collected varied from 53.73 to 76.57 (Table 10.2). For rating irrigation water, the Wilcox diagram was used, in which the %Na is plotted against EC. Figure 10.4 shows the diagram plot, which indicates that 80% of the groundwater samples fell in the "Permissible to doubtful" range and 20% in the "Doubtful to unsuitable" range. The source of Na^+ in the groundwater

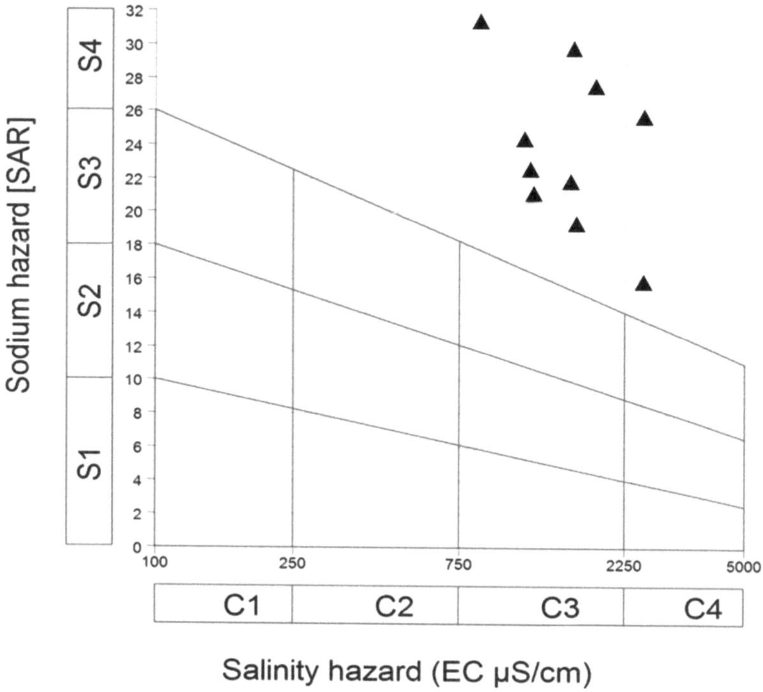

Fig. 10.3 U.S. Salinity diagram, indicating salinity hazard and sodium hazard of the ten irrigation groundwater wells sampled in the study area (triangles)

has been attributed to the weathering of feldspar and to overexploitation of groundwater.

10.3.2.4 Soluble Sodium Percentage (SSP)

SSP values preferably should be less than 60% for water to be suitable for irrigation purposes. In the present study SSP values ranged between 53.73% and 76.57%, with 50% of the water samples <60% and 50% exceeding the permissible level (U.S. Salinity Laboratory 1954) (Table 10.3).

10.3.2.5 Residual Sodium Carbonate (RSC)

The RSC index signifies the alkalinity hazard posed by water and indicates suitability of water for irrigation of clay soils (Raju 2007). The RSC values in the present study ranged from −122.43 to −35.75. Based on the U.S. Salinity Laboratory's classification level (U.S. Salinity Laboratory 1954), 100% of water samples were at suitable levels (Table 10.4).

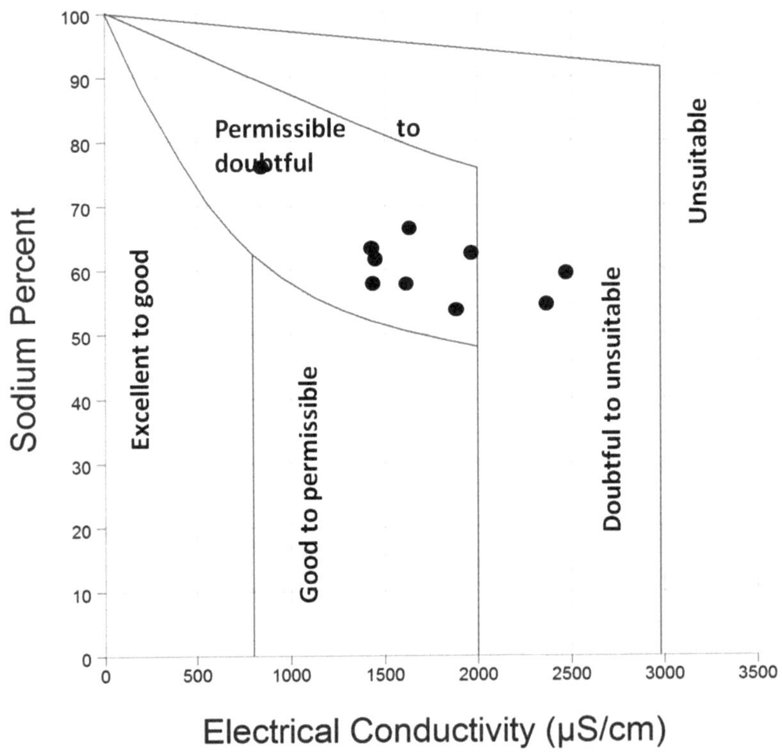

Fig. 10.4 Wilcox diagram for water quality classification of the ten irrigation groundwater wells sampled in the study area (circles)

10.3.2.6 Magnesium Hazard Ratio

MAR categorizes water into two broad classes: MAR <50 is considered suitable for irrigation, whereas MAR >50 is considered high magnesium hazard and has an adverse effect on crop yield. In this study 50% of groundwater samples were in safe level and 50% indicated water sources unsuitable for irrigation (Table 10.2).

10.3.3 Hydrochemical Facies

Figure 10.5 is a modification of the Piper diagram with a view to extend its applicability in representing water analysis in the simplest possible way. A Piper diagram is a graphical representation classifying water based on the dominant presence of cations and anions and has widespread use in assessing water type. The results are plotted on the proposed diagram to test its applicability for geochemical classification of groundwater and hydrochemical facies. The values obtained for groundwater types in the study area plotted on the Piper diagrams reveal that the major groundwater type is Na – K – Cl – SO_4 and sodium.

Table 10.3 U.S. Salinity Laboratory classification based on soluble sodium percentage value

SSP	Class	Percentage of samples
<20%	Excellent	0
20–40%	Good	0
40–60%	Permissible	50%
60–80%	Doubtful	50%
>80%	Unsuitable	0

Table 10.4 U.S. Salinity Laboratory classification based on residual sodium carbonate value

RSC	Condition	Percentage of samples
<1.25	Suitable	100%
1.25–2.5	Marginal	0
>2.5	Not suitable	0

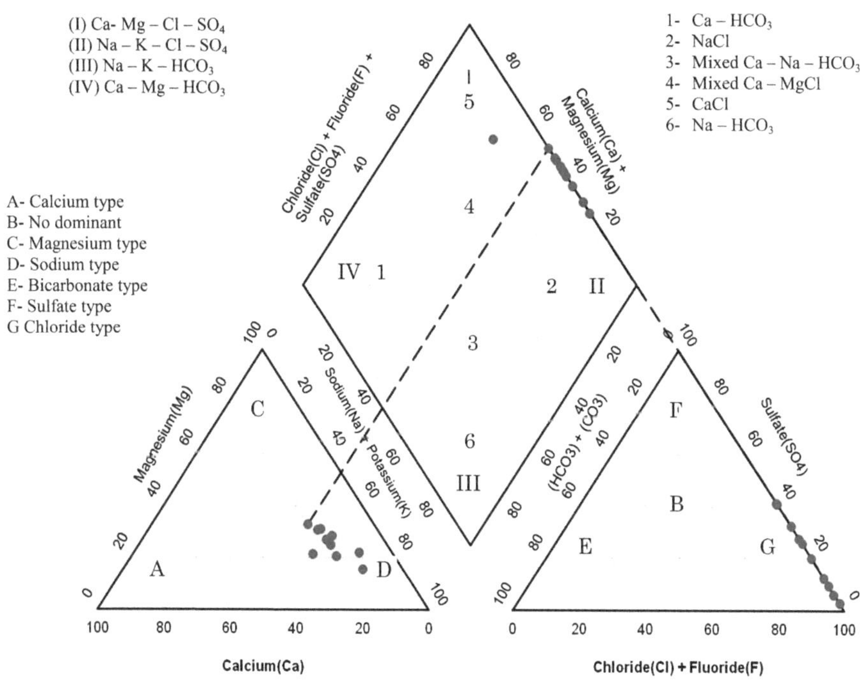

Fig. 10.5 Piper diagrams representing the hydrochemical facies of groundwater in the study area for the 30 domestic-use groundwater wells sampled (circles)

10.4 Conclusions

This study examined groundwater quality by assessing samples from 30 domestic water supply sources and 10 irrigation sources in the coastal zone of the Mekong Delta (Figs. 10.1 and 10.2). Based on our results, conclusions can be summarized as follows:

1. Chemical analyses showed that, based on WHO and Vietnamese standards, the groundwater quality had good suitability, generally to the medium level, but for drinking purposes some areas were found to require water treatment.
2. According to six computed groundwater quality indicators—sodium adsorption ratio (SAR), soluble sodium percentage (SSP), permeability index (PI), sodium percentage (%Na), residual sodium carbonate (RSC), and magnesium adsorption ratio (MAR)—the groundwater quality was found to be suitable to permissible for irrigation. However, in terms of sodium and salinity hazards, most testing locations were unsuitable for irrigation.
3. Regarding hydrochemical facies of groundwater, the groundwater samples fell under the sodium-potassium-chloride-sulfate category, and there is a dominance of sodium water type.

Acknowledgments The authors would like to acknowledge the Division of Natural Resource and Environment of Vinh Chau District, Department of Natural Resource and Environment of Soc Trang Province, and the Can Tho University College of the Environment and Natural Resources Laboratory for providing data and support.

References

APHA – American Public Health Association. (1995). *Standard methods for the examination for water and wastewater* (19th ed.pp. 211–281). Washington, DC: Byrd Prepress Springfield.

Danh V. T. (2008). Household switching behavior in the use of groundwater in the Mekong Delta (p. 115), Economy and Environment Program for Southeast Asia (EEPSEA) Research Report rr2008121.

Department of Environment in Vinh Chau. (2011). Annual environment report, pp 100–184 (in Vietnamese).

Devendra, D., Shriram, D., & Atul, K. (2014). Analysis of ground water quality parameters: A review. *Research Journal of Engineering Sciences*, *3*(5), 26–31. ISSN 2278-9472.

Doneen, L. D. (1964). *Notes on water quality in agriculture* (p. 400). Davis: Department of Water Science and Engineering, University of California, Water Science and Engineering.

DONREs – Department of Natural Resources and Environment of SocTrang. (2012). Annual report of natural resource and environment. pp. 10–184 (in Vietnamese).

Hem, J. D. (1985). Study and interpretation of the chemical characteristics of natural water (p. 2254, 3rd ed.). US Geological Survey Water-supply Paper.

IUCN- International Union for Conservation of Nature. (2011). *Groundwater in the Mekong Delta* (Discussion paper) (online). https://cmsdata.iucn.org/downloads/iucn_groundwater_in_mekong_delta_25_may_11_w.pdf

Kakonen, M. (2008). Mekong Delta at the crossroads: More control or adaptation? *Ambio, 37*(3), 205–212.

MONRes – Ministry of Natural Resources and Environment. (2015). Groundwater quality standard of Vietnam (pp. 34–84) (in Vietnamese).

Nam, N. D. G., Goto, A., & Osawa, K. (2017). Groundwater modeling for groundwater management in the coastal area of Mekong Delta. *Journal of JSIDRE, Japan, 85*, 93–103 Online ISSN: 1884-7242, Print ISSN: 1882-2789.

Nhan, D. K., Be, N. V., & Trung, N. H. (2008). *Chapter 4: Water use and competition in the Mekong Delta, Vietnam* (pp. 10–18).

Povinec, P. P., Aggarwal, P. K., Aureli, A., Burnett, W. C., Kontar, E. A., Kulkarni, K. M., Moore, W. S., Rajar, R., Taniguchi, M., Comanducci, J. F., Cusimano, G., Dulaiova, H., Gatto, L., Groening, M., Hausser, S., Levy Palomo, I., Oregioni, B., Ozorovich, Y. R., Privitera, A. M. G., & Schiavo, M. A. (2006). Characterisation of submarine groundwater discharge offshore south-eastern Sicily. *Journal of Environmental Radioactivity, 89*(1), 81–101.

Raju, N. J. (2007). *Hydro-geochemical parameters for assessment of groundwater quality in the upper Gunjanaeru River basin*, Andhara Pradesh, South India, Environmental Geology. p. 8

Sarath Prasanth, S. V., Magesh, N. S., Jitheshlal, K. V., Chandrasekar, N., & Gangadhar, K. (2012). *Apply Water Science, 2*, 165–175.

Schiavo, M. A., Havser, S., Gusimano, G., & Gatto, L. (2006). Geochemical characterization of groundwater and submarine discharge in the south-eastern Sicily. *Continental Shelf Research, 26*(7), 826–834.

Subramani, T., Elango, L., & Damodarasamy, S. R. (2005). Groundwater quality and its suitability for drinking and agriculture use in Chithar River Basin, Tamil Nadu India. *Environmental Geology, 47*(8), 1099–1110.

Tuan, L. A., Hoanh, C. T., Miller, F., & Sinh, B. T. (2007) Chapter 1: Flood and salinity management in the Mekong Delta, challenges to sustainable development in the Mekong Delta: Regional and national policy issues and research needs. pp. 6–25.

U.S. Salinity Laboratory. (1954). *Diagnosis and improvement of saline and alkali soils* (Agricultural handbook, USDA, No. 60, p. 160). Washington, DC: U.S. Government Printing Office.

World Health Organization. (2004). *Guidelines for drinking-water quality* (Vol. 1, 3rd ed.p. 547). Geneva: World Health Organization (WHO) Press.

World Health Organization. (2008). *Guidelines for drinking-water qualiy: Recommendations Incorporating first and second addenda* (Vol. 1, 3rd ed.p. 668). Geneva: World Health Organization (WHO) Press.

Chapter 11
Using the Contingent Valuation Method to Assess Communities' Willingness to Accept Compensation for Waterbird Nest Protection in the 3S Rivers, Cambodia

Phat Chandara, Seak Sophat, and Andrea H. Claassen

Abstract Anthropogenic impacts are the greatest threat to survival of plant and animal populations worldwide. The importance of involving local communities in conservation is evident, yet poverty and livelihood concerns may preclude community participation in conservation, especially in developing countries. Payment for ecosystem services (PES) may enable local participation. However, payments need to be acceptable to communities while remaining within project budgets. Contingent valuation (CV) can be useful for establishing a community's willingness to accept (WTA) compensation for participation in conservation activities and for associated opportunity costs. In the Sekong, Sesan, and Srepok (3S) River basin of Cambodia, a community-based nest protection program utilizing PES is being implemented for six waterbird species. This scheme employs local communities to prevent human exploitation (harvest of eggs and chicks) and disturbance at nesting sites. The program provides guard salaries as an incentive to local communities to report nest sites and ensure their success, rather than harvesting their contents. This study aimed to establish an appropriate level of conservation incentives for local communities to take part in conservation activities. We used CV methods to assess community WTA compensation for participating in conservation activities. Specifically, this chapter evaluates the acceptability of different payment levels by community respondents. This information will be useful for future implementation of the waterbird nest protection program in the 3S River basin, as well as for other conservation projects in Cambodia. Moreover, our approach to assessing communities' WTA compensation for participating in the conservation project can be applied to other similar participatory, community-based conservation programs.

Phat Chandara (✉) · Seak Sophat
Department of Natural Resource Management and Development, Faculty of Development Studies, Royal University of Phnom Penh, Phnom Penh, Kingdom of Cambodia

A. H. Claassen
Department of Fisheries, Wildlife, and Conservation Biology, University of Minnesota, Minneapolis, MN, USA

M. A. Stewart, P. A. Coclanis (eds.), *Water and Power*, Advances in Global Change Research 64, https://doi.org/10.1007/978-3-319-90400-9_11

Keywords Biodiversity conservation · Cambodia · Conservation incentive · Contingent valuation · Payment for ecosystem services

11.1 Introduction

Over the next few decades approximately 10–20% of all vertebrate and plant species are at risk of extinction (IUCN 2017). The greatest threat to the survival of populations of plants and animals are anthropogenic impacts, especially human exploitation for food and trade (Manel et al. 2002). Other major threats include habitat destruction, human disturbance, introduced species, loss of food resources (often exacerbated by human activities), and climate change (Purvis et al. 2000).

In recent years, the importance of involving local communities in natural resource conservation has become evident (Gruber 2010). In the past, conservation of threatened species and habitats relied primarily on centralized, mandatory regulatory mechanisms to limit access to certain areas or natural resources (Lindhjem and Mitani 2012). However, community-based conservation may potentially be more efficient and cost-effective than conventional command-and-control type approaches (Ferraro and Kiss 2002; Clements et al. 2013). Furthermore, fewer conflicts may arise between conservationists and local communities if conservation measures are voluntary rather than mandatory (Lindhjem and Mitani 2012; Bush et al. 2013). Long-term conservation outcomes for threatened species and ecosystems generally benefit from community support and direct participation (Gruber 2010). However, in order for community-based conservation to effectively protect biodiversity there need to be clear linkages between conservation goals and benefits to local communities (Ferraro and Kiss 2002). When benefits of natural resources and other ecological services are obvious, and the connections between conservation goals and local livelihoods are clear, the loss of resources may be enough to motivate and incentivize local participation in conservation.

However, approaches that involve direct incentives such as payment for ecosystem services (PES) may be advantageous if the linkages are less clear, or when poverty or sociopolitical contexts present obstacles to voluntary participation in conservation (Ferraro and Kiss 2002; Gjertsen and Niesten 2010). In developing countries such as Cambodia, rural communities are very poor and rely on natural resource-based subsistence activities such as farming, fishing, and harvest of natural resources (Ziv et al. 2010). In this context, it is unrealistic to depend on conservation models that expect voluntary behavior changes by local people (Gjertsen and Niesten 2010). On the other hand, economic incentives may be an effective tool to encourage local people to engage in conservation activities by providing clear benefits to support livelihoods (Suich 2013; Clements et al. 2013; Milne and Chervier 2014).

Bird nests are relatively easily monitored and thus well suited to participatory, community-based monitoring and conservation (Claassen et al. 2017). Moreover, nest success is an important metric of avian productivity, and improving nest success is thus a principal conservation measure for threatened bird species (Bell and Merton 2002). In Cambodia, several PES schemes have been implemented to protect nests of threatened bird species (Sok et al. 2012; Clements et al. 2013; Wright et al. 2013; Claassen et al. 2017). Some of the species targeted by these programs have little economic value but are exploited on a subsistence level by local communities who harvest eggs and chicks for food (Clements et al. 2013; Claassen et al. 2017). Payment levels for these programs were not set through a participatory community-based process, but arbitrarily by the payment providers who did not assess communities' opportunity costs of forgoing harvest of bird eggs and chicks or their willingness to accept (WTA) payments for participating in the nest protection programs.

PES approaches can be efficient and cost-effective means to conserve threatened wildlife and other natural resources (Ferraro and Gjertsen 2009; Clements et al. 2013; Lindhjem and Mitani 2012). However, it is important to set an appropriate level of compensation that is acceptable to both PES providers and recipients (Gjertsen and Niesten 2010). Payments levels need to be low enough to allow PES providers to stay within their budgets, while also using funds efficiently and effectively to reach conservation goals. Payments should also be equitable and provide enough livelihood benefits to motivate participation by local communities (Suich 2013).

Contingent valuation (CV) is an economic method of assessing the value of non-market resources (Boyle 2003). The use of CV in developing countries is widespread and has often been used to establish valuation of natural resources or environmental services arising from protected areas (Whittington 2002; Bush et al. 2013). CV methods can be used to set appropriate PES levels by assessing local community WTA compensation for participation in (or losses due to) conservation projects, while also ensuring that payments stay within project budgets (Bush et al. 2013). CV can also be used to establish economic values of ecosystem services. Although the study of ecosystem services can provide valuable insights into their benefits to local people's livelihoods, there has been less research on the non-calculable or invisible value of such resources (Berbés-Blázquez 2012). Applying CV to establish economic values is especially useful when it is otherwise difficult to establish a market value for an ecosystem service, such as in the case of threatened species that are not typically traded or sold.

In this study, we assessed community willingness to accept compensation for participation in a bird nest protection program in the Sekong, Sesan, and Srepok (3S) River basin of northeastern Cambodia. The community-based nest-guarding scheme is being implemented by the, Department of Natural Resource Management and Development of the Royal University of Phnom Penh. Six waterbird species are targeted for conservation: river tern (*Sterna aurantia*), river lapwing (*Vanellus duvaucelii*), great thick-knee (*Esacus recurvirostris*), small pratincole (*Glareola lactea*), little ringed plover (*Charadrius dubius*), and Mekong wagtail (*Motacilla samveasnae*). The scheme employs local communities to prevent human exploitation (harvest of eggs and chicks) and disturbance at nesting sites. Nest-guarding salaries are used as incentives to local villagers to report nest sites and ensure their success rather than harvesting their contents. Specifically, the aim of this study was to use CV methods to assess the minimum value of communities' WTA compensation for voluntary waterbird nest guarding along the 3S Rivers in Cambodia.

11.2 Methods

11.2.1 Study Area

The Sekong, Sesan, and Srepok (3S) Rivers are major tributaries of the Mekong River and contribute up to 20% of the Mekong's annual flow (Piman et al. 2013). The 3S River basin (13°53′N 106°40′E) covers 78,650 km^2 and supports high biodiversity, especially aquatic biodiversity (Ziv et al. 2012). Additionally, the 3S Rivers provide critical breeding habitat for ground-nesting waterbirds such as river tern, river lapwing, great thick-knee, small pratincole, and little ringed plover, which nest on sparsely vegetated, seasonally emergent sand and gravel bars (Timmins and Men 1998; Claassen 2004). Furthermore, the 3S Rivers support the livelihoods of approximately four million people who depend on its natural resources (MRC 2010; Ziv et al. 2012). In Cambodia, the 3S River basin covers three provinces: Stung Treng, Ratanakiri, and Mondulkiri. This study was conducted in nine focal villages in five communes on the Sekong and Sesan Rivers in Strung Treng and Ratanakiri Provinces. Six of the focal villages were located on the Sesan River, and three were located on the Sekong River.

11.2.2 Research Methods

In this study, we used CV methods to assess various scenarios related to communities WTA compensation payments for participating in waterbird nest protection in the 3S River basin of Cambodia. In the WTA approach, the provision point (*PP*) is the total amount of money available in the fund (Bush et al. 2013). In our case, this was the project budget available for compensating all participating community members for protecting waterbird nests and nesting sites along the 3S Rivers, Cambodia. Individuals were asked to make a bid or claim for compensation from this fund. If the sum of all bids exceeded the money available in the fund, no compensation payments were made. If the sum of bids was less than or equal to the provisioning point of WTA, individual claimants received their bid. Community nest protection agreements were then made with the individual claimants.

Following the methods of Bush et al. (2013), we defined B_j as individual bids on the compensation fund, *PP* as the total amount available in the fund, and *N* as the number of claimants. Thus,

- If $\sum_{N}^{j=1} Bj > PP$, then the sum of bids exceeded the available funds, no nest protection agreements were made, and no compensation was paid;
- If $\sum_{N}^{j=1} Bj \le PP$, the sum of bids was equal or less than the amount available, so nest protection agreements were made, compensation was paid, and people received exactly the individual amount of their bid.

During May 2015, we designed a questionnaire and conducted interviews with community members from nine villages in five communes along the Sekong and Sesan Rivers in Cambodia. The survey questionnaire contained eight questions about the respondents' socioeconomic status. Additionally, respondents were asked about their willingness to accept payment for nest guarding. Respondents were asked to answer yes or no to whether they were willing to accept a specified payment level for nest guarding. Ten levels of payment were specified, ranging from 0.5 to 5.0 U.S. dollars (USD). The maximum payment level considered was 5.0 USD to ensure that the sum of the WTA bids would not exceed the total amount available in the compensation fund.

The sample size of interviewees was based on a rule of thumb for behavioral research that recommends using a sample size of ≥10% of the total population (Roscoe 1975; Alreck and Settle 1995). In this study, we surveyed 256 community members, representing 12% of the estimated 2051 total households in the project focal villages (NCDD 2010). We obtained lists of households from village heads, and from these lists we randomly selected households to be surveyed. If no adult members of a selected household were available, we then selected a neighboring household.

We analyzed survey results by conducting a binary logistic regression to investigate how WTA was affected by level of potential payment for nest protection. WTA was the binary (dummy) response variable (1 = respondent was willing to accept the specified level of payment; 0 = respondent was not willing to accept the payment). Level of payment (bid) was the explanatory variable; 10 levels of payment were investigated, ranging from 0.5 to 5.0 USD per day in increments of 0.5 USD. We also tested the relationship between WTA and monthly income level. Monthly income was a categorical variable with five levels: <100 USD, 100–250 USD, 250–400 USD, 400–550 USD, and >550 USD. We also conducted a post hoc analysis to explore whether WTA was influenced by previous participation in the conservation program. SPSS statistical software (IBM Corp 2013) was used to conduct the logistic regression analyses.

11.3 Results and Discussion

11.3.1 Respondents' Profile

We interviewed 256 respondents from 2051 total households from 9 project focal villages in 5 communes (Sdao and Thmor Keo Communes on the Sekong River, and Ta Lat, Hat Pok, and Khaoh Pang Communes on the Sesan River). Ethnic identities of respondents were Khmer (71%), Lao (23%), and Prov (Brao; 6%). Primary sources of income included agriculture, animal husbandry, and fishing; 87% of respondents primarily depend on agriculture, although their livelihoods also rely on natural resources such as timber, nontimber forest products, and aquatic resources

Table 11.1 Monthly incomes of survey respondents (n = 256) along the 3S Rivers, Cambodia

Monthly income (USD)	No. respondents	Percentage (%)
Less than 100	153	59.8
100–250	88	34.4
250–400	12	4.7
400–550	1	0.4
550–650	1	0.4
Over 650	1	0.4

Table 11.2 Results of the binary logistic regression model of all WTA bids (n = 408) by community respondents (n = 256) on the 3S Rivers, Cambodia. Respondents were not willing to accept <4 USD per day as compensation for bird nest protection

Variable	β	SE	Wald	df	*p*-value	Exp(β)	95% CI for Exp(β)	
							Lower	Upper
Constant	5.196	1.528	11.563	1	0.001	0.006	0.0003	0.111
5 USD	3.388	1.073	9.971	1	0.002	29.615	3.615	242.607
4.50 USD	0.024	0.309	0.006	1	0.938	1.024	0.558	1.879
4 USD	0.337	0.594	0.322	1	0.570	1.401	0.437	4.491
<4 USD	–	–	–	–	–	–	–	–

along the 3S Rivers. Among respondents, 31% were uneducated (or had dropped out of primary school), 41% completed primary school, 23% completed secondary school, and 5% completed high school. Monthly incomes in the project area were low. The majority of respondents (60%) earned less than 100 USD per month, while 34% earned 100–250 USD per month (Table 11.1). Only about 1% of total respondents earned over 400 USD per month.

11.3.2 Analysis of CV Responses

Based on results from the logistic regression of all WTA bids, the only payment level variable that was significant was 5 USD (Table 11.2). However, a payment level of 4.5 USD received the greatest number of minimum WTA bids (47%), followed by 5 USD (40% of minimum bids) and 4 USD (8% of minimum bids; Fig. 11.1a). The frequency distribution of all bids (n = 408) indicated that 245 (96%) of 256 respondents were willing to accept 5 USD per day for nest guarding (Fig. 11.1b). Of those 245 respondents, 142 (55%) were also willing to accept 4.5 USD, and of those 142 respondents, 21 (8%) were also willing to accept 4 USD per day (Fig. 11.1b). None of the respondents were willing to accept less than 4 USD per day. Results of the survey indicated that 94% of respondents were sure or very sure of their decision to accept a specific payment amount as compensation for nest

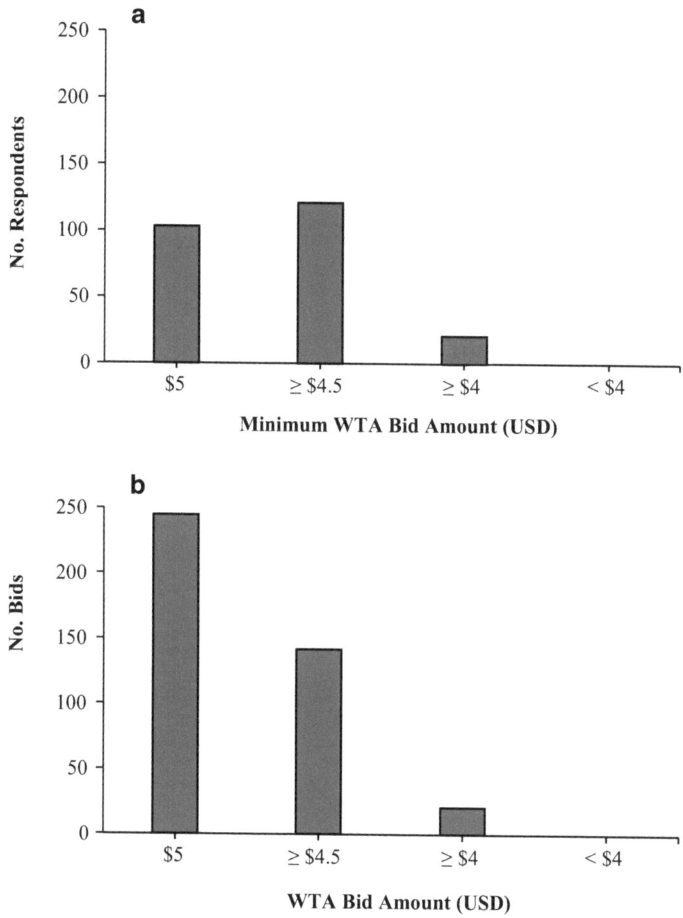

Fig. 11.1 Distribution of minimum WTA bids (**a**; n = 245) and all WTA bids (**b**; n = 408) by community respondents (n = 256) in the 3S River basin of Cambodia. No respondents were willing to accept <4 USD, and 12 respondents were not willing to accept the maximum bid level (5 USD)

guarding, and only 2% of respondents were not sure of their decisions to accept a specific payment level (Fig. 11.2).

The frequency distribution of WTA bids was consistent among monthly income levels; for all income levels, a payment level of 4.5 USD received the greatest number of minimum WTA bids, followed by 5 USD and 4 USD (Table 11.3). However, monthly income was not supported in logistic regression models of WTA.

According to the survey, a minority (4%) of respondents were not willing to accept 5 USD per day as compensation for nest guarding. These participants stated that they would not participate in the conservation program even if offered a payment level of 5 USD. Several reasons were identified during the survey for the unwillingness to accept a payment level of 5 USD per day (Table 11.4).

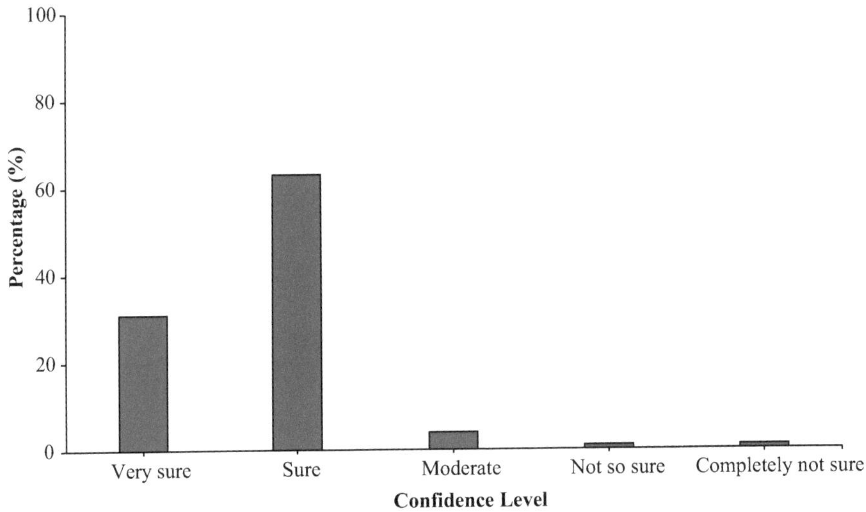

Fig. 11.2 Level of respondent confidence in their willingness to accept a specific payment amount for bird nest protection on the 3S Rivers, Cambodia

Table 11.3 Distribution of minimum WTA bids according to monthly incomes of respondents in the 3S River basin of Cambodia

	Monthly income (USD)					
Payment level (USD)	<100	100–250	250–400	400–550	550–650	>650
5	56	38	6	1	2	0
4.5	62	45	10	2	0	3
4	16	1	3	0	0	0
<4	0	0	0	0	0	0

Table 11.4 Reasons given by local villagers for unwillingness to participate in the nest protection program on the 3S Rivers, Cambodia

Reason	No. respondents
5 USD per day is too low	6
5 USD per day is too low, but I would be willing to accept 7 USD per day	3
I think the commune government should manage nest protection. I don't think there is a need for individual villager participation	1
I want longer-term financial security. Financial benefits of the program are not sustainable enough	1
I don't believe the program (and financial payments) will actually happen	1

11.4 Conclusions

Results of the CV analysis to assess communities' WTA conservation payment indicated that 5 USD is likely the most appropriate payment level for participating in the nest protection program. Individual community members were willing to accept 4, 4.5, or 5 USD per day for nest guarding (Tables 11.2 and 11.3, Fig. 11.1). None of the respondents were willing to accept less than 4 USD per day. Most starting WTA bids were 4.5 USD, although this was closely followed by bids of 5 USD (Fig. 11.1a). The frequency of all WTA bids shows that 96% of respondents would be willing to accept 5 USD as compensation for nest guarding (Fig. 11.1b). Moreover, a logistic regression analysis of all WTA bids indicates that 5 USD was the most significant payment level and had the most significant positive affect on communities' WTA compensation for nest guarding (Table 11.3). Thus, although 4.5 USD received the most starting WTA bids (Fig. 11.1a), setting the payment level at 5 USD per day would ensure acceptance by nearly all respondents from the project focal villages. Therefore, 5 USD per day would be the most appropriate level of compensation to local community members for participating in waterbird nest protection on the 3S Rivers, Cambodia.

Respondents' minimum WTA bids were influenced by previous involvement in the nest protection program. Compared to respondents who had never previously participated, respondents who had participated during the previous year (2014) tended to start their bids at 5 USD and were less willing to accept one of the lower payment alternatives offered during the bidding process. Payment was set at 5 USD in 2014, and participants were less likely to accept a lower payment than they had received previously. These results reinforce our conclusion that setting compensation at 5 USD is most appropriate, because setting payment at a lower level, such as 4.5 USD, could alienate previous participants and cause them to be unwilling to continue their involvement in the conservation program.

In our study, income level did not significantly affect WTA. A study of farmers in Madagascar found that poorer farmers were more likely to accept lower conservation payments than farmers with higher incomes (Minten 2009). Our a priori assumption was that income level would similarly influence communities' WTA compensation for nest guarding on the 3S Rivers in Cambodia. Anecdotally, it did appear that poorer respondents were more willing to accept lower payments. However, monthly income was not supported by logistic regression models of WTA. The frequency distribution of WTA bids initially appeared to suggest that fewer high income respondents were willing to accept compensation at the lower payment levels. However, this was likely an artifact of the data, as respondent data were clustered in the lower income categories. Most (60%) of our respondents were in the lowest income category of earning less than 100 USD per month (Table 11.1). We may have had too few categories for income level, resulting in loss of power of our statistical tests (van Belle 2011). Perhaps if we had split income levels into a greater number of categories we might have detected a relationship between income and WTA.

In PES schemes, it is important to balance budgetary and financial constraints of the PES providers with maintaining fair and equitable compensation to communities (Gjertsen and Niesten 2010). Furthermore, appropriate and sufficient payments to communities will ensure local participation in the conservation activities (Suich 2013). CV methods and assessments of community WTA can be used to identify the provision point where payment is acceptable to both PES providers and recipients (Bush et al. 2013). In this study, nearly all (96%) of respondents were willing to accept 5 USD per day as compensation for nest guarding (Fig. 11.1b), and the project's nest protection fund was sufficient to cover this level of payment. Even though 55% of respondents were willing to accept lower payment amounts, the higher payment amount of 5 USD per day was acceptable to a significantly greater percentage of respondents and was sufficient to enable their participation in the conservation program (Fig. 11.1). A payment level of 5 USD per day may represent an amount that is satisfactory for supporting the livelihoods of most respondents, and may provide significantly higher livelihood benefits than the lower alternative payment amounts.

During this study, sustainability of funding support was a concern to some respondents on the 3S Rivers, Cambodia. Respondents who previously participated in the nest guarding project, who were then asked about their willingness to accept a lower payment than they had received previously, expressed more concerns regarding sustainability of funding support than respondents who had not previously participated in the project. Ideally, we would have assessed WTA during the 1st year of the project to avoid causing unnecessary concern to project participants. One respondent who was unwilling to participate in the project, even at the 5 USD payment level, also had concerns about long-term financial sustainability of the project (Table 11.4). Securing the long-term funds necessary to support PES schemes can be a difficult challenge for PES providers (Wunder 2007; Milne and Chervier 2014). As such, PES schemes should be considered as components of broader long-term conservation programs aimed at achieving sustainable biological and social outcomes. For instance, PES programs can complement other strategies such as conservation education and capacity building, and developing sustainable alternative livelihood support mechanisms for communities.

Local communities along the 3S Rivers in Cambodia have expressed appreciation and support of the waterbird nest conservation project implemented by the Department of Natural Resource Management and Development of the Royal University of Phnom Penh (Phat and Seak 2015). Results of this survey indicate that the level of compensation (5 USD per day) provided by the project for nest guarding was highly accepted by local communities along the 3S Rivers, Cambodia. Results of this study will be useful for future implementation of the waterbird nest protection project in the 3S River basin, as well as for other conservation projects in Cambodia. Moreover, our approach to assessing communities' WTA compensation for participating in the conservation project can be applied to other similar participatory, community-based conservation programs.

Acknowledgements We would like to thank the Cambodian Forestry and Fisheries Administrations and the Ministry of Environment, as well as the Stung Treng and Ratanakiri provincial Forestry and Fisheries Cantonments. We would like to register our profound appreciation to local communities and district and commune authorities for administrative support and facilitation of project implementation. Also, we are grateful to our partner organizations: BirdLife International, 3S River Protection Network, Culture and Environment Preservation Association, and Wildlife Conservation Cambodia. We would like to acknowledge the dedication of our project team, including Sok Samet, Yen Run, Ouch Mara, Pheung Sophea, and Leang Sovichea. Financial support for this project was generously provided by the MacArthur Foundation and the Critical Ecosystem Partnership Fund. Finally, we express our sincere thanks to the local communities along the 3S Rivers for their support and participation in the nest protection program.

References

Alreck, P. L., & Settle, R. B. (1995). *The survey research handbook: Guidelines and strategies for conducting a survey* (2nd ed.). Chicago: Irwin.

Bell, B. D., & Merton, D. V. (2002). Critically endangered bird populations and their management. In K. Norris (Ed.), *Conserving bird biodiversity* (pp. 105–138). Cambridge: Cambridge University Press.

Berbés-Blázquez, M. (2012). A participatory assessment of ecosystem services and human wellbeing in rural Costa Rica using photo-voice. *Environmental Management, 49*(4), 862–875.

Boyle, K. J. (2003). Contingent valuation in practice. In P. A. Champ, K. J. Boyle, & T. C. Brown (Eds.), *A primer on non-market valuation* (pp. 111–169). New York: Springer.

Bush, G., Hanley, N., Moro, M., & Rondeau, D. (2013). Measuring the local costs of conservation: A provision point mechanism for eliciting willingness to accept compensation. *Land Economics, 89*(3), 490–513.

Claassen, A. H. (2004). *Abundance, distribution, and reproductive success of sandbar nesting birds below the Yali Falls hydropower dam on the Sesan River, Northeastern Cambodia*. Phnom Penh: WWF, Danida, WCS and BirdLife International.

Claassen, A. H., Sok, K., Arnold, T. W., & Cuthbert, F. J. (2017). Effectiveness of direct payments to increase reproductive success of sandbarnesting river birds in Cambodia. *Bird Conservation International, 27*(4), 495–511.

Clements, T., Rainey, H., An, D., Rours, V., Tan, S., Thong, S., Sutherland, W. J., & Milner-Gulland, E. J. (2013). An evaluation of the effectiveness of a direct payment for biodiversity conservation: The bird nest protection program in the Northern Plains of Cambodia. *Biological Conservation, 157*, 50–59.

Ferraro, P. J., & Kiss, A. (2002). Direct payments to conserve biodiversity. *Science, 298*(5599), 1718–1719.

Ferraro, P. J., & Gjertsen, H. (2009). A global review of incentive payments for sea turtle conservation. *Chelonian Conservation and Biology, 8*(1), 48–56.

Gruber, J. S. (2010). Key principles of community-based natural resource management: A synthesis and interpretation of identified effective approaches for managing the commons. *Environmental Management, 45*(1), 52–66.

Gjertsen, H., & Niesten, E. (2010). Incentive-based approaches in marine conservation: Applications for sea turtles. *Conservation and Society, 8*(1), 5–14.

IBM Corp. (2013). *IBM SPSS statistics for windows, version 22.0*. Armonk: IBM Corp.

IUCN. (2017). The IUCN red list of threatened species, version 2017.1. http://www.iucnredlist.org. Accessed 19 Aug 2017.

Lindhjem, H., & Mitani, Y. (2012). Forest owners' willingness to accept compensation for voluntary conservation: A contingent valuation approach. *Journal of Forest Economics, 18*(4), 290–302.
Manel, S., Berthier, P., & Luikart, G. (2002). Detecting wildlife poaching: Identifying the origin of individuals with Bayesian assignment tests and multilocus genotypes. *Conservation Biology, 16*(3), 650–659.
Milne, S., & Chervier, C. (2014). *A review of payments for environmental services (PES) experiences in Cambodia*. Kota Bogor: CIFOR.
Minten, B. (2009). Compensation for biodiversity conservation. *Review on Business Economic Literature, 54*(3), 362–383.
MRC. (2010). *State of the basin report 2010*. Vientiane: Mekong River Commission.
NCDD. (2010). *Commune database online*. http://db.ncdd.gov.kh/cdbonline/home/index.castle. Accessed 5 June 2017.
Phat, C., & Seak, S. (2015). *Contribution of local ecological knowledge and practices to waterbirds conservation along the Sekong and Sesan Rivers IBA, Cambodia (draft report)*. Cambodia: Department of Natural Resource Management and Development, Royal University of Phnom Penh.
Piman, T., Cochrane, T. A., Arias, M. E., Green, A., & Dat, N. D. (2013). Assessment of flow changes from hydropower development and operations in Sekong, Sesan and Srepok rivers of the Mekong basin. *Joural of Water Resources Planning and Management, 139*(6), 723–732.
Purvis, A., Gittleman, J. L., Cowlishaw, G., & Mace, G. M. (2000). Predicting extinction risk in declining species. *Proceedings of the Royal Society B-Biological Sciences, 267*(1456), 1947–1952.
Roscoe, J. T. (1975). *Fundamental research statistics for the behavioural sciences* (2nd ed.). New York: Holt Rinehart and Winston.
Sok, K., Claassen, A. H., Wright, H. L., & Ryan, G. E. (2012). Waterbird nest protection on the Mekong River: A preliminary evaluation, with notes on the recovery and release of white-shouldered ibis *Pseudibis davisoni* chicks. *Cambodian Journal of National History, 2012*, 29–41.
Suich, H. (2013). The effectiveness of economic incentives for sustaining community based natural resource management. *Land Use Policy, 31*, 441–449.
Timmins, R. J., & Men, S. (1998). *A wildlife survey of the Tonle San and Tonle Srepok River basins in northeastern Cambodia*. Phnom Penh: Fauna and Flora International Indochina Program, Wildlife Protection Office, Department of Forestry.
van Belle, G. (2011). *Statistical rules of thumb*. Hoboken: Wiley.
Whittington, D. (2002). Improving the performance of contingent valuation studies in developing countries. *Environmental Development Economics, 22*(2), 233–367.
Wright, H. L., Collar, N. J., Lake, I. R., Rours Vann, R., Sok, K., Phearun, S., & Dolman, P. M. (2013). Experimental test of a conservation intervention for a highly threatened waterbird. *Journal of Wildlife Management, 77*(8), 1610–1617.
Wunder, S. (2007). The efficiency of payments for environmental services in tropical conservation. *Conservation Biology, 21*(1), 48–58.
Ziv, G., Baran, E., So, N., Rodríguez-Iturb, I., & Levina, S. A. (2012). Trading-off fish biodiversity, food security, and hydropower in the Mekong River Basin. *Proceedings of the National Academy Sciences of the United States of America, 109*(15), 5609–5614.

Chapter 12
Assessment of Local Community Perceptions of Biodiversity Conservation in the 3S Rivers of Cambodia: Using a Knowledge, Attitudes, and Practices (KAP) Approach

Seak Sophat, Phat Chandara, and Andrea H. Claassen

Abstract The Sekong, Sesan, and Srepok (3S) Rivers in Southeast Asia provide critical habitat for wildlife and support many threatened species. The Sesan and Sekong Rivers are regarded as Important Bird Areas, which recognizes their exceptional avian diversity and critical role in supporting waterbird populations. Additionally, these rivers provide diverse natural resources and ecosystem services that support the livelihoods of millions of people of riparian countries. However, the 3S Rivers are under severe threat from hydropower development, habitat destruction, economic land concessions, extractive industries, and illegal logging. In response to these threats, a community-based conservation program was implemented for waterbirds that utilized direct payments to communities for bird nest protection. Community participation is key to effective biodiversity conservation. However, few studies have assessed local perceptions of community-based conservation programs. Therefore, we conducted a knowledge, attitudes, and practices survey of selected communities along the 3S Rivers to understand community perceptions about the nest protection program. According to the local communities, waterbird populations increased and threats decreased as a result of the program. Overall, communities had positive impressions of the program and believed the program provided them with significant livelihood benefits. The results of this research will be used as educational and diagnostic tools to assess the effectiveness of the conservation program to meet community needs and to be able to improve this and other such community-based programs in the future.

Seak Sophat (✉) · Phat Chandara
Department of Natural Resource Management and Development, Faculty of Development Studies, Royal University of Phnom Penh, Phnom Penh, Kingdom of Cambodia
e-mail: seak.sophat@rupp.edu.kh

A. H. Claassen
Department of Fisheries, Wildlife, and Conservation Biology, University of Minnesota, Minneapolis, MN, USA

M. A. Stewart, P. A. Coclanis (eds.), *Water and Power*, Advances in Global Change Research 64, https://doi.org/10.1007/978-3-319-90400-9_12

Keywords Biodiversity conservation · KAP survey · Payments for ecosystem services · 3S Rivers

12.1 Introduction

The Mekong River in Southeast Asia is one of the most biodiverse ecosystems in the world and plays an important role in supporting the livelihoods of approximately 65 million people of riparian countries in Southeast Asia (Ziv et al. 2012; Winemiller et al. 2016). Part of what makes the Mekong River so dynamic and biodiverse is its complex network of tributaries (MRC 2010). Major tributaries such as the Sekong, Sesan, and Srepok (3S) Rivers (Fig. 12.1) provide diverse sources of hydrology, biodiversity, and livelihoods for millions of people of riparian countries. The 3S Rivers cover 78,650 km^2 and are shared between Vietnam (38%), Cambodia (33%), and Laos (29%) (Piman et al. 2013). The 3S Rivers account for up to 20% of the annual flow of the Mekong River and provide natural resources such as water, fisheries, and forest products which nourish the Lower Mekong Basin and the Tonle Sap Lake (Ziv et al. 2012; Piman et al. 2013; Arias et al. 2014).

The 3S Rivers provide critical habitat for wildlife and support many threatened species. In Cambodia, over 20% of the 3S basin is currently designated as protected areas; they include a national park, wildlife sanctuaries, biodiversity conservation areas, and protected forests (Grimsditch 2012). The 3S Rivers have exceptionally high fish diversity and provide critical habitats, such as deep pools (Ziv et al. 2012). Over 213, 133, and 240 fish species have been recorded in the Sekong, Sesan, and Srepok Rivers, respectively (Baran et al. 2013). Furthermore, the Sesan and Sekong Rivers are regarded as Important Bird Areas, which recognizes their exceptional avian diversity and the fact that they contain critical habitat for threatened bird species (Seng et al. 2003). These rivers provide critical breeding habitat for regionally endangered riverine bird species such as river tern (*Sterna aurantia*), great thick-knee (*Esacus recurvirostris*), river lapwing (*Vanellus duvaucelii*), and Mekong wagtail (*Motacilla samveasnae*) (Timmins and Men 1998; Claassen 2004; Goes 2013).

In addition to their importance to supporting biodiversity, the 3S Rivers greatly contribute to human livelihoods, socioeconomic development, and culture (Ziv et al. 2012; Piman et al. 2013). Approximately four million people inhabit the areas surrounding the 3S Rivers, many of whom are from ethnic minority groups whose livelihoods are dependent on these rivers (Chinnak 2006; MRC 2010). However, the 3S Rivers are under severe threat from hydropower development, habitat destruction, economic land concessions, extractive industries, and illegal logging (Grimsditch 2012). Although biodiversity and ecosystem services along the 3S Rivers are decreasing, the area is still an important source of socioeconomic development of millions of people in diversified ethnic groups (Baird et al. 2002; Wyatt and Baird 2007; Seak 2015).

In particular, hydropower dam operations threaten waterbirds, fish, and other river-dependent wildlife, as well as the river-based livelihoods of local people (Baird et al. 2002). On the 3S Rivers, hydropower development has occurred at a

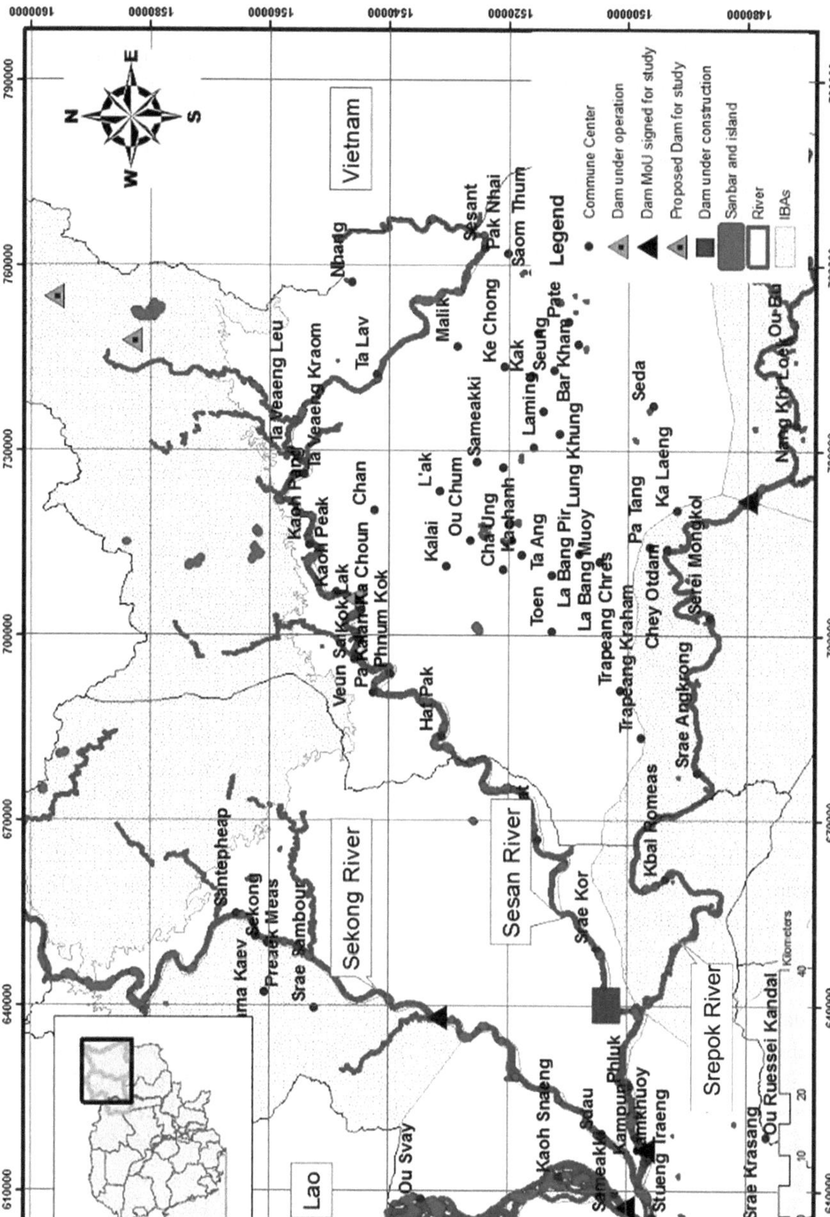

Fig. 12.1 Map of the project areas in the 3S River Basin, Cambodia

rapid pace. Due to recent economic development and high demand for energy production, over 20 hydropower dams have been built on the 3S Rivers, and more than 26 additional dams are planned for construction (Piman et al. 2013). Although most of the existing dams are located in upstream countries (i.e., Vietnam and Laos), one – the Lower Sesan 2 (LS2) Dam – was completed in September 2017 at the confluence of the Sesan and Srepok Rivers. In total, the potential for hydropower production in the 3S River basin has been estimated to be as much as 6400 MW (MRC 2010). In the 3S River basin, hydropower dams have been implicated in population declines of fish and riverine birds (Baird and Meach 2005; Claassen 2004).

Over the last couple of decades, populations of waterbirds along the Sekong and Sesan Rivers have decreased (Timmins and Men 1998; Claassen 2004; Bezuijen et al. 2008). Several reasons have been identified, such as egg harvest, animal predation, hunting, poisoning, flooding from hydropower dams, and habitat degradation (Claassen 2004; Bezuijen et al. 2008; Claassen et al. 2017). However, bird nest protection programs have been successfully implemented in other regions of Cambodia, and offer hope for also providing effective protection for threatened bird species along the Sekong and Sesan Rivers (Sok et al. 2012; Clements et al. 2013; Wright et al. 2013; Claassen et al. 2017).

In response to the situation on the 3S Rivers, in 2012 the Department of Natural Resource Management and Development (NRMD) at Royal University of Phnom Penh (RUPP) established a participatory, community-based research and conservation program for threatened waterbirds along the Sekong and Sesan Rivers in Cambodia. Since populations of waterbirds along the 3S Rivers are decreasing, and humans are the primary cause of declines, the primary focus of the project was engaging local communities to be key implementers of conservation activities. However, in a poor, developing country such as Cambodia, voluntary participation by local communities in conservation projects may not be practical (Gjertsen and Niesten 2010). Thus, we used a payment for ecosystem services (PES) approach to provide direct payments (conservation incentives) as a means to encourage participation by local communities in the project area (Ferraro and Kiss 2002). In Cambodia, several other community-based conservation projects have also utilized financial incentives (Milne and Chervier 2014; Clements et al. 2010; Wright et al. 2013; Claassen et al. 2017).

To date, few studies have explored perceptions of local communities involved in PES schemes (Petheram and Campbell 2010). However, to effectively engage communities and address both community and conservation needs, it is important to understand local perceptions of such conservation programs. Therefore, the goal of this research was to assess the perceptions of communities along the Sesan and Sekong Rivers in terms of their knowledge, attitudes, and practices regarding the bird conservation program. Results of this research will then be used as educational and diagnostic tools to assess the effectiveness of the program to meet community needs as well as achieve positive conservation outcomes, and to be able to improve this and other such community-based programs in the future.

12.2 Research Methods

During this research, qualitative, quantitative, and content analyses were employed to gather all necessary documents and information regarding socioeconomic development, current status of villagers supporting water conservation projects, and further vision to improve the effectiveness of the bird conservation program.

Knowledge, attitudes, and practices (KAP) surveys can be a useful diagnostic tool in program evaluation (Gumucio 2011). In May 2015, we conducted a KAP survey to assess the community-based conservation project. *Knowledge* was considered to be the level of understanding of local people on conservation activities and tactics for waterbird and riverine species along the 3S Rivers in Cambodia. *Attitude* was considered to be a way of being; it refers to the feelings and perceptions of local people about the conservation activities. Attitude was an intermediate variable between the local communities' knowledge of the situation of waterbird populations along the 3S Rivers and their response to this situation. *Practice* was considered to be the observable response of an individual or community to a situation and refers to how individuals or communities demonstrate their knowledge and attitudes through their actions and behaviors. In this study, we were interested in their actions and behaviors regarding the situation of waterbirds and the conservation program. We considered factors such as their participation related to conservation activities and any related activities that local communities had implemented.

The KAP survey was conducted by designing a questionnaire and conducting interviews. In total, the interview questionnaire contained 39 questions related to the respondent's background (11 questions), as well as their knowledge (13 questions), attitudes (6 questions), and practices (9 questions) regarding waterbird conservation and protection on the 3S Rivers in Cambodia. We determined sample size of survey respondents based on a rule of thumb for behavioral research that recommends using a sample size of ≥10% of the total population (Roscoe 1975; Alreck and Settle 1995). This study focused on 9 villages in 5 communes, with an estimated 2051 households (Table 12.1). We surveyed 256 community members representing 12% of the total number of households in the project focal villages. Surveyed

Table 12.1 Demographics of project focal villages in the 3S region, Cambodia (National Institute of Statistics 2010)

Province	River	Commune	Village	No. of families	Total population
Stung Treng	Sekong	Sdao	Sdao	620	1364
	Sekong	Thmor Keo	Nheang Sum	260	982
	Sesan	Ta Lat	Svay Rieng	403	1233
			Khsach Thmei	258	1300
Ratanakiri	Sesan	Hat Pok	Hat Pok	215	929
			Veun Hay	88	432
	Sesan	Kaoh Pang	Pa Tang	92	311
			Lam Av	50	286
			Pa Hay	65	224

households were randomly chosen from lists provided by village heads. If adult members of a selected household were unavailable, we selected the next nearest available household that was willing to participate in the survey.

12.3 Results and Descriptive Analysis

12.3.1 General Information About respondents

This study included nine villages in five communes, three on the Sesan River and two on the Sekong River (Table 12.1). Villagers' primary sources of income included agriculture, animal (non-chicken) husbandry, chicken husbandry, and fishing (Table 12.2). Although 87% of villagers primarily depend on agriculture, their livelihoods also rely on natural resources such as timber and nontimber forest products, and aquatic resources along the 3S Rivers.

There is a high level of ethnic diversity within the study area, including Khmer, ethnic Lao, and indigenous minorities such as Prov (Brao), Tampuan, Charay, and Kavet. Among survey respondents, 71.1% identified as Khmer, 22.7% as Lao, and 6.3% as Prov (Brao).

Based on commune data records, only 10% of adults registered in secondary school, and 30% in primary school (NCDD 2010). Of surveyed adults, the middle-aged group (35–45 years old) had the lowest level of education.

12.3.2 Waterbird Populations and Threats

Local perceptions of waterbird population trends were that 39% of respondents believed populations had increased since the conservation program began in 2012, 27% of respondents believed populations had decreased, and 34% did not know (Table 12.3). In most communes, the majority of respondents believed that waterbird populations had increased. However, respondents from Talat commune tended to believe that waterbird populations decreased. Because Talat commune is nearest to the site of the LS2 dam, survey responses from this area may indicate more significant dam-related impacts than for communities that are located farther from the dam site. Overall, 76% of total respondents from all communes stated that local participation had positively contributed to waterbird conservation, although 1% said local involvement had caused a negative impact on bird populations due to disturbance caused by community monitoring and patrol activities, and 23% of respondents did not know if community participation had been positive or negative in regards to waterbird populations.

When asked to rank project activities in order of priority for conservation, respondents listed nest protection as a "very high priority"; patrolling, awareness

Table 12.2 Sources of income of survey respondents

Source of income	No. respondents	Percentage (%)[a]
Agriculture	223	88
Animal husbandry	61	24
Chicken husbandry	86	34
Aquaculture	5	2
Fishing	56	22
Short-term/casual employment	4	2
Sales	22	9
Government employment	3	1

[a]Percentages total >100% because respondents listed more than one source of income

Table 12.3 Local perceptions (n = 256) of waterbird population trends along the 3S Rivers, Cambodia, 2012–2015

	Population increase		Population decrease		Don't know	
Commune	No. respondents	Percentage (%)	No. respondents	Percentage (%)	No. respondents	Percentage (%)
Hat Pok	15	5.9	6	2.3	23	9.0
Sdao	10	3.9	4	1.6	1	0.4
Talat	39	15.2	47	18.4	28	10.9
Thmor Keo	23	9.0	10	3.9	19	7.4
Kaoh Pang	12	4.7	2	0.8	17	6.6
Total	99	38.7	69	27.0	88	34.4

raising, and capacity building as "high priority"; and bird surveys as "medium priority". None of the project activities were ranked as low priorities.

When presented with yes/no questions about whether various potential threats and pressures were negatively impacting waterbird populations, the majority of respondents answered yes to each threat category; each threat category received 53.5–60.9% yes responses in regards to negative impacts to waterbirds (Fig. 12.2). Hunting was perceived by local communities to have the greatest impact on waterbird populations (60.9% of respondents answered "Yes"), followed by forest landgrabbing, hydropower dam development, and poisoning (59.0%, 58.6%, and 58.6% of respondents answered yes, respectively). Survey responses suggest that waterbird populations are primarily threatened by anthropogenic activities, although natural phenomena such as flooding caused by rainfall and nest predation by animals also impact waterbird populations.

Local community members perceived that impacts from most anthropogenic disturbances have decreased since the NRMD/RUPP waterbird conservation program was implemented in 2012 (Table 12.4). However, hydropower dams were the

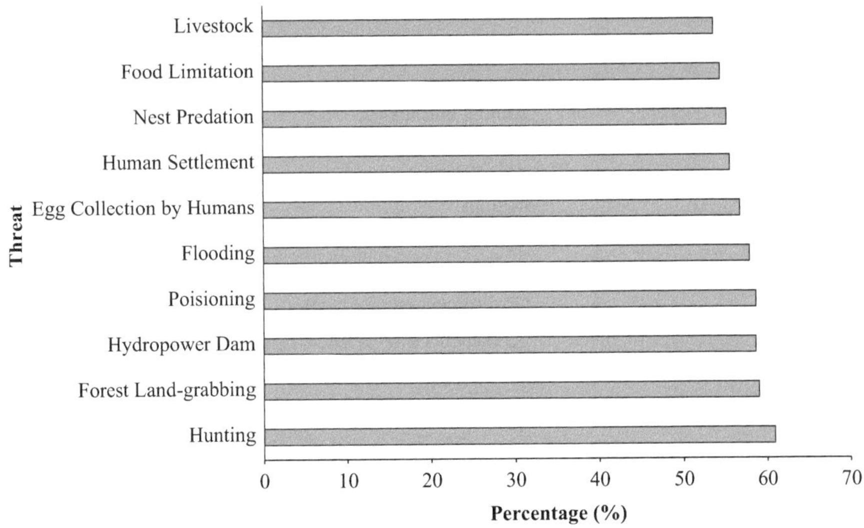

Fig. 12.2 Community perceptions of threats and pressures to waterbird populations in the 3S region of Cambodia

Table 12.4 Local perceptions of changes to the level of impacts since the conservation program began in 2012, and level of concern among respondents about the negative impacts to waterbirds in the 3S region, Cambodia

	Level of impacts			Concern about impacts		
Disturbance factors	Increased	Same	Decreased	Very worried	A little worried	Not worried
Human disturbances						
Hydropower dams	X			X		
Forest land-grabbing			X		X	
Hunting			X		X	
Poisoning			X		X	
Settlements on sandbars			X	X		
Livestock on sandbars			X	X		
Natural disturbances						
Nest predation		X			X	
Food limitation	X			X		
Flooding		X			X	

exception; according to survey respondents, negative impacts from hydropower dams have increased since the project's inception. Also, according to survey respondents, there were no decreases in the impacts from natural disturbances during the project time period; respondents indicated that food resources for waterbird species had decreased and that threats from other natural disturbance factors such as nest

Table 12.5 Threatened bird species observed by respondents (n = 256), and perceptions of which species were protected by the NRMD/RUPP conservation program

	Percentage (%) of respondents	
Bird species	Observed in past 12 months	Protected by RUPP program
River lapwing	86.7	57.0
River tern	75.8	55.5
Great thick-knee	49.6	52.3
Small pratincole	57.0	51.6
Mekong wagtail	0.0	50.4
Black-bellied tern[a]	0.0	50.4
Little ringed plover	0.0	47.7
Others	–	99.2
Giant ibis	71.1	–
Vulture	87.9	–
Little egret	–	34.4
Cattle egret	–	36.7

[a] Black-bellied tern has likely been extirpated from Cambodia; the last confirmed records were of two breeding pairs on the Sesan River in 2003 (Claassen 2004; Goes et al. 2010)

predation by animals and flooding from rainfall were the same. In general, community members were concerned about impacts from anthropogenic and natural disturbances on waterbird populations. They were primarily concerned about hydropower dams, human settlements and livestock on sandbars, and limitation of food resources for waterbirds (Table 12.4). All survey respondents expressed some level of concern regarding disturbances and pressures on waterbird populations; we did not receive any "not worried" responses in regards to the various disturbance factors (Table 12.4).

12.3.3 *Local Perceptions of the Conservation Program*

A large proportion of people in all communes had awareness about rare bird species under protection and conservation by national and international agencies; however, respondents tended not to recognize the smaller species such as little ringed plover and mekong wagtail (Table 12.5). Also, there were misunderstandings regarding what organization was protecting these species; many respondents misidentified the agency/organization responsible for bird nest protection. Several conservation programs have been implemented in the 3S basin by different agencies, including government and non-governmental organizations (NGOs). Many respondents thought that species were protected not by NRMD/RUPP but by related NGOs working closely with this conservation program, such as Birdlife International, Save Cambodia Wildlife, World Wildlife Fund, or Culture and Environment Preservation Association. Furthermore, respondents were unclear about which bird species were actually under protection. Some respondents said that some of the threatened project focal species (first seven species listed in Table 12.5) were not protected by the program, while other respondents listed species such as giant ibis and vultures that are protected by organizations other than NRMD (e.g., BirdLife International), and other respondents listed species such as egrets, which are common species and not protected by any conservation program (Table 12.5).

Despite confusion by community members regarding which organization was implementing the project, survey results indicated that 54% of all respondents were aware of the waterbird conservation program implemented by NRMD/RUPP. However, 39% stated that they had never heard of the program, and 7% believed that no such conservation program existed. Of the conservation activities conducted by the program, respondents were most aware of patrolling activities (33%), followed by awareness raising (15%), capacity building (14%), and bird surveys (13%); 24% of respondents were not aware of any of the activities being conducted by NRMD/RUPP.

Subsequent questions revealed that 72% of respondents had participated in at least one of the aforementioned conservation activities: 59% were involved in patrolling, 11% in nest and sandbar monitoring and protection, 1% in capacity building, and 1% in awareness raising activities. A follow-up question about patrolling showed that nearly all project participants had been involved in patrolling to some degree: 45% always participated in patrols, 30% were sometimes involved, 20% participated rarely, and only 4% had never participated in patrolling. The primary reasons that community members had for participating in project activities included ecological services protection (35% of responses) and protection of species for the next generation (35%). Secondary reasons for participating included being ordered to join the project by the head of the community (15% of responses) or to receive financial benefits from the project (10%); 5% of respondents said they did not know why they joined the project.

According to survey results, 39% of all respondents said that the nest protection program provided significant livelihood benefits, 21% said it provided medium benefits, 22% said benefits from the program were low, and 18% did not know. Community perceptions of benefits from the program varied according to commune; responses were most positive in Hat Pok commune, while in Kaoh Pang commune perceptions of program benefits were lowest and interviewees gave the highest percentage of "don't know" responses (Fig. 12.3). Some respondents in Talat and Thmor Keo said that livelihood support in these communities was low because the project focused more on strengthening capacity and conservation activities rather than providing community benefits.

According to the survey, 51% of respondents who were involved in the conservation project as nest guards and patrollers live approximately 2 km from the conservation sites (i.e., protected sandbars), which is about 30 min' travel time by motor boat. This short distance from home to the sandbar made it easier and more convenient for them to access conservation sites in order to effectively protect nests on sandbars and to patrol the areas around the conservation sites. However, 49% of patrollers live approximately 8 km from the sandbars. This longer distance between

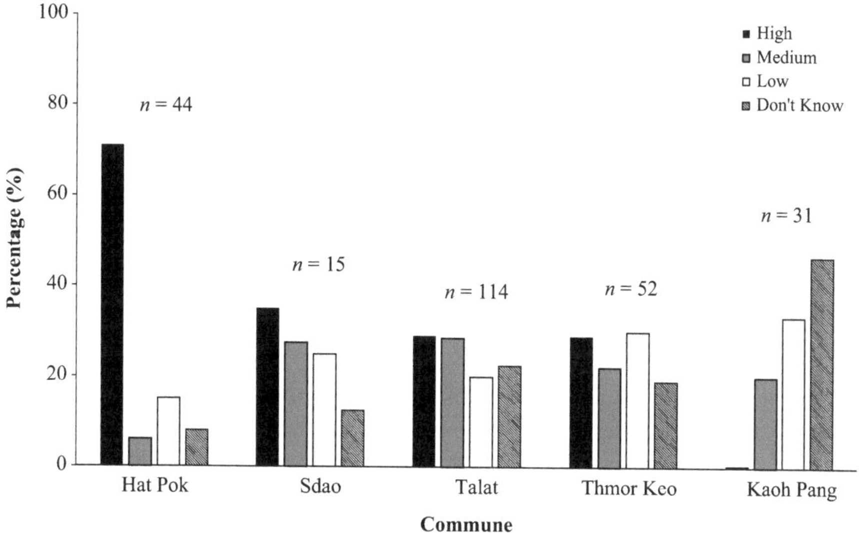

Fig. 12.3 Community perceptions of livelihood benefits from the waterbird nest protection program in the 3S River basin, Cambodia

their home and conservation sites required about 1 h of travel by motorboat, which was less convenient, so it was more difficult for them to effectively protect and patrol conservation sites.

Only 28% of respondents had not previously participated in any project-related conservation activities. Among respondents who did not participate, 68% said it was because they did not have enough time. Respondents who did not participate in project activities due to lack of time tended to be highly dependent on forest products (timber and nontimber forest products) for their livelihoods. Other reasons for not joining the project included believing that bird population declines were unavoidable (25% of responses), believing that bird populations were not declining or that declines were natural (5%), or thinking that bird conservation was the responsibility of the government rather than the community (2%).

12.4 Discussion and Conclusions

12.4.1 Waterbird Populations and Threats

Nest protection programs offer hope for providing effective protection and reducing impacts to threatened waterbird species (Sok et al. 2012; Clements et al. 2013; Wright et al. 2013; Claassen et al. 2017). Community members that we surveyed indicated that waterbird populations on the Sekong and Sesan Rivers had increased since inception of the NRMD/RUPP nest protection program in 2012. Furthermore, according to community members, the impacts from most anthropogenic disturbances have decreased since the conservation program began. In particular, our results indicate that threats from hunting, egg collection, cattle, and poisoning have been reduced after implementation of the nest protection and conservation program (Table 12.4). According to KAP survey results, 61% of all respondents considered hunting and human exploitation to be significant causes of bird population declines (Fig. 12.2). Responses indicated that hunting and other types of direct human persecution such as egg harvesting were the greatest threats to waterbirds on the Sesan and Sekong Rivers. These results mirror findings of previous studies of waterbird species on the 3S and Mekong Rivers (Claassen 2004; Sok et al. 2012; Claassen et al. 2017), as well as elsewhere in Cambodia (Clements et al. 2013; Wright et al. 2013).

Despite nest protection being one of the primary project activities, respondents generally did not consider the project to have been effective against "natural" threats such as nest predation by animals. On the Mekong River, nest protection was generally more effective against human than animal nest predation, although predation was significantly reduced by using predator exclusion fencing around nests and chicks (Claassen et al. 2017). Although our project employed some limited use of anti-predator fencing, survey responses indicate that more widespread use of anti-predator fencing may be necessary to reduce nest predation and improve nesting success of threatened waterbirds.

While results from this study suggest that the NRMD/RUPP program has been successful in reducing impacts of small-scale human activities, the project has not been successful in mitigating large-scale impacts such as from hydropower dams. Hydropower dams on the 3S Rivers have resulted in altered river flows and hydrology. Effects of current and under-construction hydropower dams are predicted to cause a 28% increase in dry (breeding) season flows and a 4% decrease in rainy (nonbreeding) season flows (Piman et al. 2013). Direct impacts of hydropower dams on threatened waterbirds include flooding nests of ground-nesting species (Claassen 2004; Claassen et al. 2017), destroying habitat (Claassen et al. 2018), and reducing populations of fish and other aquatic prey (Baird and Meach 2005; Baran et al. 2013). In the 3S region, threats and impacts from hydropower dam developments have increased since the project began. Specifically, construction of the Lower Sesan 2 (LS2) dam at the confluence of the Sesan and Srepok Rivers was ongoing during this study; the LS2 dam was finally completed in September 2017. During this survey, the majority of interviewees considered hydropower dams to be a significant threat to waterbirds on the 3S Rivers. Respondents were very worried about impacts from hydropower dams, as well as impacts to food resources (Table 12.4). Regarding food limitation, the primary concern was declines in fish populations. Fish are a critical food resource for threatened waterbird species such as river terns, and also support local human livelihoods. Fish populations in the 3S Rivers are impacted by hydropower dams, as well as by overfishing and destructive fishing practices such as explosives and electro-fishing (Baird and Meach 2005; Baran et al. 2013).

12.4.2 Local Perceptions of the Conservation Program

The majority of survey respondents from project focal communities along the Sekong and Sesan Rivers were aware of the waterbird conservation program being implemented by NRMD/RUPP and had participated in project-related conservation activities. Most communities had positive perceptions of the program and were supportive of conservation activities implemented by the program. Over half (54%) of the respondents believed that the program provided them with a high level of livelihood benefits, and only 22% considered livelihood benefits from the program to be low (Fig. 12.3).

Local perceptions about livelihood benefits from the program varied between communities. In our study, 65% of Hat Pok respondents felt that livelihood benefits were high and 14% felt they were low, whereas in Kaoh Pang commune, none of the respondents considered livelihood benefits to be high and 33% considered them to be low (Fig. 12.3). Although there were discrepancies in perceptions of livelihood benefits, all participants received the same amount of pay (US$5 per day) for project activities such as nest monitoring and patrolling. Also, the project aimed to provide comparable payment amounts to each of the focal communities, although payments varied somewhat between communities due to differences in number of participants and the geographic area covered by each community.

Discrepancies between different communities' perceptions of benefits may have resulted from language and/or cultural differences, or from differential access to education. In this study, 29% of survey respondents were from ethnic minority groups such as Lao and Prov (Brao). Language barriers may have influenced results of the KAP survey if respondents misinterpreted or did not fully understand the interview questions. Additionally, education levels of respondents may also have affected survey responses; 70% of adults in the project focal communities did not attend primary school (NCDD 2010), and 72% of survey respondents were illiterate. The large percentage of "don't know" responses may have been due to questions being too complicated or technical. Communities varied in their percentages of "don't know" responses. For example, in regards to a question about whether livelihood benefits were high, medium, or low, "don't know" responses ranged from 8% percent in Hat Pok to 47% in Kaoh Pang (Fig. 12.3). In addition to education level, language barriers may have contributed to the number of "don't know" responses. Among project focal communities, fluency in Khmer seemed to be the lowest in Kaoh Pang community. Our results suggest that future surveys should take into consideration the educational and cultural backgrounds of community respondents. Interviewers should ensure that interviewees understand the survey questions by taking measures such as devoting additional time to clarify questions or employing the help of local language translators.

Conversely, variation in satisfaction with the level of livelihood benefits may have been related to local contexts. Low perceptions of livelihood benefits may have signaled jealousy or other forms of social conflict (Clements et al. 2013). While our project did not appear to have problems with jealousy among participants over conservation incentives, other similar direct payment programs in Cambodia have experienced such problems (Clements et al. 2013; Sok et al. 2012). Nonetheless, it will be important to uphold high program standards for equality and fair distribution of benefits in order to maintain the continued support of participating communities and to ensure effectiveness of the community-based conservation program.

Interview structure was another possible factor affecting survey results. The survey questionnaire contained mostly closed-ended yes/no or limited-choice questions, rather than open-ended qualitative questions. Although easier to quantify, and requiring less training to conduct than qualitative questions, closed-ended questions can introduce bias into survey results (Chenail 2011). In particular, an acquiescence bias may have been present, whereby respondents tended to answer yes to yes/no questions (Bowling 2005; Sharman and Powell 2012). Cross-cultural differences may have contributed to acquiescence bias, as well as towards a tendency for "don't know" responses (Johnson et al. 2005). In our study, responses to yes/no questions regarding whether certain types of threats were impacting bird populations may have been affected by acquiescence bias. Using open-ended or ordinal (ranking) types of questions likely would have yielded more informative data regarding threats to waterbird populations. Moreover, limited-choice questions may have missed important information. For example, our survey question asking nonparticipants to state why they did not participate in the project had a limited, nonexhaustive set of possible responses. Other reasons for not participating likely included not knowing that the project existed, not knowing how to become involved in the project, or considering the financial incentives from the project to be too low. An open-ended question would have provided a clearer picture of why respondents did not participate in the conservation program.

Reasons why community members participated in the project may have ultimately affected their satisfaction with project benefits. For most project participants, the main reasons they gave for becoming involved in the program included wanting to protect ecosystem services and wanting to give the next generation the chance to see wildlife species. For others, financial benefits were their primary reason for joining. However, participation by 15% of respondents was not voluntarily but was because they were ordered to join the project by the head of the community. This sometimes occurred when too few people voluntarily joined the project. The lack of volunteers in some communities appeared to be linked to high levels of illegal logging, since people could make more money from illegal logging than from the conservation program. Although we did not obtain data regarding levels of illegal logging in the project area, real or perceived opportunities to engage in illegal logging likely skewed local perceptions of the value of conservation payments provided by the project. Moreover, although we did not analyze reasons for participation separately for each community, villagers who were coerced by community heads into participating in the conservation program likely had lower satisfaction with the program and its benefits compared to those whose participation was voluntary.

12.4.3 General Conclusions About the Community-Based PES Program

Results of this survey suggest that the community-based bird nest protection program generally achieved positive biological and social outcomes. Local community perceptions were that waterbird populations increased, and threats – especially small-scale anthropogenic disturbances – decreased as a result of the conservation program. Additionally, community perceptions of the program were generally positive. Most community members surveyed were aware of the program, had participated in the program, and believed that the program had improved their livelihoods. Our results indicate that local community participation in conservation activities may be a key toward protecting the remaining biodiversity and ecosystem services of the Sekong and Sesan Rivers.

However, some critical challenges need to be addressed in order for a community-based payment for ecosystem services (PES) program to achieve positive social outcomes. Specifically, there are challenges associated with raising awareness of the program, disseminating accurate information about the program, effectively encouraging local participation, and ensuring voluntary participation, fair distribution of payments, and high levels of participant satisfaction and community support. Furthermore, interview survey design should aim to maximize relevant data while also minimizing potential response bias.

Despite the challenges, our results suggest that small-scale PES projects can achieve positive environmental and social outcomes in a short time period. Other PES programs in Cambodia have also found that direct payments to individual community members can be a relatively simple way to rapidly and effectively attain positive biological and social outcomes (Clements et al. 2010, 2013; Claassen et al. 2017). However, although slower to implement and to attain results, PES programs employing a more institutional structure can potentially achieve longer-term outcomes (Clements et al. 2010). Consequently, in addition to the current conservation activities, NRMD/RUPP is working with local institutional structures to provide broader benefits to communities in the future. Thus, the nest protection program can be seen as one component of a broader community-based conservation effort to protect biodiversity and support local livelihoods in the 3S region.

Overall, our research highlights the importance of understanding local community context and perspectives when designing and implementing a community-based conservation program. Regarding the design and implementation of community-based PES programs, it is especially important to gain insights regarding motivations or obstacles to participation and to identify potential hindrances to local support such as factors affecting participant satisfaction, perception of livelihood benefits, and distribution of payments. Results of this study will be used to improve the nest protection program in the 3S region of Cambodia, and can also inform community-based conservation efforts more broadly.

Acknowledgements We would like to thank all of our partner agencies and organizations. Specifically, we would like to thank the Cambodian Forestry and Fisheries Administrations and the Ministry of Environment, as well as the Stung Treng and Ratanakiri provincial Forestry and Fisheries Cantonments. Our heartfelt gratitude goes to local communities, district and commune authorities for administrative support and facilitation of project implementation. We are grateful to our partner organizations, including BirdLife International, 3S River Protection Network, Culture and Environment Preservation Association, and Wildlife Conservation Cambodia. Furthermore, we would like to thank the members of our project team, including Sok Samet, Yen Run, Ouch Mara, Pheung Sophea, and Leang Sovichea. This project would not have been possible without the support and participation of the local communities along the 3S Rivers. Funding for this project was generously provided by the MacArthur Foundation and the Critical Ecosystem Partnership Fund.

References

Alreck, P. L., & Settle, R. B. (1995). *The survey research handbook: Guidelines and strategies for conducting a survey* (2nd ed.). Chicago: Irwin.

Arias, M. E., Piman, T., Lauri, H., Cochrane, T. A., & Kummu, M. (2014). Dams on Mekong tributaries as significant contributors of hydrological alterations to the Tonle Sap floodplain in Cambodia. *Hydrology and Earth System Science, 18*(12), 5303–5315.

Baird, I. G., & Meach, M. (2005). *Sesan river fisheries monitoring in Ratanakiri province, Northeast Cambodia: Before and after the construction of the Yali Falls dam in the central highlands of Vietnam*. Ban Lung: 3S Rivers Protection Network (3SPN), Global Association for People and the Environment (GAPE).

Baird, I. G., Baird, M., Cheath, C. M., Kim, S., Mekradee, N., Sounith, P., Nyok, P. B., Sarim, P., Savdee, R., Rushton, H., & Phen, S. (2002). *A community-based study of the downstream impacts of the Yali Falls dam along the Se San, Sre Pok and Se Kong Rivers in Stung Treng Province, Northeast Cambodia*. Cambodia: Se San Protection Network Project, Partners For Development (PFD), Non Timber Forest Products Project (NTFP), Se San District Agriculture, Fisheries and Forestry Office, Stung Treng District Office Cambodia.

Baran, E., Samadee, S., Teoh, S. J., & Tran, T .C. (2013). *Fish and fisheries in the Sekong, Sesan and Srepok basins (3S rivers, Mekong watershed), with special reference to the Sesan river*. Project report: Challenge program on water and food Mekong project MK3. Optimizing the management of a cascade of reservoirs at the catchment level. International Centre for Environmental Management (ICEM), Hanoi.

Bezuijen, M. R., Timmins, R., & Seng, T. (Eds.). (2008). *Biological surveys of the Mekong River between Kratie and Stung Treng towns, Northeast Cambodia, 2006–2007*. Phnom Penh: World Wildlife Fund Greater Mekong – Cambodia Country Programme, Cambodia Fisheries Administration, Cambodia Forestry Administration.

Bowling, A. (2005). Mode of questionnaire administration can have serious effects on data quality. *Journal of Public Health, 27*(3), 281–291.

Chenail, R. J. (2011). Interviewing the investigator: Strategies for addressing instrumentation and researcher bias concerns in qualitative research. *Qualitative Report, 16*(1), 255–262.

Chinnak, S. (2006). Natural resources for development and the destiny of marginal groups: A case study of indigenous people in Rattanakiri Province, Cambodia. In: *JCAS Symposium Series, number 25* (pp. 135–148). National Museum of Ethnology, Japan Center for Area Studies.

Claassen, A. H. (2004). *Abundance, distribution, and reproductive success of sandbar nesting birds below the Yali Falls hydropower dam on the Sesan River, Northeastern Cambodia*. Phnom Penh: World Wildlife Fund, Danida, WCS and BirdLife International.

Claassen, A. H., Sok, K., Arnold, T. W., & Cuthbert, F. J. (2017). Effectiveness of direct payments to increase reproductive success of sandbar-nesting river birds in Cambodia. *Bird Conservation International, 27*(4), 495–511.

Claassen, A. H., Forester, J. D., Arnold, T. W., & Cuthbert, F. J. (2018). Consequences of multi-scale habitat selection on reproductive success of riverine sandbar-nesting birds in Cambodia. *Avian Biology Research, 11*(2), 108–122.

Clements, T., John, A., Nielsen, K., An, D., Tan, S., & Milner-Gulland, E. J. (2010). Payments for biodiversity conservation in the context of weak institutions: Comparison of three programs from Cambodia. *Ecological Economics, 69*(6), 1283–1291.

Clements, T., Rainey, H., An, D., Rours, V., Tan, S., Thong, S., Sutherland, W. J., & Milner-Gulland, E. J. (2013). An evaluation of the effectiveness of a direct payment for biodiversity conservation: The bird nest protection program in the Northern Plains of Cambodia. *Biological Conservation, 157*, 50–59.

Ferraro, P. J., & Kiss, A. (2002). Direct payments to conserve biodiversity. *Science, 298*(5599), 1718–1719.

Gjertsen, H., & Niesten, E. (2010). Incentive-based approaches in marine conservation: Applications for sea turtles. *Conservation and Society, 8*(1), 5–14.

Goes, F. (2013). *The birds of Cambodia: An annotated checklist*. Cambodia: Centre for Biodiversity Conservation, Fauna and Flora International Cambodia Program, and Royal University of Phnom Penh.

Goes, F., Claassen, A., & Nielsen, H. (2010). Obituary to the black-bellied tern. *Cambodian Journal of National History, 2010*, 5–6.

Grimsditch, M. (2012). *3S rivers under threat: Understanding new threats and challenges from hydropower development to biodiversity and community rights in the 3S River Basin*. Ban Lung: 3S Rivers Protection Network.

Gumucio, S. (2011). *The KAP survey model: Knowledge, attitude, and practices*. Paris: Medecins du Monde.

Johnson, T., Kulesa, P., Cho, Y. I., & Shavitt, S. (2005). The relation between culture and response styles: Evidence from 19 countries. *Journal of Cross-Cultural Psychology, 36*(2), 264–277.

Milne, S., & Chervier, C. (2014). *A review of payments for environmental services (PES) experiences in Cambodia*. Indonesia: CIFOR.

MRC. (2010). *State of the basin report 2010*. Vientiane: Mekong River Commission.

National Institute of Statistics. (2010). Cambodia Demography. https://www.nis.gov.kh/index.php/en/2-uncategorised/11-cambodia-socio-economic-survey-tables. Accessed 22 June 2018.

NCDD. (2010). *Commune database online*. http://db.ncdd.gov.kh/cdbonline/home/index.castle. Accessed 5 June 2017.

Petheram, L., & Campbell, B. M. (2010). Listening to locals on payments for environmental services. *Journal of Environmental Management, 91*(5), 1139–1149.

Piman, T., Cochrane, T. A., Arias, M. E., Green, A., & Dat, N. D. (2013). Assessment of flow changes from hydropower development and operations in Sekong, Sesan and Srepok rivers of the Mekong basin. *Journal of Water Resources Planning, 139*(6), 723–732.

Roscoe, J. T. (1975). *Fundamental research statistics for the behavioural sciences* (2nd ed.). New York: Holt Rinehart and Winston.

Seak, S. (2015). Socio-economic development and transboundary issues in the Sesan, Sekong, and Srepok (3S) of the Lower Mekong Basin (LMB). *Science Verse Science Direct, 26*(2015), 1387–1407.

Seng, K. H., Pech, B., Poole, C. M., Tordoff, A. W., Davidson, P., & Delattre, E. (2003). *Directory of important bird areas in Cambodia: Key sites for conservation*. Phnom Penh: Department of Forestry and Wildlife, Department of Nature Conservation and Protection, BirdLife International Indochina Program, Wildlife Conservation Society Cambodia Program.

Sharman, S. J., & Powell, M. B. (2012). A comparison of adult witnesses' suggestibility across various types of leading questions. *Applied Cognitive Psychology, 26*(1), 48–53.

Sok, K., Claassen, A. H., Wright, H. L., & Ryan, G. E. (2012). Waterbird nest protection on the Mekong River: A preliminary evaluation, with notes on the recovery and release of white-shouldered ibis *Pseudibis davisoni* chicks. *Cambodian Journal of National History, 2012*, 29–41.

Timmins, R. J., & Men, S. (1998). *A wildlife survey of the Tonle San and Tonle Srepok River basins in Northeastern Cambodia*. Phnom Penh: Fauna and Flora International Indochina Program, Wildlife Protection Office, Department of Forestry.

Winemiller, K. O., McIntyre, P. B., Castello, L., Fluet-Chouinard, E., Giarrizzo, T., Nam, S., Baird, I. G., Darwall, W., Lujan, N. K., Harrison, I., & Stiassny, M. L. J. (2016). Balancing hydropower and biodiversity in the Amazon, Congo, and Mekong. *Science, 351*(6269), 128–129.

Wright, H. L., Collar, N. J., Lake, I. R., Rours Vann, R., Sok, K., Phearun, S., & Dolman, P. M. (2013). Experimental test of a conservation intervention for a highly threatened waterbird. *Journal of Wildlife Management, 77*(8), 1610–1617.

Wyatt, A. B., & Baird, I. G. (2007). Transboundary impact assessment in the Sesan River basin: The case of the Yali Falls dam. *International Journal of Water Resources Development, 23*(3), 427–442.

Ziv, G., Baran, E., So, N., Rodríguez-Iturb, I., & Levina, S. A. (2012). Trading-off fish biodiversity, food security, and hydropower in the Mekong River Basin. *Proceedings of the National Academy of Sciences of the United States of America, 109*(15), 5609–5614.

Part III
Governing Water: Values, Institutions, and Structures

Chapter 13
Approaching the Mekong in a Time of Turbulence

Peter A. Coclanis

Abstract Disputes regarding trans-boundary waters are almost always fraught with difficulty, so it is no surprise that disagreements over rights to the Mekong are proving problematic. Not only are the six nation-states comprising the GMS quite different from one another, but one—the PRC—is a great power. The fact that the source of the Mekong is located on the Tibetan Plateau—that is, within the territorial bounds of the PRC obviously complicates matters further regarding water rights. This paper will address issues regarding governance over the Mekong, issues becoming increasingly pressing every year because of climate change, on the one hand, and upstream dam-building/river diversion schemes, on the other. In so doing, the author will examine several approaches to/rationales for river governance—first-recourse governmental regulation, Chinese IR theories, natural law/environmental ethics, etc.—but will make the case for the efficacy of a transactions-cost approach to addressing issues of trans-boundary water rights on the Mekong. This more voluntaristic approach—which emphasizes efficiency and the accurate ascertainment and allocation, and effective enforcement of property rights regarding concerned parties, public and private—is based loosely on the work of Ronald Coase.

Keywords Mekong River · Dams · Externalities · Governance: Coase

Pointing out precarious and potentially catastrophic situations on one or another of the world's great river systems, alas, is not very difficult these days, more akin to shooting fish in a barrel than, let us say, to rocket science or, perhaps more apropos in this case, to environmental biology. For starters, one thinks of the Tigris-Euphrates system, the Nile, the Danube, La Plata, and the Colorado. In Asia, the Indus, the Brahmaputra, the Murray-Darling, the Yellow, the Yangze, the Salween, and the Mekong come immediately to mind (Chellaney 2011, 2013; Wirsing et al. 2013; Khagram 2004; McCully 2001). Here I focus on the last, the Mekong, not because

P. A. Coclanis (✉)
Global Research Institute, University of North Carolina-Chapel Hill, Chapel Hill, NC, USA
e-mail: coclanis@unc.edu

M. A. Stewart, P. A. Coclanis (eds.), *Water and Power*, Advances in Global Change Research 64, https://doi.org/10.1007/978-3-319-90400-9_13

it is the most important or the most endangered, for it is neither. Rather, I do so for other reasons.

First, because every endangered river system—like every unhappy family, according to Tolstoy—is different, and endangered/unhappy in its own way. With this in mind, close study of the particulars of this case seems both useful and essential if we are to move forward. But there are other reasons as well, relating to the fact that the problems the Mekong is currently confronting are at once extraordinarily complex and proving exceedingly difficult to address, much less to resolve. With these additional facts in mind, it seems reasonable to suggest that this might be a propitious time to think about new approaches, or, to be more precise, to approaches seldom, if ever applied to policies regarding this river. Here, I shall lay out one such approach, drawing inspiration, albeit loosely, from the insights of the late Nobel Prize-winning economist Ronald Coase in hopes that such insights might prove of some utility to policymakers in the years ahead.

At this late date it is hardly necessary to spend much time convincing people—particularly the readers of this volume—that the issues the Mekong faces are complex and difficult. Indeed, given the layered nature of such problems—one can insert nods here to peeling onions, nested Russian dolls, etc.—it may be more difficult to argue for a margin of hope regarding the Mekong, for many believe the problems are well -nigh intractable. Clearly, the diversity of problems, number of nation-states involved, profound power asymmetries that must be factored in, and dearth of effective institutional mechanisms relating to regulation/governance of transboundary waterways render optimism difficult—at least in the short run (Boer et al. 2016; Tilt 2015; Wirsing et al. 2013; Stewart and Coclanis 2011; Chellaney 2011; Molle et al. 2009; Pomeranz 2009; Osborne 2006, pp. 223–278).

Merely identifying the relevant international players is a thorny task. Most analysts talk of six such players—the members of the MRC (Laos, Thailand, Cambodia, and Vietnam) and MRC's two "affiliates"—Burma and China. The fact that one of these affiliates, China, is a super power, and largely controls access to the river complicates things further, adding a dimension of what Mark Zeitoun and Jeroen Warner famously labeled "hydro—hegemony" to the mix (Tilt 2015; Chellaney 2011, 2013; Pomeranz 2009; Zeitoun and Warner 2006). And to pile on a bit, what about the controversial status of Tibet, the source region of the Mekong? And while we are at it, what about the developed countries—players hiding in plain sight, as it were—which countries have contributed so much to CO2 emissions and to climate change, the effects of which are proving increasingly apparent and disturbing, particularly on the lower Mekong.

However complicated and contentious interstate rivalries—and they are complicated and contentious indeed—intrastate battles, which are often downplayed or even overlooked completely, may be more complex still. Different regions, sectors, and governmental units and agencies within each of the nation-states involved have discrete interests and represent varied constituencies regarding the Mekong and often clash, as they attempt to defend, promote, and support the same. Ironically, nowhere is this truer than in China, the nation-state ostensibly holding most of the cards in the "great game" being played out on the twenty-first- century Mekong (Chellaney 2011; Pomeranz 2009).

What are these issues and interests, many of which have long, tangled historical roots? Many can be subsumed under the increasingly common "nexus" rubric, which when applied to water generally denotes intersectional issues concerned with the relationship between water, on the one hand, and food production, energy, the environment, and human rights, on the other (Dodds and Bartram 2016). On the Mekong, of course, food production doesn't merely mean agriculture, but also fisheries, and questions relating to both silt flows, transportation, trade, development, inequality/economic asymmetries, and poverty reduction must also be factored in to any reasonably satisfactory approach to river governance.

And the Mekong itself must be contextualized, particularly as the problems regarding the river relate to other rivers and other problems, all connected in one way or another with the critical situation on the so-called third pole—the Tibetan/Himalayan Plateau—the greatest repository of fresh water on Earth. The troubling situation on this troubled plateau has been written about by many distinguished scholars in recent years—Michael Zhao and Orville Schell, Kenneth Pomeranz, and Brahma Chellaney, to name but a few—and it arguably can be considered *primus inter partes* among all of the world's water problems, certainly in terms of the calamitous scale of consequences if things go terribly wrong in Tibet (Zhao and Schell 2008; Pomeranz 2009; Chellaney 2011). Without getting into too many details and thereby going too far afield, suffice it to say that ten key watersheds in Asia derive their water from the Himalayas and Tibetan highlands, which watersheds are vital, indeed indispensable to life over vast areas of Asia stretching from Afghanistan to eastern China and (given water-diversion projects) from Mongolia to "megadeltas" in Pakistan, India, Bangladesh, Burma, Thailand, Cambodia, and Vietnam (Qiu 2008; Pomeranz 2009; Chellaney 2011). And, despite this volume and the committed research conducted by our contributors, it must be admitted that the problems on the Mekong, however great, are not as dire or potentially catastrophic as those involving some other rivers originating on the Tibetan Plateau, especially the Brahmaputra (Samaranayake et al. 2016; Chellaney 2011).

Not only is South Asia much more populous than the countries on the Lower Mekong, but by almost any measure—land under irrigation, groundwater pollution, aquifer depletion, the external dependency ratio for annual renewable water resources—the problems in India, Pakistan, and Bangladesh are more severe. Regarding the degree of dependency on trans-border ("external") sources of renewable water resources, for example, Bangladesh, according to the FAO, comes in at over 91% and Pakistan at over 75%, while Vietnam is at 59% and Thailand 47% (Albert 2016; Chellaney 2011, p. 27).

Alas, Bangladesh, the most dependent of these nation-states on trans-border sources of renewable water, faces a near existential threat not only from climate change and rising sea levels, but, more immediately, from changes affecting the volume of water available to it on the Brahmaputra, the most important trans-border river flowing from that part of the Himalayan Plateau under Chinese control (Albert 2016; Samaranayake et al. 2016; Chellaney 2011). Chinese dam-building and water-diversion projects on the Brahmaputra as well as dam-building and water-use practices by other countries upriver from Bangladesh threaten to cut drastically the water flowing into this desperately poor, densely populated, rice-dependent nation-state, which state is the same size as Iowa but has 170 million people living in it as opposed to three million or so in the Hawkeye State.

Nor is the Brahmaputra the only other Asian river system that bears mention before we return to the Mekong, for another very large and important system originating on the Tibetan Plateau—this one located in Southeast Asia—bears brief mention as well: The Salween (known in China as the Nu River). Although this great river, let alone the threats to it, receives little attention globally, it is of enormous importance: Indeed, in terms of the volume of its flows to areas other than China, it ranks only slightly behind the Mekong (Chellaney 2011, p. 143). The Salween too is deeply threatened by dam-building and diversion projects, but because it flows from China directly and solely into secretive Burma before emptying into the Andaman Sea (although its watershed stretches into Thailand), it is under studied and its problems often passed over quickly and superficially (Tilt 2015; Chellaney 2011, pp. 260–263).

The brief references above to the Brahmaputra and the Salween are not intended to sidetrack us, much less to downplay the problems on the Mekong. Rather, they are brought up at once to underscore the fact that there are manifold threats arising from the changing situation on the Tibetan/Himalayan Plateau and to suggest that progress regarding river governance on the Mekong must always keep this broader context in mind.

And it is to governance questions or rather to approaches to governance questions that we shall now turn, for clearly none of the above considerations can be addressed, and none of the above problems redressed unless effective governance of the Mekong can be achieved. Today, clearly, effective governance is lacking, not for lack of effort by many committed parties, certainly, but because in my view governance -design assumptions, conventions, protocols, and policies, given the "facts on the ground"—or, in this case, on the water and riverbanks—are inadequate and thus insufficient for achieving the same. This, despite the fact that a welter of governance principles and programs have been tried on and applied to the Mekong in recent decades, albeit largely unsuccessfully.

For a variety of reasons—history, geopolitics, the relative weakness of ASEAN, China's hegemonic position on the Himalayan Plateau, etc.—principles based on international protocols and rules regarding trans-border waters have foundered on the Mekong.[1] Lip service has often been paid, for example, to the International Law Association's 1966 Helsinki Rules "on the Uses of the Waters of International Rivers," which is generally seen as the first formal international statement in support of "equitable utilization" the "law of reason," and the like, but policies based rather more on conventions associated with realpolitik—the principle of prior appropriation, most notably—China's power vis à vis other nation-states on the Mekong, and its consistent assertion of, and adherence to "maximalist" positions regarding its

[1] In Europe governance protocols/mechanisms regarding transboundary waters—on the Rhine and the Danube, for example—have existed in some form for centuries. See Schiff (2017), Fitzmaurice (1996), Gerlak (2004), and Jansky et al. (2004).

rights on the river have always carried the day (IWLP 1967; Pomeranz 2009; Chellaney 2011).[2] Such positions are, of course, also largely consistent with ancient and modern Chinese conceptions of international relations. Ancient concepts such as Tianxia and the emerging movement in China today to develop Sinocentric theories of IR come to mind in this regard, and the fact that Laos, Thailand, Cambodia, and Vietnam were all "tributaries" of Imperial China at one time or another appears to buttress such concepts empirically (Cunningham-Cross and Kristensen 2014; Zhang and Buzan 2012; Noesselt 2012; Zhao 2009; Callahan 2008). With such considerations in mind, it is not surprising that China has never agreed formally to any abrogation of its riparian rights, refuses to share information with neighboring nation-states regarding its dam-building projects and plans, and has never even signed on as a member of the Mekong River Commission, established in 1996, consenting only to "affiliate" status, along with Burma (Tilt 2015, pp. 178–182; Chellaney 2011, pp. 263–272; 2013).

Conventional Western legal conceptions regarding the governance of international waters don't (and won't) fare much better when it comes to the Mekong, not only because of China's hegemonic position on the river and the gross asymmetry in power between it and the other riparian nation-states involved, but also because such conceptions are too blunt and/or too inefficient. More to the point in the case of the Mekong, according to conventional Western regulatory logic, the party or parties determined to be at fault and thus responsible for negative externalities—in this case, diminished water flow, declining fish populations, diminished biodiversity, reduced silt carriage, etc.—would either be completely prohibited from so doing by government fiat (the "command and control" approach), or via the so-called polluter pays principle forced to "solve" the negative externality problem by indemnifying either the government or the parties directly harmed, generally through fines, taxes, subsidies, etc., at levels needed to eliminate the externality problem and return a given economy to efficiency (Furner 2010, pp. 134–135; Harrison and Theeuwes 2008, pp. 58–101). Although "efficiency" was originally defined in economic terms, over time it was broadened so as to incorporate, at least in part, environmental considerations as well.

Not surprisingly, economists approaching the question of externalities have generally preferred "polluter pays" reasoning rather than regulatory mandates because such reasoning at least attempts to ground policy in economic analysis, markets, empirical measurement of prices and costs, etc. But work originating primarily, but not solely with economist Ronald Coase over a half century ago—work which has gradually developed into a core component of the influential law and economics movement—over time has succeeded in a scholarly if not necessarily in an applied sense in demonstrating that "polluter pays" reasoning (including environmental taxes famously associated with economist A. G. Pigou) can and often do lead not to efficiency, but to inefficient outcomes in their own right (Harrison and Theeuwes

[2] The "Helsinki Rules" were formally superseded by the "Berlin Rules" of 2004, which cover a broad range of water issues (domestic waterways, aquifers, etc.), while holding to the same principles regarding usage as the Helsinki Rules (IWLP 2004).

2008, pp. 79–101; Pigou 1932, Part II, Chap. xi). Why? According to Coase's revolutionary insight, because such reasoning reflects a misperception of, and, as a result, an incorrect framing of the entire class of problems conventionally subsumed under the rubric externalities.

Against the approaches described above, Coase began in the late 1950s to lay out an alternative approach to negative externalities, which are essentially "spillover effects" of an activity or activities that bring costs that affect a party or parties that did not choose to incur them. Coase's alternative approach—emphasizing rights, reciprocity, and economic efficiency—is encapsulated in what later came to be known as the Coase Theorem. Whereas the overriding assumption in both regulatory and "polluter pays" approaches is that negative environmental externalities—polluted water, dirty air, excessive noise, and, yes, reduced water flow because of dam-building—were to be addressed via the setting of governmental policies or legal rules that in essence restrained A from inflicting harm on B, Coase's operating assumptions were far different. In his view, such externalities in conceptual terms are not so simple, and it is often better and more efficient to view them in reciprocal rather than one-sided ways (Coase 1959, 1960).

That is to say, instead of asking who did what to whom, Coase argues that questions relating to negative externalities need to be reframed so as to ask, as he famously put it in 1960: "Should A be allowed to harm B or should B be allowed to harm A?" The primary goal, according to Coase, should be an economically efficient solution and, thus, "to avoid the more serious harm" (Coase 1960, p. 2). Whether A's rights or B's rights are upheld—in whole or in part—is in a sense immaterial to Coase. For Coase, efficient harm avoidance, via the assignment of rights and liabilities to the parties involved, is paramount, and to him harm avoidance works best in a consensual, bargaining context. In principle, the "contracting" approach, which invites bargaining between/among affected private parties, thereby "internalizing" solutions to externality problems, generally favors a more restricted role for government, ideally limited to bargaining facilitation and contract enforcement if and when involved parties reach an agreement (Coase 1960; Posner 1998, pp. 35–99).[3]

Coase's general approach and this theorem specifically have both generated enormous literatures over the years, and the development of two very influential scholarly movements—public choice theory and the "law and economics" tradition in legal circles—drew considerable inspiration from Coase's pioneering work, work for which he was awarded the Nobel Prize in Economic Sciences in 1991 (Posner 1998; Harrison and Theeuves 2008). Critics have identified a number of problems with Coase's approach to externalities, with some of which I wholeheartedly agree and to all of which we shall return later, but before picking at the approach, let us elaborate a bit on it and, employing both hypothetical and real examples, illustrate how it works.

---

[3] For an interesting collection of essays making cases for the efficacy of policies based on the establishment of secure and transferable property rights in water, see Gardner and Simmons (2012).

Coase himself laid out a number of hypothetical cases regarding the efficiency possibilities of his theorem, in one notable instance discussing the resolution of a conflict over negative externalities between a noisy candy manufacturer whose workplace was adjacent to a doctor's quiet office and, in another, laying out a case involving stray cattle owned by rancher A destroying crops grown on farmer B's land (Coase 1960).

David D. Friedman, Nobelist Milton Friedman's son, has also written on the utility of the Coase Theorem in a hypothetical case involving airplane noise interfering with the sleep of people living near an airport. One solution of this issue, of course, would be to regulate the amount of noise airplanes can make, but, according to Friedman, that solution may not be the most efficient one. When considered as a Coasean question of contested property rights, alternative solutions might prove more efficient such as charging airplanes for the right to make noise, or allowing airlines to pay for soundproofing homes near the airport, or to buy out nearby homeowners. It all depends, for these are empirical questions that are subject to calibrated measurement (Friedman 1990, pp. 523–524; 2014, pp. 99–100, 305–310).

Now for a real-world example from the U.S. In 2014 a group of economists working out of Boise State University in the U.S. made a strong argument regarding the successful employment of the Coasean framework in an actual case in the state of New York (Black et al. 2014). Focusing on a multi-party agreement in 1997 settling a water pollution case—the New York City Watershed Memorandum of Agreement (MOA)—the team demonstrated how the assignment of property rights and subsequent bargaining produced an economically efficient—and environmentally acceptable—outcome. The MOA came about years after the U.S. Environmental Protection Agency (EPA) sided with upstream interests and ordered New York City rather than "polluters" upstate to remove contaminants in its water that flowed downstream into the city from watersheds north of the city. Immediately after the decision, a regulatory designee of the EPA ordered New York City to remove said contaminants by building a new filtration plant.

In the view of New York City officials, the cost of building another filtration plant was prohibitive, but because in a Coasean sense property rights had been established, officials representing the city—the "loser" in said assignment and thus the responsible party for any negotiating—proposed other, less costly solutions that eventually proved acceptable to a wide range of interests: private property holders upstate, the State of New York, various and sundry environmental groups, and the EPA. Basically, instead of building a multi-billion dollar filtration plant, the city proposed to: buy some of the land upstate that was contributing to run-off and watershed degradation and to rectify problems thereon; compensate farmers for new watershed restrictions and for implementing new protections; incent farmers to employ less harmful practices; and work with government and NGOs on monitoring water quality. The agreement, however comprehensive, was still cheaper than building the filtration plant. Although not without problems, the MOU, almost 20 years later, is generally deemed a success—and an excellent example of a case using the Coasean framework in addressing externalities in a complex multi-party scenario (Black et al. 2014).

The case outlined briefly above involves internal/domestic rather than trans-border externalities and presents far fewer bargaining/contracting difficulties than would most trans-border cases. For starters, in formulating his theorem, Coase made several assumptions—economic rationality, well-defined property rights, and low transaction costs—and the last of these is obviously rendered more difficult to achieve once disputes cross borders (Coase 1960). Indeed, in many scenarios that come to mind involving negative externalities in Southeast Asia, it is not merely conceivable but perhaps even likely that at least two and maybe all three of these assumptions are lacking. In a strict sense, the Coasean frame clearly would not be applicable. But the spirit embodied in Coase's approach—especially its emphasis on property rights, reciprocity, and efficiency in addressing negative externalities—is relevant, even in trans-border cases in Southeast Asia, or so I shall argue here.[4]

Before turning to the Mekong per se, let us briefly consider another high-profile case in Southeast Asia involving negative externalities: The "haze" problem in Indonesia-Singapore-Malaysia, which has bedeviled the Straits countries for at least 40 years (Lee 2015). While this problem has proven vexing to populations in the Straits region, it is less fraught for people living in the GMS and, this being the case, should help ease us into considering problems on the Mekong from a loose Coasean perspective.

The haze problem, almost everyone agrees, is sparked mainly by forest fires in Indonesia, mainly on the islands of Sumatra and Borneo. The fires have usually been intentionally set to clear rain forests and peat land for more intensive usage, mostly notably, for agricultural units, large and small, growing palm trees for palm oil. Palm oil is quite valuable to Indonesia, which is the world's greatest producer, but the haze from the fires set obviously inflicts short-term discomfort and likely long-term harm to the health of people in the vicinity, not only in Indonesia, but in neighboring countries such as Malaysia and especially Singapore.[5] In addition to adverse effects on people's health, the haze adversely affects tourism, especially in Singapore, so negative externalities must be factored in as well. From a Coasean perspective, the appropriate question is not should the "polluter pay," but rather whose "property" rights are paramount? That is to say, does the overall social benefit of haze outweigh its social cost or vice versa? And, in either case, by how much (Coclanis and Doshi 2016)?

Once such (admittedly difficult) questions are answered, we can, *per* Coase, get down to determining how best to assign rights and liabilities. Should Indonesia (and

[4] In this regard note that John A.C. Conybeare, employing historical examples, demonstrated long ago that even in international environmental disputes wherein "the specification of property rights is insufficient to ensure socially optimal outcomes," good remedies have arisen at times via recourse to liability rules and tort remedies rather than via regulation by IO (International Organizations). In so arguing, he drew from the insights of Guido Calabresi and A. Douglas Melamed in their classic extension of Coase's original formulation of the externality problem. See Conybeare (1980) and Calabresi and Melamed (1972). In his piece, Conybeare argues—as I do here—for more openness to Coasean approaches.

[5] For an attempt to measure the negative health outcomes in Indonesia caused by Indonesian forest fires and the ensuing "haze" in one particularly bad year (1997), see Kim et al. (2017).

Indonesian interests) pay Singapore/Singaporeans to accept haze or at least some haze, or, on the other hand, should Singapore (or Singaporeans) pay Indonesia (and Indonesians) to stop setting the fires or at least reduce their incidence and effects (Sum 2015; Coclanis and Doshi 2016)?

These are extraordinarily complex issues for a variety of reasons. Transnational pollution cases are always tricky, and accurately assessing the costs of haze will be difficult, particularly because of the number and diversity of parties involved. Depending on how many—and which—parties are given formal status, transactions costs—the economic cost of negotiating an economic exchange, in this case, the exchange of *rights*—may be high. Perhaps too high to effectively internalize them via resolution by private parties. In this regard, Coase himself suggested that in cases involving smoke nuisance, where "a large number of people are involved and in which therefore the costs of handling the problem through the market or the firm may be high."… "governmental administrative regulation" might in fact lead to an improvement in economic efficiency (Lee 2013). Even then, though, it can be argued that the very act of conceptualizing and then framing such externality problems, *à la* Coase—that is to say, in a reciprocal rather than one-way manner—can help to resolve them with greater efficiency than would otherwise be the case.

To be sure, one can pursue alternative routes to addressing the haze problem, whether by taking the high road and pleading cooperative protection of the regional commons or the low road and invoking the Hobbesian jungle world of zealous competing sovereigns. One could also go the torts route and try to sue polluters (at least the big conglomerate/plantation polluters), but enforcing decisions outside of Singapore's jurisdiction would be tough. Framing the haze problem as one of reciprocal property rights that need clarification might be the way to go, even if governments rather than private parties play the lead roles. Such an approach certainly has the potential to reduce the level of tension and discord surrounding the haze.

The possibilities of an approach influenced by Coasean principles would seem especially promising if Singapore and Singaporeans could somehow overcome understandable concerns that "natural justice" requires that the "polluter pays." As a rich, developed country dealing with a problem caused at least in part by poor neighboring peasants, Singapore could, in so doing, get a jump on a solution to the haze problem and demonstrate its regional and global leadership by going the counter- intuitive reciprocity route.

In 2015 Indonesian Vice-President Jusuf Kalla raised hackles in the region by suggesting that Indonesia's neighbors shouldn't complain about 1 month of haze annually, but rather thank Indonesia for the good air quality in the region during the other 11 months of the year. However insensitive this remark—and it is insensitive indeed—in conceiving of the haze issue in reciprocal terms, it at least got some people thinking about the haze problem in new ways, offering the possibility perhaps for a new approach and a fresh start (Coclanis and Doshi 2016).

Now let us return to the Mekong with the above considerations in mind. Clearly, an improved governance arrangement on the Mekong will prove a Herculean task. How could it be otherwise, given the number of players, interests, and issues involved, the gross power asymmetries present, and the fact that, like it or not, we are already into a new riparian regime on the river? In a sense, then, the relevant, if sensitive idiom here may be water over the dam.

Again, the scenario. Six nation-states directly on the river, with the most powerful one by far, China, controlling both the headwaters—in a sense the spigot—as well as the first 40–45% of the water's course, and 21% of the entire Mekong River Basin (Pomeranz 2009; Chellaney 2011; Mekong River Commission 2005, pp. 1–10; Swain 2004, p. 117). For now, we will refrain from engaging the controversy regarding how this hydo- hegemon came in possession of these headwaters; in other words, we will not tarry over Tibet.

Five other nation-states of varied sizes and power further downstream, each with its own issues, challenges, interest groups, and position regarding river governance, governance at present only loosely dealt with—managed, if not coordinated under policy pronouncements associated with rather flimsy institutions such as the MRC and the newly established Lancang-Mekong River Cooperation group, not to mention pieties and platitudes offered by ASEAN and the UN.

Compounding the vexing situation above is the fact that even if the many thorny governance issues relating to the Mekong itself could somehow be successfully addressed, broader climate-change issues relating, for example, to rising temperatures on the Tibetan/Himalayan Plateau and rising sea levels/saline intrusion in the lower delta could well render moot any progress made regarding governance design and implementation (Boer et al. 2016; The Economist 2016; Richardson 2009; Molle et al. 2009). Establishing effective governance protocols—even identifying the relevant parties—regarding these issues is arguably more difficult still. And, again, the clock is ticking, rather loudly, what with numerous dams already in operation since the first major one—at Manwan—was completed by the Chinese in 1993, with others under construction, and still others planned on both the Mekong and its tributaries, and the effects of salt-water intrusion on the lower reaches of the Mekong already visible and being painfully felt (Osborne 2006, pp. 228–237; Chellaney 2011, pp. 263–272; Stewart and Coclanis 2011).

So what to do? First, brace ourselves for some sober assessment and blunt and bitter truth. The old Mekong—the river that informed and structured remarkable civilizations on its reaches for millennia—is gone and is not coming back. In some ways, the "old" Mekong may in fact never have existed in the first place, at least not in the timeless way some today nostalgically perceive it (Osborne 2006). Here, one could do worse than to invoke the ancient Greek philosopher Heraclitus—you can't step into the same river twice and all that—for the Mekong and life on it were a*lways* changing, but until relatively recently not very fast.

The race today to transform the river was inspired by the understandable desire by governments in the region to improve the material well-being of their populations, which desire shouldn't be scoffed at, much less condescended to, particularly since the GMS, generally speaking, contains some of the poorest people on Earth. What we must strive to do is not to eschew or avert development on the Mekong, but to situate it in a sustainable frame, and maybe, just maybe, some insights inspired by Ronald Coase can help.

Recall again Coase's guiding assumptions regarding handling negative externalities, the assignment of rights, reciprocity, and efficiency. Let us focus on what I believe to be the most original, important, and relevant of these assumptions, reciprocity—the idea that both the alleged perpetrator and the so-called victim of negative externalities have rights that once legitimated and assessed need to be prioritized according to principles of economic/social efficiency rather than on the basis of vague abstractions such as "fairness" and "equity." In so doing, ironically and somewhat counterintuitively, we might actually also get to a fair and equitable settlement, albeit by a different route and different mean, as we shall see.

For analytical purposes, let us simplify the "facts on the water" and focus on four players only, which is complicated enough, and will still allow us more directly to get at the main issues at hand. To the same end, we will forego any specifics regarding negotiating entities, whether private or governmental in nature—I am agnostic regarding a lead role for the latter—in order to zero in on the framework and terms of the debate. Thus, four players, namely, the upriver nation-states of China and Laos, which will reap most of the benefits from dam-building, and the downriver nation-states of Cambodia and Vietnam, which are being most adversely affected by the negative externalities of upriver dam-building activity. This reduction obviously simplifies the real situation considerably. Myanmar is an upriver player, but currently isn't exercising much influence on the Mekong; Thailand will benefit both from the dams it is planning on tributaries and more so by the hydro power it will purchase from Laos. Cambodia itself is building a few dams. Nonetheless, limiting our analysis in this exercise to four nation-states, as we have, will allow us to get a sense of the Coasean approach without getting bogged down in a morass of exceptions, qualifications, and details.

In so doing, what do we find? First, four very poor and underdeveloped areas. China's Yunnan Province, which the Mekong (known as the Lancang there) courses through, is the second poorest of China's 31 administrative units, the GDP per capita of the roughly 45 million people (a little over $9000 PPP) therein coming in at only 58% of the national average in 2016. The province of Tibet—source of the Lancang/Mekong, fares little better (about US $10,000 PPP), ranking 28th, with recent Han Chinese migrants to the area faring much better economically than Tibetan natives (PRC 2017).[6] The population living in the small mountainous nation-state of Laos—with a population of only about seven million—is poorer still with a GDP per capita in 2016 of only about US $5700 PPP (IMF 2017).

The two downriver nation-states are just as poor. GDP per capita in the large nation-state of Vietnam (population 95 million) was at about US $6400 PPP in 2016, and GDP per capita in 2015 for the nearly 16 million people living in Cambodia was only about US $3700 PPP, lower than in Yunnan, Tibet, Laos, Vietnam, and every other nation-state in ASEAN, including, surprisingly both Myanmar and Timor-Leste (IMF 2017). Although all of these areas are poor, one

[6]The PRC's National Bureau of Statistics' figures in the source above are for GDP by province. Population data for Yunnan, Tibet and other Chinese provinces are for 2015, and the PPP conversions use IMF data.

difference between the upriver areas of Yunnan and Laos, on the one hand, and Cambodia and Vietnam, on the other, is that the latter areas—in my view—have more economic possibilities, more degrees of economic freedom, as it were, than the mountainous areas upriver. Not for nothing, after all, were the river valleys in the latter areas historically the centers of such advanced and sophisticated civilizations as the agricultural Khmers and the fishing/seafaring Champa, while the mountainous/hilly areas upriver were much less developed, very much part of the poor (and "ungovernable") area historian Willem van Schendel and later and more famously, James Scott refer to as Zomia (Scott 2009, p. xiv).

With this context in mind, back to river governance. Although much is at issue on the Mekong, one thing is clear: China and Laos are mainly responsible for the negative externalities experienced downriver. Full stop. That said, in my view neither "regulation"—bans/prohibitions on dam-building imposed from above—or remedies via the "polluter pays" principle is likely to be effective. First of all, nobody and certainly no governmental *body* is going to tell China what to do, given the power asymmetries framing river governance. Secondly, and just as importantly, all of the concerned areas, indeed, the entire GMS desperately needs development, and hydro- power perforce will play a key part in bringing about the same. In consequence of such considerations, it seems that some other adjudicative approach is in order.

One possibility here might be that associated with Elinor Ostrom (2009 Nobelist in Economic Sciences), which argues for the "collective self-governance" of "common-pool resources." Ostrom's approach is attractive in principle, not least because of its skepticism of solutions imposed from on top by governmental or supra-governmental bodies and support instead of "polycentric," locally-determined solutions. Moreover, in some cases it is compelling in practice, but her "design principles" seem both too stringent and unrealizable for governance over the twelfth-longest river in the world, which river originates on territory controlled by a superpower, acting as hydro hegemon, before coursing along territories controlled by five other far less powerful nation-states (Ostrom 1990, Chap. 3; 2009; Poteete et al. 2010).

Granting the utility of many of Ostrom's ideas, it nonetheless seems to me that Coase's simpler approach, particularly the reciprocal framing of the problem, offers a better way forward. In reduced form, then, the question becomes should China and Laos be allowed to harm Cambodia and Vietnam (via dam-building and the negative externalities arising therefrom)? Or should Cambodia and Vietnam inflict harm on China and Laos by stopping/removing the dams and the negative externalities they cause? Viewed in this way, each of the two groups of nation-states has "rights": China and Laos to development via dam-building, and Cambodia and Vietnam to a freer-flowing Mekong. The question is transformed from one of moral culpability to one of contested rights subject to negotiation and bargaining. And with questions of guilt out of the equation, parties can negotiate over the most efficacious ways to achieve an efficient outcome, with efficiency defined broadly to incorporate sustainability as well as economic criteria.

How might such negotiations and bargaining go? Obviously, it's impossible to say, but the most efficient solution will surely comprise terms that ensure that living conditions on the lower Mekong remain economically and environmentally sustainable for the millions dependent on the river AND that China and Laos reap developmental benefits from dam-building. The devil, not surprisingly, is in the details. Two factors may lessen the "early mover' advantages of China and Laos. First, it is becoming increasingly clear that damming the Mekong may not provide the mammoth developmental surge that early proponents believed. Although Laos talks of becoming the "battery" of the GMS and wants hydropower sales to its neighbors (particularly Thailand) to be the principal source of government revenue by 2025, a number of recent studies have suggested that the hydropower potential of the Mekong's dams has been considerably overstated. According to Richard Cronin of the Stimson Center, for example, the 11 dams currently being discussed for Cambodia and Laos, taken together, only have the potential to provide 6–8% of the total electricity needs of the lower Mekong by 2030, with most of that being generated in Thailand (Cronin 2013; Cronin and Weatherby 2015; The Economist 2016).

Secondly, China has, shall we say, bigger fish to fry, which may enhance the bargaining power of the states affected by the negative externalities brought about by its upriver dam-building campaign. For example, as part of its transition to superpower status, China has increasingly been signaling that it wishes to take up more of a leadership role in global issues ranging from globalization to Asian security to climate change. One way to do so would be moderate the pressures it is bringing to bear on neighboring nation-states below it on the Mekong. The fact that China is deeply invested both in winning hearts and minds in Asia and in promoting the "one belt, one road" plan first articulated in 2013, both of which initiatives involve Cambodia and (to a lesser extent) Vietnam, give these lower Mekong states stronger hands. This will be especially true if they can convince China of the great value of the psychic income it would gain from moderating its position on the Mekong—and thereby increasing the volume and better calibrating the seasonal flow of water/silt on the river, and making provisions to ensure the viability of fisheries on the Mekong, the largest inland fishery in the world.

For their part, the states on the lower Mekong must demonstrate greater understanding of Laos' developmental dilemma, which right now, bluntly put, is either to become the battery of Southeast Asia or to continue to languish economically. Because Laos' population is so small—around seven million—it may be possible, however, for the poor nation-states of Cambodia and Vietnam to work out a plan to provide "replacement" development support to Laos in exchange for pledges to cut back on its dam-building efforts and to begin to ameliorate the damage its dam-building has already done. Here, in fact, there just may be roles for international institutions such as the ADB, the World Bank, the new China-led Asian Infrastructure Investment Bank among others. A few billions annually for 10 years could, under propitious circumstances, go a long way toward making Laos something more than a battery.

At the end of the day, thinking of the problems on the Mekong as reciprocal rather than one-way may help us to develop an approach or approaches that *might* save the river. Not by eliminating the benefits of dam-building to China and Laos, but by reducing their level to an extent sufficient to lessen, if not eliminate the negative economic and environmental externalities dam-building is causing on the lower reaches of the Mekong. However unlikely to be successful this approach seems, it stands a better chance than approaches based on traditional or new Chinese ideas about IR, than government fiats or hard-and-fast policies based on the belief that the "polluter pays," and certainly than the bromides and pieties about equity and fairness often put forth by NGOs, and international institutions and agencies. And the kicker is that even should the approach espoused above succeed to some degree, everything could nonetheless come a cropper unless something is done about the deterioration of the Tibetan/Himalayan Plateau and the rising sea floor, that is to say, about climate change more generally.

References

Albert, E. (2016). *Water clouds on the Tibetan Plateau*. New York: Council on Foreign Relations.

Black, G., Dalton, D. A., Islam, S., & Batteen, A. (2014). The Coasean framework of the New York City watershed agreement. *Cato Journal, 34*, 1–32.

Boer, B, et al. (2016). *The Mekong: A socio-legal approach to river basin development*. London: Earthscan/Routledge.

Calabresi, G., & Melamed, A. D. (1972). Property rules, liability rules, and inalienability: One view of the cathedral. *Harvard Law Review, 85*, 1089–1128.

Callahan, W. A. (2008). Chinese visions of world order: Post-hegemonic or a new hegemony? *International Studies Review, 10*, 749–761. https://doi.org/10.1111/j.1468-2486.2008.00830.x/epdf.

Chellaney, B. (2011). *Water: Asia's new battleground*. Washington, DC: Georgetown University Press.

Chellaney, B. (2013, February 7). China's hydro-hegemony. *New York Times*.

Coase, R. H. (1959). The federal communications commission. *Journal of Law and Economics, 2*, 1–40.

Coase, R. H. (1960). The problem of social cost. *Journal of Law and Economics, 3*, 1–44.

Coclanis, P.A., & Doshi, T. (2016, June 14). A fresh solution to the haze. *Straits Times* [Singapore].

Conybeare, J. A. C. (1980). International organization and the theory of property rights. *International Organization, 34*, 307–334.

Cronin, R. P. (2013). Hydropower dams on the Mekong: Old dreams, new dangers. *Asia Policy, 16* (July), 32–38.

Cronin, R. P., & Weatherby, C. (2015). *Letters from the Mekong: Time for a new narrative on Mekong hydropower*. Washington, DC: Stimson Center.

Cunningham-Cross, L., & Kristensen, P.M. (2014). *Chinese international relations theory* (Oxford bibliographies). http://www.oxfordbibliographies.com/view/document/obo-9780199743292/obo-9780199743292-0046.xml;jsessionid=47A25988513C571C0CC9201B254385CD

Dodds, F., & Bartram, J. (2016). *The water, food, energy, and climate nexus: Challenges and an agenda for action*. Oxford: Earthscan/Routledge.

Economist. (2016, February 11). *The Mekong: Requium for a river*. http://www.economist.com/news/essays/21689225-can-one-world-s-great-waterways-survive-its-development

Fitzmaurice, J. (1996). *Damming the Danube: Gabčikovo and post-communist politics in Europe*. Boulder: Westview Press.

Friedman, D. D. (1990). *Price theory: An intermediate text* (2nd ed.). Cincinnati: South-Western Publishing.

Friedman, D. D. (2014). *The machine of freedom: Guide to a radical capitalism* (3rd ed.). New York: David D. Friedman.

Furner, M. O. (2010). From 'state interference' to the 'return to the market': The rhetoric of economic regulation from the old gilded age to the new. In E. J. Balleisen & D. A. Moss (Eds.), *Government and markets: Toward a new theory of regulation* (pp. 92–192). Cambridge/New York: Cambridge University Press.

Gardner, B. D., & Simmons, R. T. (Eds.). (2012). *Aquanomics: Water markets and the environment*. New Brunswick: Transaction Publishers.

Gerlak, A. K. (2004). Strengthening river basin institutions: The global environment facility and the Danube River basin. *Water Resources Research, 40*, W08S08. https://doi.org/10.1029/2003WR002936.

Harrison, J. L., & Theeuwes, J. (2008). *Law and economics*. New York: W.W. Norton.

International Monetary Fund. (2017). *World economic outlook database*. April 2016.

IWLP [International Water Law Project]. (1967). *The Helsinki rules on the uses of waters of international rivers*. https://www.internationalwaterlaw.org/documents/intldocs/ILA/Helsinki_Rules-original_with_comments.pdf

IWLP [International Water Law Project]. (2004). *Berlin conference, water resources law*. https://www.internationalwaterlaw.org/documents/intldocs/ILA/ILA_Berlin_Rules- 2004.pdf

Jansky, L., Murakami, M., & Pachova, N. I. (2004). *The Danube: Environmental monitoring of an international river*. Tokyo/New York/Paris: United Nations University Press.

Khagram, S. (2004). *Dams and development: Transnational struggles for water and power*. Ithaca: Cornell University Press.

Kim, Y., Knowles, S., Manley, J., & Radoias, V. (2017). Long-run health consequences of air pollution: Evidence from Indonesia's forest fires of 1997. *Economics and Human Biology, 26*, 186–198.

Lee, T. B. (2013, September 4). The Coase theorem is widely cited in economics. *Washington Post* [Wonkblog].

Lee, M. K. (2015, October 2). Haze in Singapore a problem dating back 40 years. *Straits Times* [Singapore].

McCully, P. (2001). *Silenced rivers: The ecology and politics of large dams*. London: Zed Books.

Mekong River Commission. (2005). *Overview of the hydrology of the Mekong basin*. Vientiane: Mekong River Commission.

Molle, F., et al. (Eds.). (2009). *Contested waterscapes in the Mekong region: Hydropower, livelihoods and governance*. London: Earthscan/Routledge.

Noesselt, N. (2012). *Is there a 'Chinese school' of IR?* (GIGA Working Papers, Leibniz-Institut fűr Global und Regionale Studien No 188). https://www.giga-hamburg.de/en/system/files/publications/wp188_noesselt.pdf

Osborne, M. (2006). *The Mekong: Turbulent past, uncertain future* (Rev. ed.). Crows Nest: Allen & Unwin.

Ostrom, E. (1990). *Governing the commons: The evolution of institutions for collective action*. Cambridge: Cambridge University Press.

Ostrom, E. (2009). A general framework for analyzing sustainability of social-ecological systems. *Science, 325*(July 24), 419–422.

People's Republic of China. (2017). *National Bureau of Statistics*. http://data.stats.gov.cn/english/easyquery.htm?cn=E0103

Pigou, A. C. (1932). *The economics of welfare* (4th ed.). London: Macmillan.

Pomeranz, K. (2009). The great Himalayan watershed: Agrarian crisis, mega-dams and the environment. *New Left Review, 58*, 5–39.

Posner, R. A. (1998). *Economic analysis of law* (5th ed.). New York: Aspen Law & Business.

Poteete, A. R., Janssen, M. A., & Ostrom, E. (2010). *Working together: Collective action, the commons, and multiple methods in practice*. Princeton: Princeton University Press.

Qiu, J. (2008). China: The third pole. *Nature, 454*, 393–396.
Richardson, M. (2009, July 16). Dams in China turn the Mekong into a river of discord. *YaleGlobal Online*. https://yaleglobal.yale.edu/content/dams-china-turn-mekong-river-discord
Samaranayake, N., Limaye, S., & Wuthnow, J. (2016). *Water resource competition in the Brahmaputra River basin: China, India, and Bangladesh*. Arlington: CNA.
Schiff, J. S. (2017). The evolution of Rhine River governance: Historical lessons for modern transboundary water management. *Water History, 9*, 279–294.
Scott, J. S. (2009). *The art of not being governed: An anarchist history of upland Southeast Asia*. New Haven: Yale University Press.
Stewart, M., & Coclanis, P. A. (2011). *Environmental change and agricultural sustainability in the Mekong Delta*. New York: Springer.
Sum, E. S. K. (2015, September 18). Pay Indonesia farmers to stop burning? *Today* [Singapore].
Swain, A. (2004). *Managing water conflict: Asia, Africa and the Middle East*. London/New York: Routledge.
Tilt, B. (2015). *Dams and development in China: The moral economy of water and power*. New York: Columbia University Press.
Wirsing, R. G., et al. (2013). *International conflict over water resources in Himalayan Asia*. Basingsoke: Palgrave Macmillan.
Zeitoun, M., & Warner, J. (2006). Hydro-hegemony – A framework for analysis of trans-boundary water conflicts. *Water Policy, 8*, 435–460.
Zhang, Y., & Buzan, B. (2012). The tributary system as international society in theory and practice. *Chinese Journal of International Politics, 5*, 3–36.
Zhao, M., & Schell, O. (2008). Tibet: Plateau in peril. *World Policy Journal, 25*, 171–180.
Zhao, T. (2009). A political world philosophy in terms of all-under-heaven (Tian-xia). *Diogenes, 221*, 5–18. https://doi.org/10.1177/0392192109102149.

Chapter 14
Trust Crisis and Building Trust in Transboundary Water Cooperation Along the Lancang-Mekong River

Li Zhang

Abstract As China and the lower Mekong countries (Myanmar, Laos, Thailand, Cambodia, and Vietnam) strengthen their efforts regarding the development of water resources of the Lancang-Mekong River, the two parties are undergoing a growing trust crisis that is becoming a key obstacle to improving transboundary water cooperation. This chapter presents a theoretical and analytical framework of international trust for the purpose of analysis and resolution and then discusses three cases of mistrust between China and the lower Mekong countries involving the Mekong River drought, waterway security, and hydrological data sharing. In these cases, insufficient communication and information channels, neglecting trust mechanisms for maintaining cooperation, and choosing national interests over international trust are the main root causes of trust crisis in transboundary water cooperation. This chapter concludes by drawing from the literature on international trust building to propose several ways to build trust for transboundary water cooperation: easing the fundamental contradiction between national interests and international trust, establishing a sense of trust maintenance under the Lancang-Mekong River Cooperation Mechanism, and expanding communication and information channels for both parties to better understand each other and solve water issues together.

Keywords International trust · Trust crisis · Mutual trust building · Transboundary water cooperation · Lancang-Mekong River

L. Zhang (✉)
Center for China's Neighbor Diplomacy Studies and Institute of Southeast Asian Studies, Yunnan University, Kunming, China
e-mail: zhang_li@ynu.edu.cn; zhl612@vip.qq.com

M. A. Stewart, P. A. Coclanis (eds.), *Water and Power*, Advances in Global Change Research 64, https://doi.org/10.1007/978-3-319-90400-9_14

14.1 Introduction

The Lancang-Mekong River[1] is a natural link between China and the lower Mekong countries (Myanmar, Laos, Thailand, Cambodia, and Vietnam). It is an important juncture for deepening the development of water resources, transportation channels, and environmental protection between the two parties. In November 2015, the Lancang-Mekong Cooperation (LMC) mechanism was formally established, and water cooperation was listed as one of five priority directions. In March 2016, at the first LMC leaders' meeting, China expressed its willingness to jointly set up the Lancang-Mekong Water Cooperation Center with the lower Mekong countries to share information and jointly protect the ecological resources along the river (Li 2016). But in recent years, there have been many disputes between China and the lower Mekong countries regarding the development of the river. One of the significant reasons is the lack of mutual trust. To be sure, this lack of trust also exists among the lower Mekong countries themselves, but this chapter focuses on the lack of mutual trust specifically between China and the lower Mekong countries. Mistrust between these two parties ultimately has led to difficulties in settling problems relating to drought, waterway security, information sharing, and other issues. International trust theory will help analyze the trust crisis in the Mekong River transboundary water cooperation and identify its root causes. Such a theory could also help find ways to reduce problems, promote the development of water cooperation under the LMC mechanism, and construct a community with a shared future for humanity for all Lancang-Mekong countries.

14.2 Basic Theory and Analytical Framework of International Trust

In the 1990s, international relations disciplines borrowed and developed the concept of "trust" from sociology to analyze the direction of international affairs and to explore the relationship between trust and cooperation (Kydd 2005; Rathbun 2011; Cross et al. 2013). Because the concept of international trust has not yet formed a complete theoretical system, this chapter employs existing work on international trust to build an analytical framework for solving trust issues of transboundary water cooperation in the LMC.

[1] This river is usually called the Lancang-Mekong River, while it is called the Lancang River in China. For the purpose of simplicity, hereinafter it is referred to as the Mekong River.

14.2.1 Basic Theoretical Literature on International Trust

Based on their different theoretical backgrounds and orientations, international relations scholars have explained the theory of international trust from different perspectives and applied it in various ways to the study of international relations.

Deborah Welch Larson (1997) presents three alternative explanations of trust and distrust in international relations: rational choice, domestic structures, and social psychology. She suggests that distrust may lead policy makers to dismiss the other side's cooperative signals or proposals. Paul R. Brewer et al. (2004) argue that citizens base their opinions about world affairs in part on generalized beliefs about how much their nation can trust other nations. They find that social trust, political trust, partisanship, and age influence this form of trust, which we call "international trust."

Andrew H. Kydd (2005) thinks that trust is a belief that the other side prefers mutual cooperation to exploiting one's own cooperation and that there are four main implications of the trust theory. First, cooperation requires a certain degree of trust between states. Second, though conflict between trustworthy states is possible, when we see conflict it is a sign that one or both states are likely to be untrustworthy. Third, in multilateral settings, hegemony—the presence of a very powerful state—can promote cooperation, but only if the hegemon is relatively trustworthy. Fourth, if two parties are genuinely trustworthy, they will usually be able to reassure each other of this fact and eventually cooperate with each other.

Aaron M. Hoffman (2002) believes the following four points. First, trust refers to an attitude involving a willingness to place the fate of one's interests under the control of others. Second, trusting relationships are behavioral manifestations of trust. Third, the intensity and scope of trust and trusting relationships are capable of variation. Fourth, trusting others involves making predictions about their future actions. Ken Booth and Nicholas J. Wheeler (2008) argue that trust will exist when two or more actors interpret the mutual attitude and behavior of each other and believe that the other side can be trusted both at the present and in the future and will not act in the way that will harm the interests and values of each other.

Yin Jiwu (2016) believes that the types of international trust can be divided into five basic categories: affective trust and cognitive trust, declared trust and cost trust, institutional trust and culture trust, reciprocity trust and fair trust, and homogeneous trust and reputation trust. The direction and asymmetry of international trust determine the origin and dimensions of trust.

14.2.2 Analytical Framework of International Trust

The study of international trust in international relations is more about the question of what international trust is, how it manifests, and what its origin and consequences are. It is not concerned much with the emergence and resolution of trust crises in detail, nor does it discuss why trust issues arise in an established cooperation relationship or mechanism. Thus, the three main issues discussed here are (a) the relationship between trust information and trust judgment obtained from both direct and indirect channels; (b) the relationship between international trust and national interests; and (c) the relationship between international trust and cooperation and both international and regional mechanisms. On this basis, this chapter tries to build an analytical framework regarding international trust.

Information feedback and integrity will affect trust judgment through direct and indirect channels. Before establishing a trust relationship with another country or an international organization, state actors usually obtain trust information through either direct channels, observing the existing behavior of the target country or organization and judging if there has been any treachery to determine whether or not to trust and cooperate with it; or through indirect channels, such as third-party neutral institutions and other participating countries, evaluating international behavior as the standard to judge whether the other country or international organization is trustworthy and if it is ready to cooperate. Similarly, the channels for interrupting the international trust process are also divided into two categories: the target country either (a) directly breaks an agreement and takes actions that are totally ignorant or dismissive of the feelings and interests of the partner country, or (b) uses the media to spread negative news to influence the development of trust relations.

In the process of developing foreign relations by state actors, especially establishing cooperative mechanisms, national interests and international trust should be mutually reinforcing and equally important. Both are starting points (and root causes) of cooperation and constructing cooperative mechanisms. But the difference between international and national interests is that national interest is the trigger point, while international trust is the maintenance point. Interests trigger the desire to cooperate, and only with trust can countries increase the pace, deepen the scope, and expand the scale of cooperation so that cooperative mechanisms can bring about sustainable and in-depth development.

International trust should be an important prerequisite for the establishment of cooperation and cooperative mechanisms, and it is also a necessary condition for developing sustained cooperation and maintaining mechanisms. Trust is important for any a long-term, stable cooperation or mechanism. If there is no trust among different parties, and they cooperate only because of self-interest and the desire to avoid responsibilities, then issues will inevitably arise and the cooperative mechanism may eventually collapse. When indirect factors affect trust building, this imperfect cooperative mechanism will become more vulnerable. In addition, sustained operation of the cooperative mechanism also depends on continuous maintenance of the trust relationship. If international trust factors are abandoned after the establishment of bilateral relations and related mechanisms, the cooperative mechanism will be limited, and the capacity and the original intention of the governance of the relevant regional and global issues will be lost.

Thus, the analytical framework of international trust includes three parts. First, trust information and trust judgment are obtained through direct and indirect channels. The integrity, correctness, and continuity of trust information will affect the establishment and development of the basis of trust. Second, international trust and

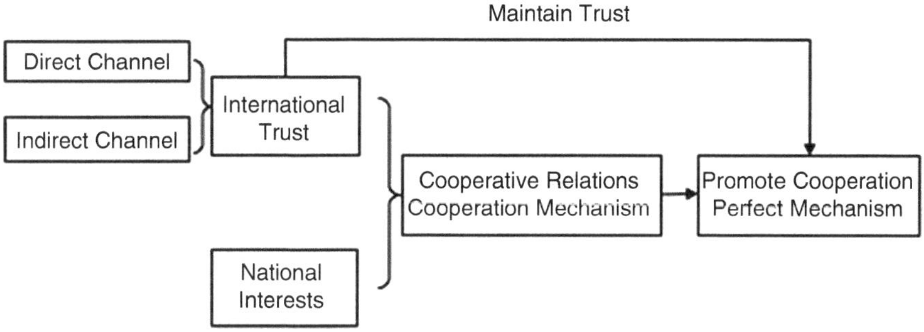

Fig. 14.1 Analytical framework of international trust. (Source: Author)

national interests are the starting points for a stable, sustainable development of cooperation and cooperative mechanisms. Third, international trust is an indispensable factor not only in the construction but also the maintenance of cooperation and cooperative mechanisms (Fig. 14.1).

14.3 Cases of Trust Crisis in Transboundary Water Cooperation Between China and the Lower Mekong Countries

In recent years, efforts by China and the lower Mekong countries to develop the Mekong River are being strengthened. The scope of mutual cooperation between China and the lower Mekong countries concerns many areas, such as hydropower development, shipping cooperation, and water management, etc.. But international mutual trust between them has not been strengthened, despite the increasing number of cooperation projects. On the contrary, several trust crises have arisen that affect cooperation and development. Due to limited space, this section analyzes three of these cases, involving the process of solving the Mekong River drought, constructing waterway security along the Mekong, and sharing hydrological data and other information related to the river.

14.3.1 Case 1: Solving the Mekong River Drought

In March 2016, China and the lower Mekong countries were affected by drought. During the period when affected parties attempted to deal with the drought, China and the lower Mekong countries questioned each other, and China tended to bear the brunt of accusations. The drought settlement process is a case study in the state of trust relations between China and the lower Mekong countries.

Located farthest downstream, Vietnam confronted the most severe drought it had faced in 90 years. Around 139,000 hectares of rice-cultivated area was damaged, and 575,000 people faced drinking water difficulties (Hunt 2016). At first, Vietnam asked Thailand and Laos to help, but neither of them did. Then at the request of the Vietnamese government, the Chinese government made a positive response when China also suffered drought. From March 15 to April 10, 2016, China increased the daily discharge capacity of the Jinghong hydropower reservoir by 2000 m^3/s to release more water downstream. A total of 12.65 billion cubic meters of water was discharged from the reservoir from March to May 2016. These releases amounted to between 40% and 89% of flows along various sections of the Mekong River. The emergency water supplement increased the overall water level or discharge along the Mekong mainstream to 0.18–1.53 m or 602–1010 m^3/s. If these emergency releases had not occurred, flows would have been 47% lower at Jinghong, 44% lower at Chiang Saen (Thailand), 38% lower at Nong Khai (Thailand), and 22% lower at Stung Treng (Cambodia). This additional flow also alleviated salinity intrusion in the Mekong Delta (Mekong River Commission 2017). China's emergency water supplement won praise from Vietnam, Cambodia, and Myanmar.

Scholars from Vietnam pointed out that it is short-sighted to prevent dam construction to reduce the occurrence of drought. Cambodia and Vietnam also have hydropower dams of their own on the Mekong River, so the reasons for the opposition to China and Laos continuing to build dams are untenable. Besides, China's efforts to save the Mekong River and to solve all kinds of related issues in the future will still be indispensable (Giang 2016). And according to the Mekong River Commission's new report, six Chinese dams that have been built in the mainstream on the upper Mekong basin since 1993 do affect water flows in the lower Mekong Basin, generally reducing the flow during the wet season and increasing it during the dry season (Mekong River Commission 2017) (Figs. 14.2 and 14.3).

But at the same time, there were many doubts about the intention of releasing water by China. A social scientist working for the International Water Management Institute in Laos suggested that China released the water from Jinghong hydropower reservoir in Yunnan province because of strategic considerations relating to the "One Belt and One Road Initiative", intending to strengthen China's infrastructure cooperation and investment in the lower Mekong River region and to ensure its own influence in investment and politics in the region. An opinion from Radio Free Asia stated that water released from the dam has had a very limited effect on the water level in Vietnam, because the released water will be absorbed by the dry soil before flowing to Vietnam (Radio Free Asia 2016). Duong Van Ni, a professor at Can Tho University, pointed out that China released 2.3 billion cubic meters of water, but little change has been noticed in the level of the Mekong downstream. Le Anh Tuan, vice director of Can Tho University's Research Institute for Climate Change, said that all the water behind the Jinghong would hardly constitute a drop in the bucket. According to his analysis published in the *Saigon Times* online, the Jinghong hydro-

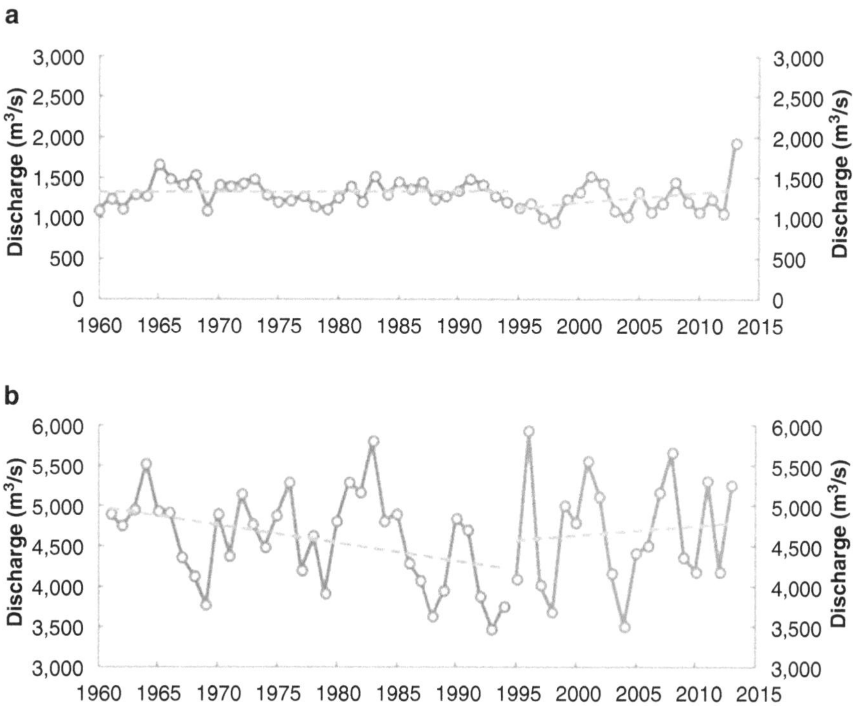

Fig. 14.2 Trends of annual dry season flows at (**a**) Chiang Saen, Thailand, and (**b**) Kratie, Cambodia, for 1960–2013. Colors indicate flows before (green) and after (blue) construction of Chinese dams in the upper Mekong basin. (Source: Mekong River Commission 2017) (Color figure online)

power reservoir would be empty after about 30 h if China released all the water Vietnam was requesting. China also suffered from the drought that beleaguered the Vietnamese, and many countries along the river did not have enough water, so they had no confidence in China's release of water (Nguyen 2016).

According to the above accounts, the drought occurred for a number of natural and man-made reasons. However, some scholars, researchers, journalists, and people from the lower Mekong countries always believe that droughts are caused by Chinese dams built in the mainstream on the upper Mekong basin. They do not understand or trust China. Meanwhile, China sacrificed its own benefits to release water but still did not gain the understanding and trust of the lower Mekong countries. Thus, trust between China and the lower Mekong countries is lacking.

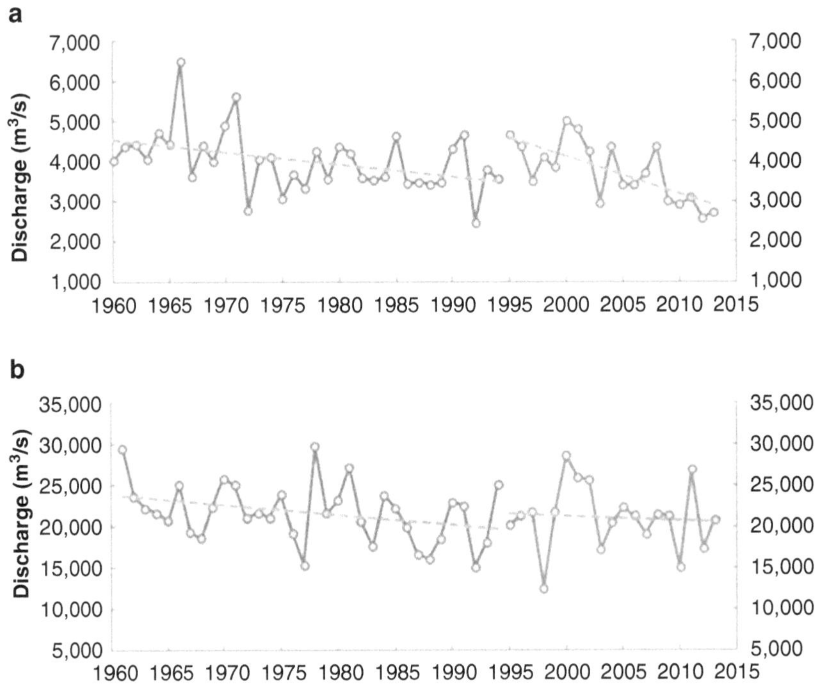

Fig. 14.3 Trends of annual wet season flows at (**a**) Chiang Saen, Thailand, and (**b**) Kratie, Cambodia, for 1960–2013. Colors indicate flows before (green) and after (blue) construction of Chinese dams in the upper Mekong basin. (Source: Mekong River Commission 2017) (Color figure online)

14.3.2 Case 2: Constructing Waterway Security

Issues of waterway security on the Mekong River impact both China and the lower Mekong countries. However, since the 2011 Mekong River Massacre, the lower Mekong countries have been worried about China's participation in co-governing the region's water resources.

In August 2011, travelers were robbed in the Golden Triangle waters, forcing Mekong River passenger transport to be suspended. In October 2011, two Chinese cargo ships, *Hua Ping* and *Yu Xing 8*, were attacked on a stretch of the Mekong River in the Golden Triangle region on the borders of Myanmar and Thailand. All 13 crew members on these two ships were killed and dumped into the Mekong River. After this so-called Mekong River Massacre, Lancang-Mekong River navigation was suspended for the first time in 10 years.

In December of the same year, a joint patrol was set up by China, Laos, Myanmar, and Thailand to restore freight traffic on the Mekong River. But in January 2012, two successive armed attacks occurred: on January 4, an unlawful armed marshal launched rockets into Myanmar patrol boats and Chinese cargo ships near a port in

Myanmar; and on January 14, the Chinese ship *Sheng Tai 11* on its way from Chiang Saen, Thailand, to the port of Xishuangbanna, Yunnan Province, was shot from the Laos side by people of unknown identity.

Before 2011, China had been working with the lower Mekong countries to co-develop the economic function of the waterway. Since the massacre, however, the Chinese government has been under pressure from the relatives of the victims and other people in China. The Chinese government was also aware of the importance of the security function of the Mekong River. Then in 2011 China started the joint patrol on the Mekong River with Myanmar, Laos, and Thailand. In 2013, the Safe Mekong Operation Project was jointly launched by China, Thailand, Laos, and Myanmar, which minimized and seriously controlled the movements of drug-trafficking syndicates and smugglers, resulting in a decline in illegal drug seizures. The achievements included the arrest of 3398 alleged drug traffickers and the seizure of 25,884,580 methamphetamine pills, 8153 kg of cocaine, 20 tons of methylene chloride, 1556 kg of heroin, 179 kg of opium, and 10,300 kg of other chemical substances (Xinhua News 2015). In October 2015, the Ministerial Meeting on Law Enforcement and Security Cooperation along the Mekong River was held in Beijing. A joint declaration was issued by China, Myanmar, Laos, and Thailand, which set a goal of transferring the existing cooperation mechanism on law enforcement and security along the Mekong River to a regional organization for cooperation.

There were 70 instances of joint patrols on the Mekong River from December 2011 to May 2018. However, although China, Laos, Myanmar, and Thailand tried to carry out joint patrol enforcement, some security issues still exist. More important, during the process of handling the Mekong River Massacre and security cooperation, there was still a serious trust crisis between the lower Mekong countries and China.

First, some lower Mekong countries believe that China took excessive steps to prosecute the traffickers. In addition, although China paid much of the cost to form the joint patrol and Safe Mekong Operation Project, some lower Mekong countries believe that its fundamental purpose is to demonstrate China's power to the lower Mekong countries, rather than to increase safety on the Mekong River (Santasombat 2016). In addition, although Laos, Myanmar, Thailand, and China issued a joint statement on security cooperation in the Mekong River basin, so far, Laos, Myanmar, and Thailand have not signed the Agreement on Law Enforcement Security Cooperation in the Mekong River Basin. This shows that the lower Mekong countries are still suspicious of cooperating with China on waterway safety. They have concerns about China's motives (Bao 2014).

Second, China believes that it has provided the most public goods and funds in the process of joint patrol and that waterway security construction also contributes to the development of shipping trade in the entire Mekong River basin. But the lower Mekong countries think China's policy involves more unilateralism than multilateralism, and cooperation between them in waterway security is limited. As a result, China has expressed its concern and regret that the lower Mekong countries cannot engage in more in-depth security cooperation.

14.3.3 *Case 3: Sharing Hydrological Data and Other Information*

During the process of environmental assessment and hydrological data sharing, China and the lower Mekong countries are both displaying a lack of confidence in the mutual information and data provided. First, there is a trust crisis in the accuracy of hydrological data. The lower Mekong countries believe that China's relevant environmental assessment data are not entirely correct or comprehensive. The lower Mekong countries believe that, after the outbreak of drought in the Mekong River area in 2010, the drying up of the Mekong River was attributable to the closure of four Chinese dams in the upper reaches of the river—reportedly to conserve water for electricity generation in China (Walker 2010).

Since then, China has actively negotiated and published relevant data showing that it accounted for only 13.5% of the total amount of Mekong River water resources and that its basin area ranked only third among all the six countries along the river (Qin 2010). But some of the lower Mekong countries did not recognize this. Four years later, at the Second Mekong River Commission Summit held in Vietnam in 2014, Vietnamese journalists openly questioned Chinese officials regarding the same issue. But when Chinese officials asked reporters for specific data and examples to prove their allegations, the reporters couldn't provide them. After the summit, the Vietnamese officials apologized to the Chinese officials.

Second, there is a trust crisis in water information sharing. The lower Mekong countries have expressed doubt about China's limited publication of hydrological data. Studies have pointed out that China has promised to disclose more hydrological data, but they also pointed out that transboundary river hydrological data were related to national security issues, and that China has its own right to develop the part of the Lancang River within its territory.

China's moves caused a lack of trust among the lower Mekong countries regarding China's hydropower development and power plant construction (Magee 2014; Voice of America 2014). In fact, China and the Mekong River Commission signed an agreement on data sharing in 2002 and renewed it in 2008. Since then, China has provided data relating to precipitation and water levels collected from the two monitoring stations located on the Mekong mainstream at Yunjinghong and on a tributary at Manan. During 2010, the region was hit by a severe drought due to extremely low rainfall in the dry season and a particularly early end to the wet season in 2009. China was able to provide the Mekong River Commission with hydrological data for emergency use. China has expressed its willingness to share the data during the dry season should similar emergency conditions occur (Mekong River Commission 2013).

In November 2015, the LMC mechanism was established, with China, Myanmar, Laos, Thailand, Cambodia, and Vietnam as members and water resources cooperation as one of the five priority directions of this mechanism (Ministry of Foreign Affairs of the People's Republic of China 2016). On March 23, 2016, at the first LMC leaders' meeting, Premier Li Keqiang expressed the willingness of the Chinese government to jointly set up the Lancang-Mekong Water Cooperation Center with the lower Mekong countries to strengthen technical cooperation, talent, and information exchange and to promote green, coordinated, and sustainable development (Li 2016). That same month there was a severe drought, increasing pressure on the Mekong River. Again, China shared hydrological data with the lower Mekong countries.

But some lower Mekong countries thought that the information that China shared with them was not enough. At the same time, China also has justifications for not disclosing the relevant increase in water consumption, as well as other information, with the lower Mekong countries. For example, agricultural water has a direct impact on the river, and Vietnam uses a large amount of agricultural water. With the increase in the area planted in rice in Vietnam, the demand for agricultural water in the Mekong River is also increasing, which is bound to affect the river and may impede the pace of agricultural development. However, in the event of drought, Vietnam did not give relevant agricultural water data and information on the impact of the river, but only condemned the upstream countries.

14.4 Root Causes of Trust Crisis in Transboundary Water Cooperation Along the Mekong River

The three cases above demonstrate the need for more common interest in transboundary water cooperation for China and the lower Mekong countries, and there are other serious trust issues as well, concerning less common interests such as dam construction and fishery development. This section examines the root causes of the trust crisis in the context of an analytical framework of international trust.

14.4.1 China and the Lower Mekong Countries Lack Sufficient and Direct Information Channels

According to the international relations literature on trust building between nations, whether state actors can receive comprehensive and objective information through direct and indirect channels can have a significant impact on the establishment and maintenance of trust. The change of the attitude of the lower Mekong countries regarding China's transboundary water cooperation is based mainly on the analysis of China's water resources development and on information provided by nongovernmental organizations, media, and foreign countries. Conversely, China's attitude toward the lower Mekong countries mainly originates from its observation of the Mekong River water development situation and water conditions and from passively receiving a lot of negative news about the development of its own water resources. Therefore, in terms of information access, the indirect channels to obtain the trust information are far more important than direct channels for China and the lower Mekong countries.

In addition, the limited nature of indirect information channels and the often untruthful or unreliable information provided have led to deepened mutual misunderstanding, raising mutual suspicion and exacerbating the trust crisis. For instance, when the drought happened in the Mekong Basin in 2016, China and the lower Mekong countries did not communicate with each other immediately and therefore could not easily understand the other party's situation. On the contrary, they tried to guess the other party's intention and action by assessing international news media reports and observing the other party's behaviors and reactions. Sometimes, they blamed each other because of false or misleading news. These factors have a negative impact on trust between China and the lower Mekong countries. Besides, for the lower Mekong countries, China not only is located upstream but also is a very large economy. The lack of more effective and direct mutual communication makes the lower Mekong countries overly worried.

14.4.2 China and the Lower Mekong Countries Both Believe That National Interests Override International Trust

According to the analytical framework of international trust, national trust and international interest are the starting points for building rational and lasting cooperation and cooperative mechanisms. Although China and the lower Mekong countries have mutual trust in their cooperation on transboundary water resources development, the status of trust is not equal to the status of interest. Is this because the trust of the lower Mekong countries toward China has not reached the level of trust of China toward the lower Mekong countries (or vice versa)?

From the existing degree of transboundary water cooperation, it seems that China and the lower Mekong countries are finding it difficult to balance the principles of water sovereignty (national independent development rights) and harmless right (common maintenance rights). Water sovereignty emphasizes the direct expression of national interests, while harmless right emphasizes trust in other countries. However, in the course of specific operations, both China and the lower Mekong countries are always worried that the other party (and among the lower Mekong countries themselves) will put water sovereignty over harmless right. For example, regarding drought, the downstream countries are always worried about China will control water, and China is also worried that unjustified accusations will bring obstacles to the normal development of water resources. In the case of waterway security, the lower Mekong countries worry that their own water sovereignty will be affected by China and always see a "Chinese color" in geopolitics. Meanwhile, China worries that waterway security issues will bring much more danger to the shipping trade and crew safety.

Therefore, both China and the lower Mekong countries tend to choose national interests rather than international trust in transboundary water cooperation. This causes many more issues, such as the interruption of new cooperation projects. In addition, some of the trust crisis in cooperation that may not be serious in itself may be used by countries as leverage for bargaining in water cooperation negotiations, making the trust crisis even worse.

14.4.3 China and the Lower Mekong Countries Lack Trust Maintenance in Water Cooperation

The analytical literature on international trust emphasizes that trust needs to be maintained and strengthened after bilateral cooperation or cooperative mechanisms are established in order to ensure the effectiveness of cooperation and to reduce trust crises. China and the lower Mekong countries began cooperation regarding water development, waterway security, and precipitation and water level data and established cooperative relations under the LMC mechanism. But what China and the lower Mekong countries did in terms of trust maintenance was not enough. They tended to suddenly launch a public opinion offensive or emergent response, or they lacked prior communication concerning the behavior of the other party when receiving a large amount of one-sided, indirect information.

For example, in the 2016 drought, if China had previously shared information regarding the drought on its own areas or if the lower Mekong countries had communicated and interacted directly with China after the outbreak of drought, they would have understood each other better, rather than becoming easily susceptible to other one-sided news and reports. The same thing happened with waterway security and information sharing. If China and the lower Mekong countries could communicate with each other regularly and often on the relevant topics and discuss the difficulties and needs of both sides, this would reduce mutual doubts and concerns, deepen cooperation, and change attitudes regarding joint water development.

14.5 Ways to Build Mutual Trust Between China and the Lower Mekong Countries in Transboundary Water Cooperation

If the trust crisis between China and the lower Mekong countries in transboundary water cooperation is not effectively resolved, it will affect water resources development and bilateral cooperation projects. If the trust crisis further intensifies, it could possibly even develop from competition for water rights and interests into an all-out "water war."

14.5.1 Easing the Fundamental Contradiction Between National Interests and International Trust

According to the analytical framework of international trust, national interests and international trust are starting points for developing cooperation and establishing cooperative mechanisms. Easing contradictions between the two points will directly affect future cooperation and stability of cooperative mechanisms. The key to the contradiction between national interests and international trust in transboundary water cooperation between China and the lower Mekong countries is the dispute between water sovereignty and harmless right. Water sovereignty emphasizes the self-development of the Mekong River water resources, and harmless right emphasizes the common maintenance of the Mekong River, which leads to the emergence of structural contradictions.

The key to the dispute is not to abandon either set of rights but to treat the dialectical relationship between the two points objectively and to consider social and ecological responsibilities in the development of transboundary water resources. First, both parties need to view the stakes between water sovereignty and harmless right from a comprehensive and dialectical perspective. China and the lower Mekong countries need to be aware that each member country both has rights and needs to respect the others' rights, rather than just taking one side as a weapon or as an excuse to avoid responsibility.

Second, both parties need to assume social and ecological responsibility. The key to the struggle between water sovereignty and harmless right is to deal with the relationship between development and protection. China and the lower Mekong countries need to identify their own interests and to integrate protection of the interests with other countries in the process of developing the Mekong River, in order to fulfill social and ecological responsibilities during the development process. Meanwhile, China and the lower Mekong countries must all take responsibility, stop attributing blame, and fully embody sincerity in order to finally solve the contradiction between water sovereignty and harmless right and contribute to the resolution of the trust crisis.

14.5.2 Establishing Trust Maintenance in Transboundary Water Cooperation

The analytical framework of international trust shows that the effectiveness of cooperation will be further enhanced through continuous trust maintenance, which is conducive to the resolution of differences and trust crises. Therefore, China and the lower Mekong countries must first establish in their collective consciousness a community with a shared future for humanity to effectively encourage all the countries to maintain the basis of trust and cooperation. A community with a shared future for humanity for all Lancang-Mekong countries connotes a community of shared interests, shared responsibilities, and shared future in transboundary water cooperation, emphasizing the common, rational, scientific, and sustainable development of Mekong River water resources and shared responsibility in realizing this development.

Therefore, the collective consciousness of establishing a community with a shared future for humanity for all Lancang-Mekong countries requires the joint efforts of China and the lower Mekong countries. First, China should combine its own water security and water development interests with those of the lower Mekong countries. New water cooperation mechanisms and related agencies can be set up within the LMC mechanism to effectively explore and solve mutual cooperation issues between the two parties. Second, the two parties should be prepared to consult with and compromise with the other side in envisioning and establishing a community with a shared future for humanity to maximize the benefit of all parties.

14.5.3 Expanding Communication and Information Channels

According to the analytical framework of international trust, state actors gain trust information through direct and indirect channels in the process of trusting other countries. Therefore, China and the lower Mekong countries should actively create and expand channels to understand each other, which will help the two parties to deepen their trust.

First, China and the lower Mekong countries should actively break through the misunderstanding of water cooperation in bilateral political exchanges. Both parties should treat the water issue as one of the important considerations in the construction of bilateral political trust to fully express their sincerity, take action to fix the existing problems in a timely manner, and increase bilateral water cooperation trust and will. Second, China and the lower Mekong countries should hold exchange meetings under the LMC in the Lancang-Mekong Water Cooperation Center, discussing hydrological information, water development, and issues of waterway exploration, while taking advantage of capital, technology, and talent to enhance the mutual water cooperation project from all aspects, under the principle of mutual benefit on the specific use of water, river protection, waterway management, and aquatic products management. In addition, to refute false talk regarding contradictions between China and the lower Mekong countries on water cooperation, disputes should be solved ultimately through mutual consultation to consolidate water cooperative relations.

14.6 Conclusion

The international trust crisis in transboundary water cooperation is an important but easily overlooked factor affecting the development of cooperation between China and the lower Mekong countries. Despite the rising level of transboundary water cooperation between both parties, mutual trust has not grown commensurately. In the future, China and the lower Mekong countries should make full use of the platform of the LMC mechanism to continue to strengthen communication, expand communication channels, and combine their own national interests and the needs of other countries to enhance trust. At the same time, the two parties should also recognize that the solution of the trust issue is a long-term task and that both sides need to commit to solving practical issues so as to ensure sustainable development of the Mekong River basin, to form a community with a shared future for humanity for all Lancang-Mekong countries.

References

Bao, G. J. (2014). The factor of trust in the Mekong River security cooperation and China's strategic choice. *Asia & Africa Review*, 37–38. 包广将：《湄公河安全合作中的信任元素与中国的战略选择》，《亚非纵横》，2014年第3期，第37－38页.

Booth, K., & Wheeler, N. J. (2008). *The security dilemma: Fear, cooperation and trust in world politics* (p. 230). New York: Palgrave Macmillan.

Brewer, P. R., et al. (2004). International trust and public opinion about world affairs. *American Journal of Political Science, 48*(1), 93.

Cross, S., et al. (2013). *Shaping south east Europe's security community for the twenty-first century: Trust, partnership, integration*. New York: Palgrave Macmillan.

Giang, N. K. (2016). New rule-based order needed to save the Mekong. *East Asia Forum*. http://www.eastasiaforum.org/2016/03/29/new-rulebased-order-needed-to-save-the-mekong. Accessed 5 Sept 2016.

Hoffman, A. M. (2002). A conceptualization of trust in international relations. *European Journal of International Relations, 8*(3), 376–379.

Hunt, L. (2016). *Drought fans tensions along the Mekong, The Diplomat*. http://thediplomat.com/2016/03/drought-fans-tensions-along-themekong. Accessed 3 Sept 2016.

Kydd, A. H. (2005). *Trust and mistrust in international relations*. Princeton: Princeton University Press.

Larson, D. W. (1997). Trust and missed opportunities in international relations. *Political Psychology, 18*(3), 701.

Li, K. Q. (2016). *Li Keqiang's speech at the first Lancang-Mekong Cooperation Leaders' Meeting (full text), Ministry of Foreign Affairs of the People's Republic of China*. http://www.fmprc.gov.cn/web/ziliao_674904/zyjh_674906/t1350120.shtml. Accessed 3 Sept 2016. 《李克强 在澜沧江-湄公河合作首次领导人会议上的讲话(全文)》, 外交部. http://www.fmprc.gov.cn/web/ziliao_674904/zyjh_674906/t1350120.shtml

Magee, D. (2014). China fails to build trust with Mekong neighbours. In: I. Hilton (Ed.), *The uncertain future of the Mekong River*. http://www.thethirdpole.net/wp-content/uploads/2014/03/mekong__new14-2.pdf. Accessed 5 Sept 2016.

Mekong River Commission. (2013). *Mekong River Commission and China boost water data exchange, Mekong River Commission*. http://www.mrcmekong.org/news-and-events/news/mekong-river-commission-and-china-boost-water-data-exchange. Accessed 3 Sept 2016.

Mekong River Commission. (2017). *The effects of Chinese dams on water flows in the lower Mekong Basin, Mekong River Commission*. http://www.mrcmekong.org/news-and-events/news/the-effects-of-chinese-dams-on-water-flows-in-the-lower-mekong-basin. Accessed 4 Sept 2016.

Ministry of Foreign Affairs of the People's Republic of China. (2016). *Five features of Lancang-Mekong River Cooperation, Ministry of Foreign Affairs of the People's Republic of China*. http://www.fmprc.gov.cn/mfa_eng/zxxx_662805/t1349239.shtml. Accessed 6 Sept 2016.

Nguyen, N. (2016). China's water release unlikely to slake Vietnam's thirst. *Radio Free Asia*. http://www.rfa.org/english/news/vietnam/chinaswater-unlikely-to-slake-vietnams-thirst-03232016154743.html. Accessed 4 Sept 2016.

Qin, G. (2010). China has been showing a responsible attitude in the development and utilization of the Mekong River water resources. *Xinhua News Agency*. http://news.xinhuanet.com/world/2010-03/30/c_128105.htm. Accessed 5 Sept 2016. 《秦刚:中方一直本着负责任的态度开发利用湄公河上游水资源》， 新华网, http://news.xinhuanet.com/world/2010-03/30/c_128105.htm

Radio Free Asia. (2016). China's water release unlikely to slake Vietnam's thirst. *Radio Free Asia*. http://www.rfa.org/english/news/vietnam/chinas-water-unlikely-to-slake-vietnams-thirst-03232016154743.html. Accessed 6 Sept 2016.

Rathbun, B. C. (2011). From vicious to virtuous circle: Moralistic trust, diffuse reciprocity, and the American security commitment to Europe. *European Journal of International Relations, 18*(2), 375–401.

Santasombat, Y. (2016, April 8). *The Impact of China's rise on Southeast Asia*. Presented at the Friday Forum, Center for Southeast Asian Studies, University of Wisconsin-Madison.

Voice of America (Chinese Version). (2014). Disputes occur on building dams on the transboundary river and experts call on China to strengthen project transparency. *Voice of America* (Chinese Version). http://www.voachinese.com/Content/china-dams-controversy-20140219/1855097.html. Accessed 5 Sept 2016. 《跨境河流上兴建水坝引发争议 专家呼吁中国加强工程透明度》， 美国之音. http://www.voachinese.com/content/china-dams-controversy-20140219/1855097.html

Walker, A. (2010). Damned Chinese: Mighty mekong a memory. *New Mandala*. http://www.new-mandala.org/damned-chinese-mighty-mekong-amemory. Accessed 2 Sept 2016.

Xinhua News. (2015). Safe Mekong operation project scores remarkable anti-drug achievements. *Xinhua News*. http://news.xinhuanet.com/english/2015-04/07/c_134130862.htm. Accessed 5 Sept 2016.

Yin, J.W. (2016). The origin of international trust: A comparative analysis of typology. *Teaching and Research, 98*. 尹继武:《国际信任的起源: 一项类型学的比较分析》,《教学与研究》, 2016年第3期, 第98页.

Chapter 15
Common Challenges of Smallholders in ASEAN: Lacking Access to Land, Water, Market, and State

Danny Marks

Abstract This chapter focuses on smallholder farmers in Cambodia, Lao PDR, Myanmar, and Vietnam (CLMV), examining the relationship between their access to and uses of the environment, changing regimes at regional and national levels, and environmental problems. It argues that environmental and food security problems are also always governance problems and that they cannot be solved unless their political-economic dimensions are addressed. Although the vast majority of the food supply in Southeast Asia comes from smallholders, many face not only accelerating climate change but also poverty, large-scale land-grabbing, and limits to markets, technology, credit, and water. Women smallholders especially face legal and social hurdles, including access to land, credit, and education. This chapter offers case studies of how the ASEAN Economic Community, which aims to form a single market to better integrate in the global economy, could actually worsen the situation for smallholders in CLMV. Noting that smallholders' level of resilience is significantly affected by political economy variables, such as degree of access to power, the effects of public policies, and lack of state support, it then offers recommendations of what ASEAN, national governments, and civil society could do to help smallholders. Suggestions include harmonizing regulatory frameworks; reducing non-tariff barriers; ensuring access to, management of, and ownership of land; reviving the agrarian economy; implementing pro-women policies; and helping smallholders to better access markets and receive fairer prices. Moreover, future studies of agricultural and environmental problems should account for how power geometries affect the creation of these problems.

Keywords ASEAN economic community · Smallholders · Political economy · Trade policy · Land-grabbing · Women farmers

---

D. Marks (✉)
Department of Asian and International Studies, City University of Hong Kong, Kowloon Tong, Hong Kong
e-mail: danny.marks@cityu.edu.hk

M. A. Stewart, P. A. Coclanis (eds.), *Water and Power*, Advances in Global Change Research 64, https://doi.org/10.1007/978-3-319-90400-9_15

15.1 Introduction

As Southeast Asia has become more integrated, particularly with the advent of the ASEAN Economic Community (AEC) in 2015, it is worth remembering that for every five plates of food eaten in the region, about four of them are provided by the region's smallholder farmers (FAO 2012). They form the backbone of most Southeast Asian countries' economies. However, today many of them are snared in poverty and face numerous challenges that prevent them from escaping this trap. Challenges faced by women smallholders, who constitute 45% of economically active women in the region (Rao 2011), are even greater. While national governments have taken some significant steps in reducing the poverty of smallholders, more must be done.

Some of these challenges have persisted for the last few decades. Smallholders, particularly in Cambodia, Lao PDR, Myanmar, and Vietnam (CLMV), still have difficulties accessing both regional and global markets. They still have limited means to produce their crops, lacking technology, access to credit, and, in many areas, access to water. Women smallholders still face a number of both legal and social hurdles, including access to land, credit, and education.

At the same time, some of these challenges are changing or magnifying. Starting in 2015, countries in the region together formed the AEC, which they hoped would enable the region to form a single market and production base and become better integrated in the global economy. However, if proper safeguards are not put in place, the AEC could reify or actually worsen the situation of some smallholders, particularly those in CLMV. In addition, the impacts of climate change are accelerating, meaning that farmers will likely have to endure more intense and prolonged droughts and floods, which could jeopardize their crops. And perhaps the largest problem they face remains the same and, in some places, is becoming worse: limited access to and security over land. Large-scale land-grabbing is still occurring and, in some countries, such as Myanmar, is likely to be done more often.

While smallholders continue to be productive, innovative, and resilient to these numerous shocks and problems, governments can do more to help protect their interests and rights. The intergovernmental Association of Southeast Asian Nations (ASEAN) can assist them in numerous ways, such as by harmonizing regulatory frameworks and helping reduce non-tariff barriers (NTBs). Specific policies of each national government will vary from country to country, but in general, each should ensure that smallholders can access, manage, and own land; revive the agrarian economy; implement pro-women policies; and enable smallholders not only to better access markets but also to receive a fairer price for their products.

This chapter examines the relationship between smallholders' access to and uses of the environment, changing environmental regimes at both the regional and national level, and environmental problems. It argues that environmental and food security problems are always governance problems too. It begins by giving an overview of smallholders in the region and describing the common issues faced by smallholders in Southeast Asia. It then describes how trade is affecting smallholders in the region and projects how the AEC could affect them. Next, the report provides case studies of these issues in the four countries in Southeast Asia: Cambodia, Thailand, Vietnam, and Myanmar. It then offers recommendations of what both ASEAN and national governments in Southeast Asia could do to help smallholders address these problems. Last, it concludes by suggesting how civil society organizations can best help smallholders. One limitation of this study is that it focuses mostly on the lower Mekong countries and does not include much information about Indonesia, Malaysia, and the Philippines. Nonetheless, many of the challenges discussed below are faced by smallholders in these countries as well.

15.2 Situation of Smallholders in Southeast Asia and Common Challenges

In this chapter, smallholders are defined as farmers who own 3 ha of land or less and depend mostly on household members for labor. While the exact percentage of farmers in Southeast Asia who are smallholders is unknown, they undoubtedly constitute the vast majority. Wide diversity among smallholders in the region exists. They have differing income levels, ranging from US$200 to $1100 per farmer per year (Haggblade and Boughton 2013); use different levels of technology and inputs; and live in countries with divergent systems of government, ranging from autocracy to democracy, and different levels of governmental support.

Yet at the same time, many commonalities between smallholders exist. First, they make major contributions to their country's economy, including exports (see Table 15.1). Second, most of them produce paddy rice. Third, they are members of the lowest rung of society: they have the highest rates of poverty, receive the least support from the government, and are often discriminated against by the upper

Table 15.1 Economic statistics of ASEAN countries

Country	Population 2012 (millions)	Nominal GDP 2012 (billion US$)	Average Real GDP Growth, 2008–2012	Agriculture value (% GDP, 2012)	Exports (% GDP, 2012)	Imports (% GDP, 2012)
Brunei	0.4	16.95	0.7	0.7	81.4	31.2
Cambodia	14.9	14.06	5.4	36.7	54.1	59.5
Indonesia	246.9	878.04	5.9	12.8	31.6	29.4
Lao PDR	6.6	9.30	8.0	30.8	38.0	44.3
Malaysia	29.2	303.53	4.2	10.1	87.5	75.5
Myanmar	52.8	51.44	4.90	48.4	0.2	0.1
Philippines	96.7	250.27	4.7	12.6	26.1	31.8
Singapore	5.3	274.70	4.4	0.0	200.7	178.5
Thailand	66.8	365.56	2.9	12.2	78.0	75.3
Vietnam	88.8	141.67	5.9	21.3	89.8	90.2

Source: Daite (2015)

echelons of their societies. Fourth, the agricultural sectors in each country are highly unequal. A small number of large agribusiness companies reap most of the profits, and their owners live comfortably, while smallholders make a pittance compared to them.[1] In CLMV, many smallholders barely eke out an existence, the majority being subsistence producers. Fifth, despite their marginalized status, they achieve higher levels of labor productivity than large farms do (Shenggen et al. 2013). In countries that have successfully reduced poverty in the region, such as Thailand and Vietnam, productivity gains by smallholders significantly drove this reduction. Last, they face a set of common challenges inhibiting them from improving their productivity and food security and moving away from poverty: a lack of access to land, threat of land-grabbing, increased threat of climate change, limited social protections, lack of access to markets, limited bargaining power, and a significant gender gap.

15.2.1 Lack of Access to Land and Land-Grabbing

The biggest problem smallholders face throughout the region is lack of access to sufficient land and high land insecurity. Smallholders either own a limited amount of land, 3 ha or less, which makes it difficult for them to boost their incomes considerably; rent land (which is a small percentage); or in many cases, especially in the lesser developed countries, live on land they do not legally own, thereby making them legally landless. Various scenarios have developed that have led them to become landless:

1. They have been living on this land for decades but do not have legal security over this land, often using it under customary tenure and, due to changes in land laws, are at risk of being evicted due to changes in ownership.
2. They have sold their land due to debts or could never afford land and occupy either state or public land, which they do not own.
3. The land they had been living on has been transferred to the state, such as many forested areas in Thailand.
4. They have already been evicted due to their land being confiscated by the state or private actors who are backed by the state. Rarely do smallholders receive compensation set at market value for their land.

This last scenario has become more prevalent during the last 10–15 years as land-grabbing has become larger scale, systematized, and legalized. The amount of land confiscated from smallholders and given to private foreign and domestic investors and state-own enterprises totals millions of hectares in a number of countries (Amanor 2017). The land-grabbing has been facilitated by national governments changing their land laws to become more investor friendly. Foreign investors have

[1] For example, Dhanin Chearavanont, chairman of the agriculture conglomerate Charoen Pokphand is the wealthiest Thai, with assets of US$11.4 billion, and Doan Nguyen Duc, chair of Hoang Anh Gia Lai Group, has assets close to US$1 billion. See *Nation* (2014) and VietnamNet Bridge (2013).

rarely abided by any codes of conduct or ensured that smallholders are fairly benefitting from their projects. In some instances, their practices have breached the laws in both the host country and their own countries.

Smallholders have not been submissive victims but instead have challenged these land seizures. They have protested, refused to leave their land, sued companies, and, in some instances voted for parties promising to stop land concessions if elected. In response, they have been severely repressed, been arrested, have had their properties burnt down, and, on a few occasions, murdered.[2] Although victories have so far been few, smallholders have refused to back down and are becoming more politically mobilized and empowered.

15.2.2 Increased Threat of Climate Change but Limited Social Protection

According to Germanwatch's Climate Risk Index (Germanwatch 2016), of the ten countries affected most by climate-related disasters from 1996 to 2015, four are in this region: Myanmar (2nd), Philippines (5th), Vietnam (8th), and Thailand (10th). The region as a whole is highly vulnerable to the impacts of climate change, which are magnifying as greenhouse gases accumulate. The crop yields of smallholders will be badly affected by these impacts, particularly the rise in number and intensity of floods, droughts, and heat waves and the increase in rain variation. In a number of countries, particularly CLMV, smallholders have limited access to irrigation, so they cannot access additional water if there is insufficient rainfall for their crops. They are also especially vulnerable because many of them are in debt or have little savings, so one season of poor yields could devastate them financially.

The state therefore needs to help farmers who are more exposed to risks of food insecurity during such crises by providing various forms of social safety nets, such as feeding programs, food-for-work programs, and food price subsidies. Countries in the region have created a number of such programs, but they are poorly targeted and prone to a high rate of leakage to nonpoor households. Further, protectionist trade policies, such as Indonesia's import ban on rice in 2004, can undermine the benefits of these safety nets. In this case, the Indonesian government subsidized the price of rice for 25% of poor households' monthly consumption, but the ban plus the low elasticity of rice demand led to higher prices of rice and increased poverty in the country during 2005–2006 (Desker et al. 2013). Moreover, many countries lack comprehensive social protection policies beyond safety nets that are necessary to help smallholders, such as access to cheap or free health care, vocational training programs, adequate minimum wages to supplement their incomes, and comprehensive education in rural areas.

[2] For example, in 2012 two land right activists were killed in southern Thailand (see Human Rights Defenders 2012) and one activist was killed in Cambodia (see Diakonia 2012).

15.2.3 Lack of Access to Markets and Limited Bargaining Power

Further exacerbating their situation is that farmers receive limited income for their products because they lack access to markets and have limited bargaining power. In CLMV, producers, especially smallholders, have difficulty accessing the regional and global markets. Cross-border trade has been increasingly inhibited by NTBs, particularly from Thailand's neighbors to Thailand (Zola 2009). Many products lack required certifications and do not meet food safety standards that would enable them to be exported regionally and globally. For example, Vietnamese poultry is not yet allowed into the European Union and the United States due to these countries' stringent food safety requirements (Tuoitrenews 2013). Smallholders often have limited market information and may lack production technologies needed to meet these standards. Interregional agricultural trade is hampered by the lack of a common agriculture policy, which has been agreed upon in the European Union. Food safety standards and regulations have yet to be harmonized in the region (Vdassani 2013).

Within countries, farmers also struggle to receive a fair price for their products, obtaining only a small percentage of their final retail prices. Limited processing facilities, poor logistics, and shoddy infrastructure limit their products' life span and market reach. Further, because many are in debt, they feel compelled to sell their products immediately to middlemen, thereby reducing their profits. Those in rural areas, particularly remotes ones, often have little choice but to sell to traders and cannot dictate the terms of trade. Also, many farmers engaged in contract farming with agribusinesses are pressured to accept unequal profit and risk sharing in these schemes. Farmers' organizations can help farmers improve their bargaining position and lobby for effective market institutions. However, these organizations are poorly developed in many countries, and governments often have an ambivalent stance toward them.

15.2.4 Gender Gap

While the gender gap has decreased in Southeast Asia during the past few decades, particularly in countries such as Cambodia and Vietnam, women smallholders still face many inequities and constraints that hurt them and their families. Since most smallholders in the region are women, women are confronted with all of these aforementioned issues. Land grabs are particularly damaging for them because they suffer from losing their homes, livelihoods, and kinship ties. Evictions also increase their vulnerability to acts of violence (Kusakabe 2015).

In addition, women face a set of problems due to the remaining gender gap. First, inequity over rights and access to land still exists. In Cambodia, for example, women are more likely to be landless or have considerably smaller plots of land. Further, a joint land title in Cambodia between a husband and wife does not necessarily bestow legal rights to the woman because customary practices sometimes do not fully recognize women's ownership rights. In addition, in a number of countries, rural husbands still dominate the process of decision making regarding land because of the traditional mind-set and education gap (STAR Kampuchea 2013). Second, when smallholder farms move to mechanization, women often are left behind because they normally stay at home and sometimes are unable to access these new technologies. Third, women have more difficulties accessing credit. Banks tend to discriminate against them, forcing them to turn to microfinance institutions or loan sharks, which have higher interest rates (ActionAid Cambodia program officer, personal communication, February 2014). Fourth, a labor gap still exists that hurts women's off-farm incomes. This gap has emerged from difference in wages, access to training, and types of available employment (Minh and Maerten 2012; Huyer 2016).

15.3 Trade, the AEC, and ASEAN

15.3.1 Unequal Impacts of Regional and Global Trade

How trade impacts smallholders in ASEAN is complicated, and the impacts are multiple, both positive and negative. Given this chapter's limited length, only a few are discussed here. One commonality is that smallholders have been unable to benefit much from the opening of their countries' borders to both regional and international markets. Intra-ASEAN agriculture trade is still small, consisting of only 7% of the total in 2010, because in this sector each country produces a wide range of agricultural products rather than specializing in a few (Desker et al. 2013). Numerous NTBs, other protectionist measures, and the lack of a harmonization of standards and regulations have also hindered regional trade, particularly cross-border trade. NTBs add more to trade costs than do tariffs, which account for just 6% of the total cost (Anthony et al. 2013). Nonetheless, the amount of interregional agricultural trade has risen. For example, Vietnam's agricultural exports to ASEAN totaled almost US$4 billion in 2012, expanding 400% since 2000 (Tinh and Long 2013).

Globally, ASEAN's overall exports are rising, more than doubling from 3.5% in 1980 to 7.6% in 2015 (ASEAN Secretariat 2016). However, agricultural exports account for only 7.6% of ASEAN's total exports, suggesting that this sector's integration with the international market is limited and that much production is geared toward meeting domestic needs (Bano et al. 2013). Research conducted by the Southeast Asia Council on Food Security and Fair Trade found that smallholders have difficulties taking advantage of the opportunities arising from expanding ASEAN markets due to limited market information, bargaining power, access to credit, and postharvesting technology. Instead, processors, middlemen, and exporters have accrued most of the benefits from this expansion (Asian Farmers' Association for Sustainable Rural Development 2007).

In some cases, trade liberalization has actually harmed smallholders in the region. According to the Asian Farmers' Association, the agricultural trade rules of the World Trade Organization has led to the dumping of cheap agricultural imports from outside countries, thereby displacing smallholders. For example, Thai farmers had trouble selling garlic and longan after the China-Thailand free trade agreement was enacted because these Chinese products was imported at zero tariffs and became cheaper than the Thai ones (Noi 2010). Also, Vietnamese and Thai vegetables are cheaper than local produce in Cambodia, making it difficult to develop a domestic vegetable sector (Cambodian Institute for Research and Rural Development official, personal communication, February 2014).

Preferential trade schemes, such as Everything but Arms, which enables least developed countries, including Cambodia, Laos, and Myanmar, to import certain goods without duties or quotas to the European Union, have been designed with the intention to help smallholders access foreign markets, but in some cases they have actually encouraged land-grabbing. For example, in Cambodia, companies have cleared over 100,000 ha of land, most of it seized from farmers living there, to build sugar plantations (Hodal 2013).

15.3.2 The AEC: Reifying Differences in the Region?

The year 2015 marked the beginning of the AEC, whose blueprint was adopted in 2007. Through a set of norms, rules, and institutions, the AEC project seeks to transform ASEAN into a competitive and globally integrated single market consisting of a free flow of goods, investment, capital, and labor (Juego 2014).

The Heinrich Böll Foundation argues that the AEC will lead to the region's elite acquiring more power and wealth but that the rest, particularly smallholders in the lesser developed countries, will lose out. By eliminating tariffs and NTBs, the AEC removes important tools that lesser developed countries can use to strengthen their economies. Economic development throughout the world reveals that the use of tariffs and industrial policies has enabled a number of countries to grow rapidly by encouraging the development of local capacity and domestic industries. Further, by creating policies seeking to attract foreign capital, particularly foreign direct investment, the AEC is creating a system that gives highest priority to profit margins, capital flows, and market forces rather than the interests of smallholders. To attract investment, governments are discouraged from regulating the behavior of foreign and multinational companies, which could lead them to increasingly participate in land-grabbing and other predatory behavior. The lesser developed countries could compete for a race-to-the-bottom to attract foreign companies by lowering labor costs and environmental standards. If this happens, smallholders will suffer because their off-farm labor options will become worse and the local environment will become degraded (Juego 2014).

The AEC could therefore reify existing regional differences. The richer countries, such as Singapore and Malaysia, could continue to provide higher-earning activities such as marketing, processing, and packaging, while the lesser developed countries will have the role of providing raw materials, unprocessed agricultural goods, and low-wage labor. If this happens, the AEC would help sustain the region's unequal growth (Focus on the Global South official, personal communication, February 2014).

15.3.3 ASEAN's Limited Role So Far in Helping Smallholders

So far ASEAN has had a limited role in helping smallholders. While ASEAN has developed a number of platforms, such as the ASEAN Integrated Food Security Framework, an ASEAN Working Group on Climate Change, and the ASEAN Multisectoral Framework on Climate Change: Agriculture and Forestry toward Food Security, so far these platforms have had limited impact because they have most been "talk shops" without much funding, manpower, or policy collaboration behind them. These frameworks also do not explicitly enable civil society engagement. Another issue is that smallholders have had limited opportunities to raise their concerns during ASEAN meetings, particularly during the ASEAN Ministerial Meetings on Agriculture and Forestry. Further, ASEAN has yet to develop any regional codes on the problem of agricultural land grabbing in the region. The good news is that these platforms already exist to serve as foundations for increased collaboration, and there is a successful precedent. The implementation of the ASEAN Agreement for Disaster Management and Emergency Response during Cyclone Nargis demonstrated the benefits of working across governments and with civil society organizations to tackle regional challenges (ASEAN Secretariat 2010).

15.4 Country Case Studies

15.4.1 Cambodia

Agriculture comprises about 37% of Cambodia's GDP, and over 70% of the nation's working population works in agriculture, the vast majority of which are smallholders. Around 80% of farmers grow rice. Accordingly, the Cambodian government has prioritized agricultural and rural development in its national development strategy in order to reduce poverty and grow the economy. Positively, since 2000, Cambodia has successfully become rice self-sufficient nationally, although pockets of deficits still exist, and has reduced child mortality and malnutrition.

Despite these improvements, most Cambodian smallholders are mostly subsistence based and impoverished, and malnutrition stunts the growth of about 40% of the population (Asian Development Bank 2013). These smallholders are confronted with a number of obstacles preventing them from escaping poverty. First, they either own little land or are landless. Many who had owned land have been evicted due to economic land concessions (ELCs). Second, they cannot grow much rice or other products because little of the land is irrigated. Third, and related to the two previous challenges, they earn little or no profit from their products. Fourth, due to climate change, they must endure more droughts and floods but receive limited assistance from the government.

15.4.1.1 Land

Approximately 21% of the country's total land is arable. Similar to many other countries in the region, the pattern of landholding in Cambodia is highly unequal. The average size of cultivated land per household is less than 1 ha, and about a quarter of rural farmers do not own any land (Oldenburg and Neef 2014). Without much land, even if smallholders increase their production, it is difficult for them to produce enough food. At the other end of the spectrum, the government has granted a small group of well-connected Cambodian and foreign investors land concessions of thousands of hectares that enables them to use that land for up to 99 years (ADHOC 2013).

In recent years, Cambodia has experienced a dramatic spike in land-grabbing. As of 2013, the government has allocated 73% of the arable land to ELCs—about 15% of Cambodia's total land (Global Witness 2013). Much of this land was occupied by smallholders who had been sustaining their livelihoods there but lacked formal land titles. In most instances, companies that had received ELCs evicted these smallholders, sometimes forcibly, often giving them little compensation and no new land. In the instances where they were resettled, this new land given to them was often infertile. Rural landlessness, mostly triggered by forced evictions, surged from 13% in 1997 to around 25% in 2010 (UN Capital Development Fund 2010). In 2011, 42% of rural households were either "land poor" (owning less than 0.5 ha) or landless (Asian Development Bank 2014).

Many of these land concessions were granted outside of the existing legal code. Cambodia's Land Law of 2001 created new property rights categories, such as state public land, which are mostly forested areas, and state private land. However, the government did not clearly distinguish the difference between state public and state private land. Most state private land has been given to the grantees of ELCs. Further, many items in the Land Law have never been enforced. Just as with many other laws, a huge gap exists between the Land Law's policy and its implementation, largely due to weak governance and collusion between the state and the private sector (Oldenburg and Neef 2014).

Numerous communities have protested against the government and sued the companies that evicted them. However, courts have yet to rule in their favor, and none of the land has been returned to local communities. The police have also jailed villagers, companies have burnt down their homes, and some of their leaders have been murdered (ADHOC 2013). In 2012, due to mounting pressure, Prime Minister Hun Sen announced a moratorium on ELCs, although 33 were still granted afterward, and initiated Directive 001, which was supposed to speed up land titling throughout the country (Oldenburg and Neef 2014). However, field research in Ratanakiri Province revealed that the program worsened the land security of numerous villagers, including indigenous populations. Researchers found that authorities had withheld titles from villagers, did not grant land title documents, barred villagers from access to their land, and transferred land to powerful local elite (Focus on the Global South 2013). A local nongovernmental organization (NGO) stated that in Kampong Thom Province, in order to measure land for the program, powerful interests had cleared forests from which smallholders used to forage for nontimber forest products (personal communication, February 2014). In 2013 the prime minister suspended the program a few months before the July election, but in November he announced that it would resume shortly (Human Rights Watch 2014). In December, after local communities in Kratie Province protested, government officials promised to quickly deliver 1500 land titles to them as part of this program (Naren 2013).

Consequently, the land situation for smallholders remains grim. According to a local NGO, many of these affected and disgruntled villagers expressed their displeasured by voting for the opposition party, Cambodia National Rescue Party, in 2013 and by joining the party's protests. These protests peaked at around 100,000 people in December 2013, and affected groups throughout the country are becoming more empowered. According to the same NGO (personal communication, February 2014), people had lived in a climate of fear before the election, but during the election campaign youths throughout the country came out and shattered this climate.

15.4.1.2 Water

Another big problem is that the majority of smallholders lack adequate access to water. As of 2013, only about 22% of rice farming was irrigated, and current irrigation systems mostly benefit well-off farmers rather than smallholders (National Institute of Statistics 2015). Although the country has over 2000 canals, according to a local NGO, only about 10% of these canals can be used for a lengthy period. The rest can be used for only a short time or when water begins to recede. Consequently, the amount of canal water that can be used regularly is meager. In addition, the country's poor water governance also limits irrigation's effectiveness. A study revealed farmer water user communities, local bodies responsible for water management, often do not collect enough fees to maintain irrigation systems and rarely notify farmers before they release water or coordinate farmers' water demand (NGO Forum on Cambodia 2011). Water management is further enfeebled by the lack of feedback mechanisms and coordination between different levels of government (Cambodia Development Resource Institute 2011).

The remaining farmers must depend upon rainfall for agriculture. A lack of irrigation means that most producers can grow only a single, rain-fed rice crop annually. Over 80% of rice production occurs during the wet season (Kimsay 2013). This lack of investment in irrigation severely hampers many smallholders' rice yields and other agricultural production. Local NGOs have been lobbying the government to spend more on irrigation. In the 2016 national budget, less than 2% was devoted to agriculture, despite most Cambodians working in this sector (Sokheng and Turton 2015).

15.4.1.3 Low Productivity and Lack of Income

Lack of access to land and water severely hinders smallholders' productivity and hence their income. A number of factors prevent them from improving their income. First, they have limited access to technologies. Few of them own a buffalo, the primary source of draft power, or have access to electricity (only 26% of the population) or credit for agricultural production (World Bank 2011). Second, the costs of inputs, such as gasoline, fertilizer, and pesticide, are higher because Cambodia imports all of them. Third, the price of rice is volatile, and when prices are low, farmers cannot repay their loans and can fall into debt. Unlike its Thai counterpart, the Cambodian government does not guarantee a minimum price for rice or other products. Fourth, the government provides inadequate extension services due to its limited budget. Provincial and district offices have limited staff, capacity, and transportation to assist farmers. In some districts, there are no staff. Overall, these numerous factors dampen productivity: the average national rice yields are approximately 3.3 tons per hectare, the lowest in the region (Chun 2014).

In addition, smallholders have limited access to markets, particularly those abroad. Farmers grow a number of varieties of rice, but only a few are accepted for export. Also, the country's limited storage facilities means the quality of the rice can deteriorate quickly unless sold immediately. Sometimes neighboring countries refuse to buy rice at the border, citing its poor quality. Often borrowing a lot of money to produce the rice, smallholders directly and immediately sell to private collectors after they harvest their crops so that they can immediately repay their debts. But doing this lowers the price they receive (ActionAid Cambodia office official, personal communication, February 2014).

15.4.1.4 Increased Threat of Disasters but Limited Social Safety Nets

Two local NGOs both stated that climate change is a serious current and future issue affecting smallholders because it is increasing the magnitude of both droughts and floods, and this trend is expected to accelerate during the next decades. Since most farmers rely upon rain to grow their crops, such disasters can devastate their production and consequently drive them into debt or deeper in it. In 2016, 40,000 Battambang rice-farming families suffered heavy losses from a severe drought (Sokhorng 2016). This same year, over 17,000 ha of rice field were affected by floods (Kunmaka 2016). The government has developed a social protection policy to help farmers during droughts and floods by giving them food relief. However, these interventions can be improved because they do not reach certain groups in need of assistance. Further, government programs are poorly targeted, causing a high rate of leakage to nonpoor households (Council for Agriculture and Rural

Development, World Food Programme, and World Bank 2009). According to a local NGO official (personal communication, February 2014), disaster relief has been politicized: the government gives preferential treatment to those who voted for the ruling party. Due to the limited capacity of the state, NGOs often play a more important role during times of crises. Outside of times of crises, farmers receive limited support from the state, such as health care, education, and vocational training.

15.4.2 Myanmar

Agriculture plays a central role in Myanmar: in 2016 it accounted for 38% of the country's GDP, employed more than 60% of the population, and generated 25% of exports (Zorya 2016). The majority of farmers are smallholders. Despite its strategic location and high resource base, this sector is the least developed in the entire region. Income per capita in agriculture is almost US$200 a year, much less than those of their regional peers (Haggblade and Boughton 2013). This low level of income means that those living in rural areas experience high rates of poverty and food insecurity. Crop production comprises about 80% of farmers' income, and paddy rice accounts for about half of all planted area; 40% of smallholders produced crops primarily for sale, while the remaining 60% are subsistence based (Haggblade and Boughton 2013).

For the past few decades, the government severely stunted the growth of this sector, exercising excessive controls on production and marketing. Forced government procurement at below market rates gave farmers little incentive to expand production. Almost nonexistent extension services, lack of access to credit, and limited investment in infrastructure have also hurt smallholders. During the past few years, the government positively has liberalized the sector and expressed motivation to help smallholders, recently passing the Law on Enhancing the Economic Welfare of Farmers. Farmers, however, still face a host of problems. The biggest ones are lack of access to land and land insecurity, climate change, water insecurity, and the government's limited investment in agriculture.

15.4.2.1 Land

Approximately half of all rural household do not have the right to cultivate their land, forcing many of these households to work casually. Landlessness is over 70% in the Mekong Delta region. For those who do own land, most own only a couple of hectares or less (see Table 15.2).

Smallholders' access to land has declined over the past decade because, as of March 2012, the government had allocated 3.5 million hectares of land to local agribusiness companies, almost all of which have close ties with the military. On much of this land, smallholder farmers had been living in these areas previously and have been forcibly evicted or received scant compensation, and in some cases

Table 15.2 Percentage of land ownership by rural area in Myanmar as of 2011

Amount of land owned (hectare)	Delta/Coastal	Dry zone	Hilly areas
0	72	43	26
0–2	7	37	63
2–4	9	12	9
>4	12	8	2
Total	100	100	100

Source: USAID Burma (2013)

arrested for protesting. In July 2012, after receiving thousands of land complaints from farmers, Parliament established the Land Investigation Commission, which subsequently released a report stating that about half of the nation's land-grabs were due to military involvement and recommended that the military return unused farmland and compensate farmers. The military initially promised to return 250,000 ha but later changed its stance, saying that it would return only 18,000 ha (Woods 2013).

This trend of land-grabbing could unfortunately continue for the foreseeable future if the government's plan to develop the country is realized. According to the Ministry of Agriculture and Irrigation's master plan, the government is now seeking to woo foreign investors to convert 10 million acres of "wasteland," land that is currently occupied and used by millions of smallholders, into "productive" land through private commercial agricultural and industrial production, such as textile factories and palm oil, rubber, and cassava plantations. In 2012 the government passed two laws that favor large foreign investors and set the legal foundation for giving large-scale agricultural and industrial concessions to foreign investors. The Vacant, Fallow, and Virgin Land Law gives the Agricultural Ministry sole authority to manage farmland management and reallocation and has a clause that the ministry cannot be sued for any wrongdoings. Very few farmers possess statutory land use rights under this law. The Foreign Investment Law gives investors the right to use land for up to 70 years, a large increase compared to the previous law, and removes limits on foreign ownership (Woods 2013).

During the past couple of years, farmers' grassroots networks have newly emerged as political actors. They have united over common grievances, particularly land-grabbing, but other issues as well, such as access to credit, markets, and lack of freedom to crop. In response to these networks' demands and with the upcoming 2014 election in mind, the government passed the aforementioned Law on Enhancing the Economic Welfare of Farmers in late 2013. Originally the bill set a minimum guarantee for the price of rice, which was popular with farmers but could have been disastrous to the government's budget. In its final form, Parliament revoked this provision. The bill is still supposed to help farmers receive a fair price for the products and sell them freely. However, the details of how the government will do this are still unclear. Further, to the frustration of smallholders, the bill includes agribusiness as "farmers" and does not address land disputes (Win and Kean 2013).

15.4.2.2 Limited Water Supply

While the Myanmar is blessed with ample water resources, only 10% of them are utilized, with most of it for irrigation. As of 2015, only about 15% of the land was irrigated (Zorya 2016). This lack of irrigation means that most farmers can grow crops only during the wet season, must rely upon rain as the sole source of water, and are more vulnerable to the effects of floods and droughts.

15.4.2.3 Climate Change

Studies indicate that the country's agriculture sector is highly vulnerable to the impacts of climate change, predicting that there will be more floods and droughts and that rainfall will become more sporadic. This volatility and increase in disasters will negatively affect production, particularly of rice, and could also contribute to price volatility for locally produced goods. For example, in 2015, monsoon rains damaged 527,000 ha of crops, with some areas losing their entire crop for that season (FAO 2015).

15.4.2.4 Limited Investment in Agriculture

Smallholders sorely need the government to invest more in this sector. Of its limited investment so far, most of the money has been directed toward state-owned enterprises. So far, extension services have been virtually nonexistent. Farmers instead must rely upon each other and other private actors. Little has also been invested in research, only 20% as much as some of its regional counterparts. Last, the government has done little to help farmers increase their access to credit and to improve archaic infrastructure in rural areas (USAID Burma 2013).

15.4.3 Thailand

As of 2016, around 40% of Thais worked in agriculture, about two-thirds of these as rice farmers. The agricultural sector, which contributed 11% to Thailand's GDP, comprised mostly smallholders, totaling about 7.5 million households, with an average land holding of 3.7 ha per household (Sondergaard et al. 2016). In 2010, the percentage of cultivated land under smallholders was 76%. Compared to most countries in the region, Thailand has achieved a high level of development in the agricultural sector. This is largely due to upgrades in agricultural infrastructure, particularly large-scale irrigation and rural roads, as well as the provision of credit to farmers by state-owned banks and the widespread adoption of green technology in previous decades. As a result of this growth, Thailand has positioned itself as one of the top ten suppliers in the world food trade, ranking in the top three globally in terms of rice, cassava, and sugarcane exports. This growth has also led to significant drops in rural poverty and child malnutrition (Leturque and Wiggins 2010).

However, despite the country's relative success, changes are still needed in this highly unequal sector.[3] While large agribusinesses reap the majority of the profits, smallholders are under a high level of duress due to a number of problems. The biggest problem is that a vast number of Thai farmers face a high degree of land insecurity: they either are landless or do not have enough land, despite the country having enough land for all Thais. Further, many farmers have unmanageable levels of debt, which according to the Department of Mental Health is the largest factor behind a spate of suicides of 13 farmers in 2014 (Sarnsamak 2014). In addition, most models of contract farming being used are unfair to smallholders. Last, there is concern that large agribusinesses in Thailand will replicate their practices in neighboring countries, thereby preying upon smallholders there.

15.4.3.1 Land

Despite the government's efforts beginning in the 1970s to address high levels of tenure insecurity and landlessness, these problems remain today. In 2017, it was estimated that there were 9.6 million landless or nearly landless Thais, about 14% of the population (Chantanusornsiri 2017). Landless farmers live on privately owned land, rent their farmland, or live on state-owned lands. For example, 1.2 million Thais live in "protected forest areas" without any official land tenure documentation. Many of them had lived in these areas before the government claimed the land as its own. One negative consequence of them living in these areas is that sometimes the government charges them with trespassing or illegally occupying the land. In 2001, almost 300 farmers in the north were arrested under such charges. Overall, almost half (48%) of farm households own less than 1.6 ha of land (Lubanski 2012). The percentage of farmers who own land fell dramatically from 44% in 2004 to 15% in 2011 (Macan-Markar 2016). At the same time, 10% of the population own about 90% of total private land. Further, 70% of privately owned land has been either abandoned, vacated, or not put to productive use, often being held for speculative purposes (Lubanski 2012). The Land Institute Foundation declared that the annual cost of this underutilized land is approximately 127.4 million baht (Leonard and Ayutthaya 2002). These figures clearly suggest the huge inequality in land ownership in Thailand.

How has this situation arisen? To briefly summarize, in the 1980s most of the country's land was still unregistered, so in response, with encouragement and support from the World Bank, the government sought to systematically hand out land titles. However, the land titling program that it initiated was riddled with corruption,

[3] Thailand's Gini coefficient is 0.51, the highest in Southeast Asia. Further, the poor are overrepresented in the agriculture sector. Poor families are heavily concentrated in this sector: almost half of the poor work in agriculture. About 90% of the poor reside in rural areas. See Bird et al. (2011).

such as many title deeds being issued under false names, and other problems, such as communal lands being arbitrarily broken up. The program unintentionally transferred large swatches of rural land from the poor to the wealthy, who have often left the land idle and sparked rural land conflicts. Moreover, the program did not address the issue of numerous smallholders illegally living on state land (Leonard and Ayutthaya 2002). Little has changed since the program's completion despite numerous protests and court cases largely because those in power in the government are often wealthy landowners themselves and because of land governance is highly fragmented: 16 agencies are involved in managing land (ADB consultant, personal communication, February 2014). Historically land ownership has been skewed toward the elite in a feudalistic hierarchy.

Positively, in the past few years, the government has legally accepted communal land titles and created a Community Land Title Office. While this concept is not new to Thailand, its legal status is. In 2011 two communities, Khlong Yong in Nakhon Pathom and Ban Mae Aoa Pasang in Lamphun, were granted legal permission to manage their land communally. According to a report, the Khlong Yong community was successful because it remained united, had charismatic leadership, and received media attention. Less positively, 433 other communities have applied for communal land titles, of which 53 have been approved by the Community Land Title Office. However, none of them have yet received permission from locally responsible government agencies (Lubanski 2012). Besides handing out these additional communal land titles, local activists are also advocating that the government help smallholders by implementing a progressive property tax, creating a large land bank, and establishing a fund to help poor farmers pay for legal fees.

15.4.3.2 Indebtedness

A survey found that the average household debt of rice growers in Thailand stood at 103,047 baht (US$ 3200) in 2011, rising 6% from 2010 (see Fig. 15.1). The majority of farmers also revealed that they would have to repay their formal debts by borrowing money from loan sharks. The reasons behind this increase in debt are manifold. One persisting trend is an increase in the cost of inputs, which are

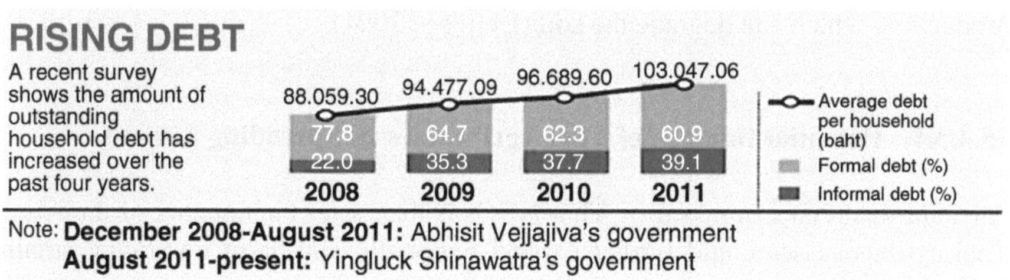

Fig. 15.1 Household debt per year in Thailand from 2008 to 2011. (Source: Arunmas and Khachornwutthinan 2012)

controlled by large agribusinesses. In the central region, many farmers rent land, hence increasing their costs (local Thai NGO official, personal communication, February 2014). Further, the government has struggled to pay farmers due to legal stipulations that it cannot increase its debt as a caretaker government. But even before its status change, the rice-pledging scheme, initiated in 2011, according to the International Monetary Fund, was cumbersome, complicated, and full of problems, including allegations of widespread corruption (Sattaburuth and Pongpao 2012). Consequently, instead of participating in the program, many farmers sold their rice to middlemen at lower prices. In addition, many cheap vegetables and fruits from China are being dumped into the market and hurting farmers who cannot compete price-wise (Noi 2010). Overall, farmers are not receiving enough money for their products. An analysis shows that smallholders receive only 24% of the export value of rice, while the rest goes to exporters, traders, and millers (Delforge 2004). Since rice is easy to grow but not very profitable, many farmers engage in off-farm activities to supplement their income. In 2011, they earned only 45% of their income from agriculture (Isvilanonda 2012). Thus, it seems that instead of helping farmers increase their access to credit, what is needed is helping them reduce their debt.

15.4.3.3 Contract Farming

In the region, Thailand has the greatest experience with contract farming, using it with the widest range of crops (FAO 2008). As the models of contract farming vary according to company and crop, it can be both beneficial, such as garlic in Chiang Mai, and harmful (Walker 2007). According to the Alternative Agriculture Network, in the fishery and hog-raising industries, this relationship is highly unequal. Large agribusinesses control the production of inputs, use of land, distribution systems, and marketing, while farmers earn little income from their labor while bearing large risks (Purugganan 2012). Further, contracts often strongly favor the interests of the companies. For example, Frito Lay's contracts stipulate that farmers must sell their product only to the company, but the company is not required to buy the product from the farmers. Many schemes also can increase farmers' level of debt because farmers must borrow in order to invest in new technologies and infrastructure. Last, contracting companies often dictate that farmers heavily use chemicals to maximize production, which can degrade the soil (Delforge 2004).

15.4.3.4 Potential Damage of Thai Agribusinesses Spreading Abroad

An additional concern raised by Thai NGOs is that, after the opening of the AEC, Thai agribusinesses could further expand regionally and harm traditional smallholder systems by using genetically modified organisms, introducing unequal contract farming schemes, and perhaps partaking in land-grabs in neighboring countries (local Thai NGO official, personal communication, February 2014). The agriculture conglomerate Charoen Pokphand has already gained control of the poultry industry in Laos and gained a foothold in the corn for animal feed sector in Myanmar through using contract farming schemes. Local Burmese NGOs have already raised concerns about the Charoen Pokphand model. "If this model is replicated across the country then negative consequences for production sustainability will likely result," said Tobias Jackson of the Food Security Working Group (quoted in Boot 2013).

The Thai company Khan Kaen Sugar Industry has already been involved in land-grabbing in Koh Kong Province in Cambodia. The company owns 70% of a Cambodian sugar company that was granted a land concession to build a sugar plantation. To clear the land, the Cambodian company violently evicted over 4000 villagers from the land and involuntarily relocated them. The villagers said that they were not consulted beforehand and that the land transfer was illegal. The company has denied any wrongdoing (O'Toole 2011).

15.4.4 Vietnam

In 2014 the Vietnamese agricultural sector contributed over 17% to the country's GDP and employed nearly half of the workforce (Oxford Business Group 2016). Rice and coffee are the country's largest agricultural exports, followed by pepper, starches, cassava, and rubber. The agriculture sector has grown tremendously over the last few decades, catapulting many out of poverty. In the 1980s the country was a net food importer, but amazingly, only two decades later the country was the one of the world's leading agriculture exporters. During these years, its annual growth rates of numerous products skyrocketed: 50% annually for rice and over 100% for coffee. Most of this growth occurred in small farms, with plantations contributing only modestly; for example, rubber grew a modest 16% annually. Thus, Vietnamese smallholders, about nine million households, contributed greatly to the country's massive poverty reduction. The poverty rate is now below 15%, much less than in 1979 when it was 58% (Pincus 2012).

This does not mean, however, that smallholders are prospering. During the past few years, as the government has switched to pursuing a more pro-industry and pro-foreign investment economic model, growth has mostly been captured by the richest urban households, while rural incomes have hardly improved. Many smallholders are still below or hover barely above the poverty line and are encountering a number of challenges thwarting them from improving their livelihoods further. In 2010 the poverty level was 27% in rural areas, and 16% suffered from food shortages for almost five months per year (Minh and Maerten 2012). Just as in most other countries in the region, inadequate access to land and land insecurity are their greatest challenge. In addition, Vietnamese smallholders struggle to earn profits, for numerous reasons, and are increasingly suffering from natural disasters. Another issue of concern is that a few Vietnamese companies have been involved in land grabs in neighboring countries.

15.4.4.1 Land

Due to the country's high population and limited amount of arable land, the average land endowment of each agricultural household is one of the lowest in the world: less than half a hectare per household, with most farmers owning less than 2 ha. Since 70% of the population live in rural areas, land use is highly intensive. Land ownership is also highly unequal. According to official statistics, 664 state-owned enterprises have the right to use over 6.8 million hectares, while 326,000 minority households lack sufficient arable land (Wells-Dang 2013).

During the past decade, land rights have become highly contested. The main driver of contestation is the seizure of nearly 1 million hectares of farmers' agricultural land by the government for what it deems economic development purposes, ranging from building infrastructure to private projects, such as industrial and housing estates and golf courses. Most disputes have arisen over the clearing of the land for private projects, often on the periphery of urban areas, from which farmers receive sometimes only 10% of its value in compensation. At least two million Vietnamese were displaced or hurt in some manner by land conversions from 2006 to 2011 (Communist Review 2011). During this period, farmers filed one million complaints and petitions and protested often. As one villager decried, "It used to be that leaders took the land of the rich and allocated it to us. Now they take our land and give it to the rich." A survey revealed that only 10% of those have lost their land feel that they received adequate compensation (Wells-Dang 2013).

Widespread discontent over these land allocations led to numerous calls to revise the 2003 Land Law, which favored the interests of investors and urban residents over those of smallholders. Whereas farmers' leases for agricultural land are only 20 years, and they cannot change the use of their land for other purposes, leases for commercial projects are 50 years and for real estate projects are unlimited, and these leaseholders can do whatever they want with the land. Further, government agencies have set the price of land as low as 30% of market value to encourage land conversion and investment. In addition, women encounter disparities in terms of their access to land because of gaps in the law (Wells-Dang 2013).

In 2014, the National Assembly approved a revision of the Land Law. While the revision protects smallholders more than before, the changes are minor. They included an extension from 20 to 50 years for leases for agricultural land, a mechanism for public consultations for land use planning and changes (although what form this will take is still unclear), and a weakening of provincial governments' power over land allocation. But the core problems that originally led to land disputes remain. State agencies can acquire land without giving adequate compensation, smallholders still could be uninformed of these plans, and incentives for corruption remain rife (Oxfam Vietnam officer, personal communication, February 2014).

15.4.4.2 Limited Profits

While farmers' income levels have improved drastically in the past few decades, they remain low, and some groups have been left behind. One major reason that their incomes are low is because, while farmers have skillfully improved their amount of production, their products are still of an inferior quality compared to those in other countries, such as rice in Thailand and coffee in Colombia. According to the World Bank, the prices of Vietnamese rice and coffee, two of the country's biggest products, are 15% lower than those of its competitors (Overseas Development Institute 2013). Moreover, some products have been rejected in some foreign markets due to food quality requirements. For example, Vietnamese pork, beef, and chicken are still not allowed to enter the United States and the European Union.

The state has not allocated enough funding and other resources to build the capacity of farmers so that they can improve their products' quality and their products can receive certification. The government spent only 6% of its budget in 2010 on agriculture, most of it devoted to irrigation and drainage (FAO 2013a). Further, seeking to gain publicity, local authorities often prefer large-scale projects rather than helping smallholders.

Overall, smallholders have a disadvantaged position in the market and little bargaining power. State-owned enterprises monopolize input supply, postharvest processing, and marketing of products. In many areas, farmers are forced to rely on private dealers to buy inputs and to sell their outputs. According to Oxfam, the price margins for some products bought by private dealers in some areas can be between 20% and 50% (Minh and Maerten 2012). Also, farmers' organizations are weak, partially because of the government's skeptical attitude toward them. Local NGOs are concerned that, without subsidies and increased investments, smallholders will be unable to compete with farmers from other countries, such as China. Additional liberalization of Vietnamese trade could therefore make smallholders more vulnerable (ActionAid Vietnam officer, personal communication, February 2014).

Another problem is high inflation, particularly of inputs. From 2008 to 2013, the price of inputs has increased more than the price of rice: fertilizer rose by 250% but rice increased only 120%, thereby denting farmers' purchasing power (Minh and Maerten 2012). Many of these inputs are imported, which further raises their prices. A recent jump in the price of petroleum has benefited Vietnam's crude oil exports, but the agricultural sector has experienced losses because they use a large amount of petroleum in production and processing.

15.4.4.3 Increased Impact of Climate Change with Limited Government Assistance

Vietnam has been ranked by the World Bank as one of the five countries most vulnerable to climate change globally. To smallholders, climate change means increased droughts, floods, and heat waves, which could obliterate their harvests. For example, a tropical storm in October 2017 killed 230,000 livestock and damaged more than 40,000 ha of crops (Nguyen and Tostevin 2017). Because of climate change, the country risks losing 7.2 million tonnes of rice yield, or 3% of its agricultural land, by the second half of this century (Vietnamnews.Vn 2017).

During such times, government assistance is needed. However, Vietnam's social protection is fragmented and prone to mistargeting and leakage. For example, unregistered migrants, some ethnic groups, and informal workers do not receive any coverage, while higher-income groups benefit more than the poorest ones (Vietnamnews.Vn 2017). Needed social protection programs, such as cash transfers, have not yet been initiated. Further, free health insurance is limited to state employees, and vocational training opportunities exist but are not linked well to market opportunities (Oxfam Vietnam officer, personal communication, February 2014).

15.4.4.4 Regional Land-Grabbing by Vietnamese Companies

In recent years some Vietnamese companies, such as the joint-stock company Hoang Anh Gia Lai Group (HAGL) and the state-owned enterprise Vietnam Rubber Group, have entered into joint ventures with local partners in Laos and Cambodia to build rubber plantations. The impacts of these projects have been twofold. On the one hand, they contribute to the countries' economic development by increasing its rubber exports and by improving its infrastructure. For example, HAGL built a US$40 million airport in Attapeu Province in Laos (Tuoitre News 2012). On the other hand, as Global Witness has reported, these projects have raised social concerns in terms of the relationship between the joint ventures and local communities, because the latter have been expelled from their land and lost access to the forests, including spirit forests and burial grounds, which were formerly theirs (Global Witness 2013). There is a worry that as the regions' borders become more open, the expansion of these practices may lead to increasing tensions and problems for smallholders.

15.5 Recommendations

Smallholders are the backbone of ASEAN economies and constitute the majority of agricultural workers. Southeast Asian governments and ASEAN must not buy into pervasive anti-smallholder mythology and instead realize that in land-constrained countries, which they all are, growth models based on broad-based small-farmer-led growth rather than large-scale commercial farming have been more successful worldwide in boosting productivity, reducing poverty, and sustaining nationwide growth. Therefore, with support from ASEAN, governments should promote agricultural development strategies based on such a model. Civil society also needs to play an important role helping them.

ASEAN should consider the following measures:

- Adopt the Food and Agriculture Organization (FAO) "Voluntary Guidelines for the Responsible Governance of Tenure of Land, Fisheries and Forests in the Context of National Food Security." The guidelines state, "States should consider promoting a range of production and investment models that do not result in the large-scale transfer of tenure rights to investors and should encourage partnerships with local tenure right holders." FAO has hosted regional workshops on these guidelines to raise awareness of them.[4] ASEAN's adoption of them would encourage member states to pursue alternatives to direct large-scale land acquisitions.
- Harmonize regulatory framework and standards regarding food safety and nutrition, such as phytosanitary, food inspection, and laboratory standards. It should follow the model of the EU Common Agriculture Policy. To have every country agree upon common standards will require difficult negotiations and trade-offs but will boost trade and enable smallholders to better access regional markets. It is important also that these standards are pro-poor and that smallholders can easily meet them.
- Help establish regional social safety nets to help smallholders during times of crises. One example would be to create an insurance mechanism that pools together collective risks of agricultural production in the region. It would use weather-based indices and give farmers insurance when they are struck with major agricultural disasters. Countries that are net consumers of agricultural products should help subsidize this scheme (Corona 2013). This tool will become increasingly helpful as climate change accelerates.

While exact government policies that are needed to help smallholders in each Southeast Asian country vary, the ones below are a few important broader recommendations. They are geared especially toward the lesser developed countries in ASEAN.

National ASEAN governments should consider the following measures.

- Land governance:
  - Support and distribute community land titles. The practice of communal land ownership has existed throughout Southeast Asia since people have been farming. Despite this long history, only Cambodia, Laos, Philippines, and Thailand in the region currently provide legal recognition of this form of ownership, and overall only a handful of titles have been granted.

---

[4] For example, FAO hosted a workshop in Bangkok on the guidelines in August 2013. See FAO (2013b).

- Create land banks to help landless or nearly landless farmers buy land with low interest rates and minimal down payments.
- Publish information about land concessions and acquisitions, including proposals, the bidding processes, active concessions and acquisitions, and the location of state land and protected areas. Doing so would make the sector more transparent.
- Conduct impact assessments of proposed concessions, publicize them before they are granted, and consult people about resettlement. Judgments related to concessions should be based on the interest of the overall population.

• Water and climate change:
  - Expand the amount of irrigated land. Especially in CLMV, limited access to water severely hurts farmers' production and increases their vulnerability to droughts. Governments therefore need to scale up their investments.
  - Increase the amount of funding devoted to agriculture research. This research should focus on technologies that are greener, more affordable, and more suitable to smallholders, particularly those that will help them adapt to climate change. Such technologies will help increase the quality of farmers' products and their productivity.

• Market access and trade:
  - Promote farmer cooperatives. Governments should assist the formation of cooperatives by teaching smallholders cooperative organization skills, providing them extension services, and creating schemes to help them access credit and loans. Cooperatives are beneficial because they increase smallholders' bargaining power, lower transaction costs in obtaining loans, and enable them to have better access to information.
  - Create legal frameworks to govern contract farming to curb exploitation of smallholders. Governments should monitor and evaluate the performance of contractors, such as the cost of their inputs and the timeliness of disbursing payments to farmers, and should threaten to punish companies that do not honor their obligations.
  - Make sure smallholders' opinions are included when negotiating trade agreements. Governments should disclose the terms of the trade negotiations early so that meaningful discussions can be had, hold public hearings and consultations with smallholders, and disseminate the terms of the agreement in non-technical language.

• Address the gender gap:
  - Make sure land laws are gender-equal in terms of ownership provisions.
  - Provide technologies suitable for women to use.
  - Encourage the emergence of women's organizations, such as unions and cooperatives.

- Civil society should consider the following measures:
  - Advocate to governments, donors, and companies of the value of supporting smallholders. They should do this by releasing studies and documentaries, disseminating reports to the media, and lobbying officials during meetings.
  - Spread best practices and lessons of successes in other countries, such as Brazil's social protection programs, and successes in regional countries' agriculture sectors, such as those in Thailand, to government agencies. National and local government officials, such as those in Myanmar and Vietnam, are interested in learning about what has been successful elsewhere. International NGOs can help by spreading these ideas to officials and local civil society organizations (CSOs) and teaching them how to implement them.
  - Monitor foreign agribusiness companies investing in nearby countries. If these companies commit wrongdoings, such as land-grabbing, CSOs should publicly reprimand them, inform foreign buyers of their malpractices, and demand that their governments punish them.
  - Continue to press at the national level for governments to expand the distribution of communal land titles while at the local level help communities map their land. While doing the latter is not a silver bullet, it has numerous benefits: it helps these communities assert contrary claims to land, provides interim recognition of these lands, enables them to have better knowledge of the community's territorial assets, builds community ownership and confidence, and helps them resolve local land and resource conflicts (Adler et al. 2009).
  - Support and help empower groups who have been opposing land-grabbing. Doing this could perhaps mean empowering farmer organizations, teaching them about advocacy strategies successful elsewhere, connecting them to media outlets, and assisting them in organizing their own political parties and platforms.

15.6 Conclusion

This chapter has shown that the challenges facing smallholders in Southeast Asia stem from governance shortcomings. Their level of resilience is significantly affected by political economy variables, such as degree of access to power, the effects of public policies, and lack of state support. Consequently, echoing what Blaikie and Brookfield (1987) argue, this chapter suggests that the environmental problems discussed here cannot be solved unless these political-economic dimensions are addressed. Moreover, as this chapter asserts, any future studies of agricultural and environmental problems in the region must take into account how power geometries affect the creation of these problems.

References

ADHOC. (2013). *A turning point? Land, housing and natural resources rights in Cambodia in 2012*. Phnom Penh: ADHOC. http://www.adhoc-cambodia.org/wp-content/uploads/2013/02/ADHOC-A-Turning-Point-Land-Housing-and-Natural-Resources-Rights-in-2012.pdf

Adler, D., Ironside, J., & Ratanak, M. (2009). *Mapping and tenure sin Cambodia's indigenous communities*. Phnom Penh: Heinrich Böll Stiftung. http://www.academia.edu/download/31038135/Mapping_and_Tenure_Security_in_Cambodias_Indigenous_Communities_final.pdf

Amanor, K. (2017). The global land grab: Beyond the hype. *The Journal of Peasant Studies, 44*(2), 506–509. https://doi.org/10.1080/03066150.2017.1287670.

Anthony, M. C., Teng, P. S., & Chng, B. (2013). *ASEAN economic community (AEC) 2015: Opportunities and challenges for food security* (NTS Policy Brief No. PO13-04). RSIS Centre for Non-Traditional Security Studies.

Arunmas, P. & Khachornwutthinan, P. (2012). Rice farmers' debts surge. *Bangkok Post*, August 17, 2012. http://www.bangkokpost.com/print/307965/

ASEAN Secretariat. (2010). *A humanitarian call: The ASEAN response to cyclone Nargis*. Jakarta: ASEAN Secretariat. http://www.aseansec.org/wp-content/uploads/2013/07/1.pdf

ASEAN Secretariat. (2016). *ASEAN economic community chartbook 2016*. Jakarta: ASEAN Secretariat. http://www.aseanstats.org/wp-content/uploads/2016/11/AEC-Chartbook-2016-1.pdf

Asian Development Bank. (2013, April). *Asian development outlook (ADO) 2013: Asia's energy challenge*. https://www.adb.org/publications/asian-development-outlook-2013-asias-energy-challenge

Asian Development Bank. (2014). *Cambodia country poverty analysis 2014*. Manila. https://www.adb.org/sites/default/files/institutional-document/151706/cambodia-country-poverty-analysis-2014.pdf

Asian Farmers' Association for Sustainable Rural Development. (2007, October 26). *Towards a fair trade in an integrating ASEAN: Perspectives of small men and women farmers in AFA – Asian farmers association for sustainable rural development*. http://asianfarmers.org/?p=335

Bano, S., Takahashi, Y., & Scrimgeour, F. (2013). ASEAN-New Zealand trade relations and trade potential: Evidence and analysis. *Journal of Economic Integration, 28*(1), 144–182. https://doi.org/10.2307/41825679.

Bird, K., Hattel, K., Sasaki, E., Attapich, L. (2011, November). *Poverty, income inequality, and microfinance in Thailand*. Manila: Asian Development Bank. http://hdl.handle.net/11540/1311

Blaikie, P., & Brookfield, H. (1987). *Land degradation and society*. London: Methuen http://www.cabdirect.org/abstracts/19876703924.html.

Boot, W. (2013). CP workers' woes highlight concerns about big agriculture in Burma. *The Irrawaddy*, June 27, 2013. https://www.irrawaddy.com/business/cp-workers-woes-highlight-concerns-about-big-agriculture-in-burma.html

Cambodia Development Resource Institute. (2011). *Improving the governance of water resources in Cambodia: A stakeholder analysis*.

Chantanusornsiri, W. (2017, July 20). New property tax bill offers chance to narrow income gap. *Bangkok Post*. https://www.pressreader.com/thailand/bangkok-post/20170720/281797104049654

Chun, N. (2014). *Improving rice production and commercialization in Cambodia: Findings from the farm investment climate assessment*. Manila: Asian Development Bank.

Communist Review. (2011, January). *Impacts of the present land policy and management on rural development*. http://english.tapchicongsan.org.vn/Home/Theory-and-Reality/2011/199/Impacts-of-the-present-land-policy-and-management-on-rural-development.aspx

Corona, S. (2013). *The case for a Southeast Asian agriculture risk pool*. Geneva Association. http://www.asiacapitalre.com/Documents/ACR-CH/media_centre/our_views/2013/pdf/AIR%20-%20The%20case%20for%20a%20Southeast%20Asian%20agriculture%20risk%20pool.pdf

Council for Agriculture and Rural Development, World Food Programme, and World Bank. (2009). *Safety nets in Cambodia: Concept note and inventory*. http://www.foodsecurity.gov.kh/sites/default/files/docs/WB-Cambodia-Social-Safety-Net-Concept-Eng.pdf

Daite, R. B. (2015). The ASEAN economic community 2015 and regional trade: Some prospects for ASEAN agriculture. *Asian Journal of Agriculture and Development, 12*(2), 17–32.

Delforge, I. (2004). *Thailand: From the kitchen of the world to food sovereignty*. Focus on the Global South. https://focusweb.org/node/506

Desker, B., Caballero-Anthony, M., & Teng, P. (2013). *Thought/issues paper on ASEAN food security: Towards a more comprehensive framework* (ERIA Discussion Paper Series ERIA-DP-2013-20). Singapore: Rajaratnam School of International Studies. http://www.eria.org/ERIA-DP-2013-20.pdf

Diakonia. (2012, April 27). *Cambodia: Activist Chut Wutty killed*. http://www.diakonia.se/en/Info/News–Publications/News-archive-2012/Cambodia-Activist-Chut-Wutty-killed/

FAO (Food and Agriculture Organization). (2008). *Overview of contract farming in Thailand: Lessons learned*. http://www.fao.org/uploads/media/ADBI%20contract%20farming%20thailand.pdf

FAO (Food and Agriculture Organization). (2012). *Smallholders and family farmers*. http://www.fao.org/fileadmin/templates/nr/sustainability_pathways/docs/Factsheet_SMALLHOLDERS.pdf

FAO (Food and Agriculture Organization). (2013a). *Report of workshop on private corporate sector investment in agriculture in Southeast Asia*. http://www.fao.org/fileadmin/user_upload/tcsp/docs/1_Final%20Report%20Bandung.pdf

FAO (Food and Agriculture Organization). (2013b). *Technical workshop for the Asia region raising awareness of the voluntary guidelines on the responsible governance of tenure of land, fisheries and forests in the context of national food security*. http://www.fao.org/fileadmin/templates/rap/files/meetings/2013/131209_report-awareness.pdf

FAO (Food and Agriculture Organization). (2015). *Myanmar floods: Huge impact on agricultural livelihoods, more international support urgently needed: FAO in emergencies*. http://www.fao.org/fileadmin/user_upload/emergencies/docs/Final%20fact%20sheet%201%20-%20web%20version.pdf

Focus on the Global South. (2013). *Moving forward: Study on the impacts of the implementation of order 01BB in selected communities in rural Cambodia*. https://focusweb.org/system/files/Student%20Volunteers%20Report_%20ENG.pdf

Germanwatch. (2016). *Global climate risk index 2017*. https://germanwatch.org/en/12978

Global Witness. (2013). *Rubber Barons*. https://www.globalwitness.org/en/campaigns/land-deals/rubberbarons/

Haggblade, S., & Boughton, D. (2013). *A strategic agricultural sector and food security diagnostic for Myanmar* (IDWP 131). Prepared for USAID/Burma. http://ageconsearch.umn.edu/record/161372

Hodal, K. (2013, July 9). Cambodia's sugar rush leaves farmers feeling bitter at 'Land Grab.' *The Guardian*, sec. World News. http://www.theguardian.com/world/2013/jul/09/cambodia-sugar-land-grab-claims

Human Rights Defenders. (2012, December 12). *Re: Thailand—Killing of human rights defenders and members of Southern Peasants Federation of Thailand Ms. Montha Chukaew and Ms. Pranee Boonrat*. International Federation for Human Rights. http://www.fidh.org/en/asia/thailand/Re-Thailand-Killing-of-human-12664

Human Rights Watch. (2014). *Country summary: Cambodia*. http://www.hrw.org/sites/default/files/related_material/cambodia_7.pdf

Huyer, S. (2016). Closing the gender gap in agriculture. *Gender, Technology and Development, 20*(2), 105–116. https://doi.org/10.1177/0971852416643872.

Isvilanonda, S. (2012). Success factors for smallholder rice production in Thailand. In *Agri Benchmark Conference*. Pretoria, South Africa.

Juego, B. (2014). *The ASEAN economic community project: Accumulating capital, dispossessing the commons.* Heinrich Böll Stiftung. http://vbn.aau.dk/ws/files/207086805/Bonn_Juego_2014_The_ASEAN_Economic_Community_Project_Development_Debates_in_Asia_pp._12_19.pdf

Kimsay, H. (2013, May 20). Agriculture needs more loans, irrigation systems, experts say. *Phnom Penh Post.* http://www.phnompenhpost.com/business/agriculture-needs-more-loans-irrigation-systems-experts-say

Kunmaka, M. (2016, October 28). Rice crop damaged by floods. *Khmer Times.* http://www.khmertimeskh.com/news/31368/rice-crop-damaged-by-floods/

Kusakabe, K. (2015). *Gender analysis of economic land concessions in Cambodia and in Northern Laos: Case of rubber plantations* (LDPI Conference Paper No. Vol. 43).

Leonard, R., & Narintarakul Na Ayutthaya, K. (2002). Taking land from the poor, giving land to the rich. *Watershed, 8*(2), 14–25.

Leturque, H., & Wiggins, S. (2010). *Thailand's progress in agriculture: Transition and sustained productivity growth.* London: Overseas Development Institute.

Lubanski, J. (2012). *Land is life: A policy advocacy case study of the Northern Thailand land reform movement.* MA Thesis, SIT Graduate Institute. http://digitalcollections.sit.edu/capstones/2565

Macan-Markar, M. (2016, March 19). Debt fills Thailand's rice bowl. *Nikkei Asian Review.* https://asia.nikkei.com/Politics-Economy/Economy/Debt-fills-Thailand-s-rice-bowl

Minh, L. N., & Maerten, B. 2012. *Growing a better future in Viet Nam: Expanding rights, voices and choices for small-scale farmers.* Policy and Practice. Hanoi: Oxfam. http://policy-practice.oxfam.org.uk/publications/growing-a-better-future-in-viet-nam-expanding-rights-voices-and-choices-for-sma-250351

Naren, K. (2013, December 31). After protest, government speeds up land titles for Kratie Families. *Cambodia Daily.* https://www.cambodiadaily.com/archives/after-protest-government-speeds-up-land-titles-for-kratie-families-49768/

Nation. (2014, March 5). *Eleven Thais make Forbes' billionaires list.* http://www.nationmultimedia.com/business/Eleven-Thais-make-Forbes-billionaires-list-30228374.html

National Institute of Statistics. (2015). *Census of agriculture in Cambodia 2013.* Phnom Penh. http://www.fao.org/fileadmin/templates/ess/ess_test_folder/World_Census_Agriculture/Country_info_2010/Reports/Reports_5/KHM_ENG_REP_2013.pdf

NGO Forum on Cambodia. (2011). *Learning from irrigation projects in Cambodia: Benefits and challenges.* Phnom Penh.

Nguyen, M., & Tostevin, M. (2017, November 5). Death toll from typhoon in Vietnam rises to 27. *Reuters.* https://www.reuters.com/article/us-asia-storm-vietnam/floods-swamp-vietnam-as-typhoon-death-toll-reaches-27-idUSKBN1D5042

Noi, G. S. (2010, February 17). ASEAN Jittery about trade pact with China. *Strait Times.*

O'Toole, J. (2011, September 5). Bittersweet harvest. *Bangkok Post.* http://www.bangkokpost.com/business/economics/255018/bittersweet-harvest

Oldenburg, C., & Neef, A. (2014). Reversing land grabs or aggravating tenure insecurity? Competing perspectives on economic land concessions and land titling in Cambodia. *Law and Development Review, 7*(1), 49–77. https://doi.org/10.1515/ldr-2014-0014.

Overseas Development Institute. (2013). *Agriculture: Business and development exchange summary.* http://www.odi.org.uk/sites/odi.org.uk/files/odi-assets/projects-documents/269.pdf

Oxford Business Group. (2016, December 4). Viet Nam's agricultural sector arrives at a crossroads. *VietNamNet.* http://english.vietnamnet.vn/fms/business/167997/viet-nam-s-agricultural-sector-arrives-at-a-crossroads.html

Pincus, J. R. (2012). *Vietnam's reforms: The road to market leninism.* The Legatum Institute. https://lif.blob.core.windows.net/lif/docs/default-source/country-growth-reports/prosperity-in-depth-vietnam-39-s-reforms–the-road-to-market-leninism.pdf?sfvrsn=0

Purugganan, J. (2012, November 26). *Questioning ASEAN's path to development: Grassroots leaders raise concerns on trade and investment policies, projects.* Focus on the Global South. https://focusweb.org/content/questioning-asean%E2%80%99s-path-development-grassroots-leaders-raise-concerns-trade-and-investment

Rao, N. (2011). *Women's access to land: An Asian perspective* (Expert paper 3). UN Women. http://www.un.org/womenwatch/daw/csw/csw56/egm/Rao-EP-3-EGM-RW-30Sep-2011.pdf
Sarnsamak, P. (2014. February 19). Rice-scheme woes not major factor in suicides, Agency Says. *The Nation*. http://www.nationmultimedia.com/national/Rice-scheme-woes-not-major-factor-in-suicides-agen-30227201.html
Sattabururh, A., & Pongpao, S. (2012, June 12). Flood compo protesters threaten roadblock. *Bangkok Post*. http://www.bangkokpost.com/news/politics/297627/flood-compo-protesters-threaten-roadblock
Shenggen, F., Joanna, B., Michiel, K., & Alex, H. (2013). *From subsistence to profit: Transforming smallholder farms*. Washington, DC: Intl Food Policy Res Inst.
Sokheng, V., & Turton, S. (2015, October 22). Budget figures released. *Phnom Penh Post*. http://www.phnompenhpost.com/national/budget-figures-released
Sokhorng, C. (2016, August 25). Battambang rice-farming families fear floods. *Phnom Penh Post*. http://www.phnompenhpost.com/national/battambang-rice-farming-families-fear-floods
Sondergaard, L. M., Luo, X., Jithitikulchai, T., Poggi, C., Lathapipat, D., Kuriakose, S., Martin, S., Eduardo, M., Reungsri, T., & Mohib, S. A. (2016). *Thailand getting back on track*. Bangkok: World Bank. https://openknowledge.worldbank.org/bitstream/handle/10986/25740/Thailand-SCD-final-public-11162016.pdf?sequence=1
STAR Kampuchea. (2013). *Shadow report on women's land rights in Cambodia*. Phnom Penh.
Tinh, L. D., & Long, H. H. (2013). *Vietnam in ASEAN and ASEAN in Vietnam* (Asia-Pacific Bulletin 242). Washington, DC: East-West Center. http://scholarspace.manoa.hawaii.edu/handle/10125/30982
Tuoitre News. (2012, February). *HAGL starts work on $40 Mln Airport in Laos*.
Tuoitre News. (2013, November 2). Challenges await Vietnamese firms after TPP. http://tuoitrenews.vn/news/business/20131102/challenges-await-vietnamese-firms-after-tpp/29176.html
UN Capital Development Fund. (2010). *Local development outlook: Cambodia*. New York. http://trove.nla.gov.au/version/170674793
USAID Burma. (2013). *Background Paper No.4: Food and nutrition security in Burma*. http://www.usaidbest.org/docs/burmaBackground.pdf
Vdassani. (2013, June 14). *Trade opportunities between ASEAN and SAARC*. http://southasia.ifpri.info/2013/06/14/trade-opportunities-between-asean-and-saarc/
VietnamNet Bridge. (2013, March 10). *How many billionaires are there in Vietnam?*. http://english.vietnamnet.vn/fms/business/68185/how-many-billionaires-are-there-in-vietnam-.html
Vietnamnews.Vn. (2017, June 2017). *Climate change forecast to shrink rice yield*. http://vietnamnews.vn/environment/378426/climate-change-forecast-to-shrink-rice-yield.html
Walker, A. (2007, October 17). Contract farming. *New Mandala (blog)*. http://www.newmandala.org/contract-farming/
Wells-Dang, A. (2013). *Promoting land rights in Vietnam: A multi-sector advocacy coalition approach*. Washington DC.
Win, S. P., & Kean, T. (2013, November 3). Revised farmer protection law quietly comes into effect. *The Myanmar Times*. https://www.mmtimes.com/in-depth/8688-revised-farmer-protection-law-quietly-comes-into-effect.html
Woods, K. (2013). *The politics of the emerging agro-industrial complex in Asia's 'Final Frontier': The war on food sovereignty in Burma*. https://www.tni.org/en/briefing/politics-emerging-agro-industrial-complex-asias-final-frontier
World Bank. (2011, February 25). *Cambodia: Villagers enjoy cheaper, reliable electricity*. http://www.worldbank.org/en/news/feature/2011/02/25/cambodia-villagers-enjoy-cheaper-reliable-electricity
Zola, A. (2009). *Scoping study on cross-border agribusiness in Lao PDR: Focus on Champasak Province*. World Bank. http://documents.worldbank.org/curated/en/782861468263761056/Trade-development-facility-scoping-study-on-cross-border-agribusiness-in-Lao-PDR-focus-on-Champasak-province
Zorya, S. (2016, May 27). *Unleashing Myanmar's agricultural potential*. http://blogs.worldbank.org/eastasiapacific/unleashing-myanmar-agricultural-potential

Chapter 16
Negotiating Water Institutions in the Đồng-Nai River Basin, Vietnam: Unstable Balance Between Conservatism and Innovation

Huynh Thi Phuong Linh and Olivier Tessier

Abstract In a context of increasing struggles over access to water and environmental challenges, this chapter analyzes institutional efforts to support integrated water resources management in the framework of the Phước-Hòa Water Resources project (funded by Asian Development Bank [ADB] and Agence Française de Dévelopment [AFD] in 2006–2018) that concurrently aimed to increase the Dầu-Tiếng Reservoir's capacity and water infrastructure. The chapter investigates the institutional development process for management of the Dầu-Tiếng water resources system in a context of multiple uses and increasing environmental challenges to water quality and quantity, especially under the current obvious and severe effect of salinity intrusion. The chapter then sheds light on local dynamics surrounding the efforts to apply participatory irrigation management (PIM) and to increase agricultural productivity in Tân-Biên irrigated areas developed under the Phước-Hòa project. Through the case of the Phước-Hòa project, the chapter emphasizes the difficulty of defining the right unit for action in a context of shifting environmental and social conditions and the need to allow more flexibility and time in building water institutions. This chapter also seeks to contribute an empirical case study to the large and increasingly important literature on water governance in Southeast Asia and other developing areas. The chapter builds on the results of a multidisciplinary research project on local water governance in the Đồng-Nai basin. The research project's results, while contributing to the pool of scientific knowledge, will help the AFD reflect on the factors driving and limiting the legacy of the development project.

Keywords Local water governance · Multiple-used water system · Institutional process · Agricultural development · Participatory Irrigation Management · Dong Nai river basin

Huynh Thi Phuong Linh (✉) · O. Tessier
Ecole Française d'Extrême-Orient, Hồ-Chí-Minh City, Vietnam
e-mail: hl@vd-office.org

M. A. Stewart, P. A. Coclanis (eds.), *Water and Power*, Advances in Global Change Research 64, https://doi.org/10.1007/978-3-319-90400-9_16

16.1 Introduction to the Complex and Multifunctional Đồng-Nai Basin

Đồng-Nai basin (Fig. 16.1), with a catchment of 36,481 km^2, is the biggest basin entirely within the boundaries of Vietnam and among the three largest river basins in Vietnam. It contains the principal area of urban concentration and industrial development of the country, including the Hồ-Chí-Minh megacity, Lâm-Đồng, Đắk-Nông, Bình-Phước, Bình-Dương, Tây-Ninh, Đồng-Nai, Bình-Thuận, Long-An, and Bà-Rịa–Vũng-Tàu. The basin population is about 19 million, constituting 21% of the national population, of which about 48% live in rural areas (data 2015, Nguyen Vu Huy 2016). The area accommodates a productive economy (constituting 48% of the national economy), which in turn places huge and complex demands on the water supply, including domestic water and water for industrial processes, for hydropower production, for irrigation, and for combating the intrusion of salinity. Because of its key social, economic, and political position in the country and the complex character of water governance, integrated water governance in the Đồng-Nai River basin becomes important and urgent.

The Đồng-Nai basin is made up by six smaller river basins: Vàm-Cỏ-Tây, Vàm-Cỏ-Đông, Sài-Gòn, Bé, Đồng-Nai, and La-Ngà (Fig. 16.1). The flow in the basin relies on a rainfall regime,[1] so it varies in space and time (Đỗ Đức Dũng et al. 2014). There are two seasons: the rainy season, from May/June to November, and the dry season. The rainy season correlates to the monsoon season (*mùa*) and provides most of the annual rainfall. The basin is closely linked to smaller rivers in the coastal area. The low-lying delta is strongly influenced by tides from the East Sea.[2] The basin has faced the influence of salinity, with a peak influence often falling in the last month of the dry season (April or May), when the outflow from the basin is lowest.

The basin possesses different elevations along its reach, from mountainous to delta to coastal areas. It is also characterized by a complex typology and geology, with diverse soil types, composed mainly of ferralsol/oxisol (about 3 million hectares, 51% natural area), acrisols (about 1.2 million hectares), and another 8 soil types. About 26% of the basin soil is unfavorable for farming: soil that is sandy or salty, thionic fluvisols or acid sulfate soil, and leptosols (gravelly or stony). The forestland in the region has decreased to 2 million hectares in 2014 (Đỗ Đức Dũng and Nguyễn Ngọc Anh 2015).

[1] Precipitation is 2000 mm on average and varies significantly across areas. While the estuarine area of Đồng-Nai River, the Vàm-Cỏ-Đông basin, and lower Đa Nhim have lower rainfall (1500–1700 mm annually), rainfall is 2500–3000 mm at the midstream of Đồng-Nai River, the upper Bé River, and the upper and midstream La-Ngà River.

[2] Due to the large amplitudes of the tide (up to 3.5–4.0 m along the coast) and the low riverbed slopes, the tide enters deeply into rivers and canals. The tide effect can reach up to the foot of the Trị An waterfall of Đồng-Nai River, 132.8 km from the sea; to the Dầu-Tiếng dam site on the Sài-Gòn River, 184.4 km from the sea; and to the Cambodia border on the Vàm-Cỏ-Đông River, 208 km from the sea (Corderi 2011).

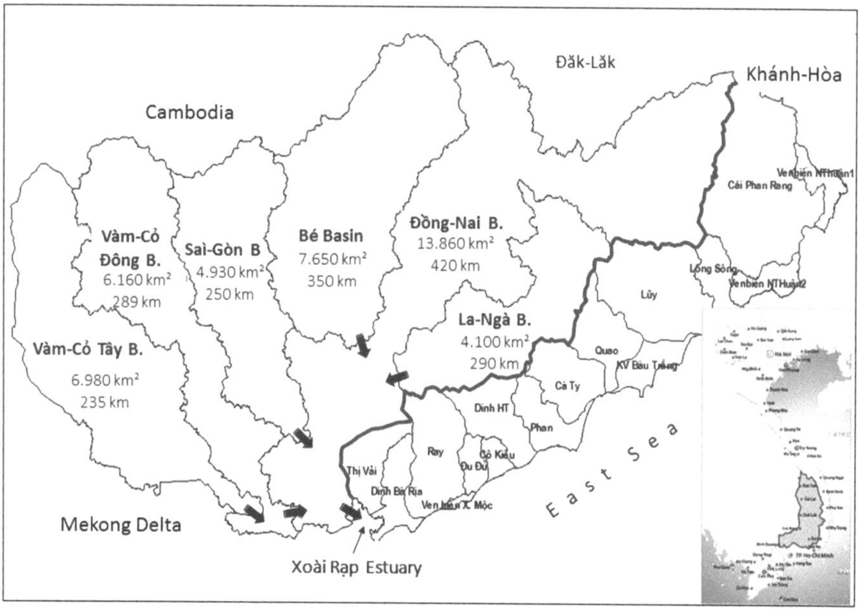

Fig. 16.1 The location of the sub-basins in the Đồng-Nai River basin and the adjacent coastal basins. (Source: Southern Institute for Water Resources Planning 2016)

Water development in the Đồng-Nai basin in the last four decades has been dominated by infrastructure investments, especially those devoted to the development of hydropower. Because of the favorable condition for cascades, the basin has been subjected to exploitation to produce energy for a long time, beginning with the first Đơn Dương/Đa Nhim project in the 1960s funded by the Japanese. By 2015, the basin already hosted 12 hydropower projects[3] (Fig. 16.2, Table 16.1). Many of the projects recently are under heated debate for their huge impact on the environment. For instance, projects ĐN6 and ĐN6A have been stopped because of the risk they posed to the Cát-Tiên National Forest. The Đồng-Nai 8 project has been replaced by five small projects (planned): Tà Lài, Phú Tân 1 and 2, Thanh-Sơn, and Ngọc-Định. Artificial reservoirs are also common in the small rivers and streams, such as Da-Tôn, suối Mây, suối Cả, Gia-Ui, and Cầu-Mới.

The seemingly endless endeavor to control nature has shaped the waterscape of the region. Since the establishment of the Dầu-Tiếng hydraulic system in the late 1980s, water has been stored in Dầu-Tiếng Reservoir in the upper Sài-Gòn River and redistributed for the basin's multiple uses. Since the development of Dầu-Tiếng, "reservoir water" has become a notion in everyday life in Tây-Ninh to distinguish it from river water or groundwater. Also, since then, the government of Vietnam and its development partners have striven to improve the capacity and management of

[3] Three hydropower projects are on Bé River, 7 on Đồng-Nai River, including the biggest reservoir Trị An—14,890 km² with 2.76 billion m³; and two on La-Ngà River.

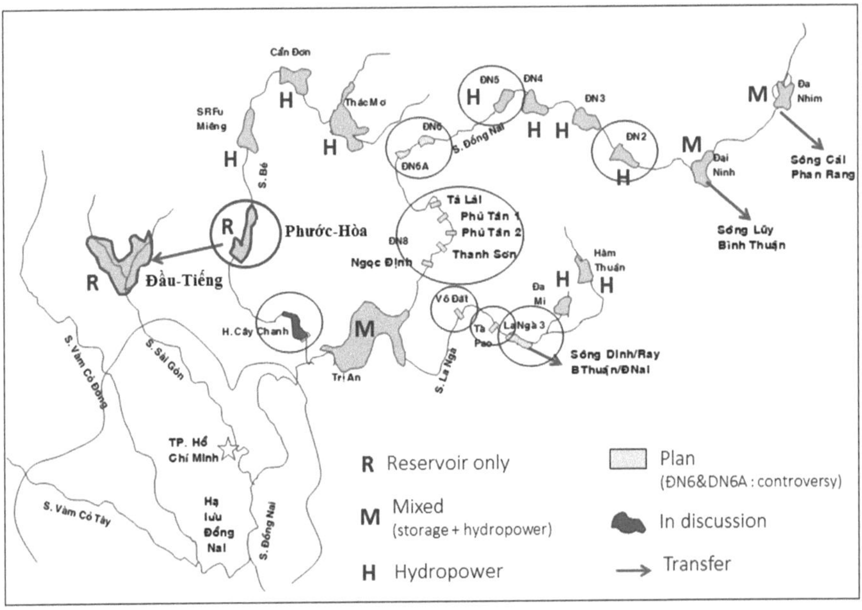

Fig. 16.2 Water projects on the Đồng-Nai River basin. (Source: Southern Institute for Water Resources Planning 2016)—In yellow color: planned projects, red color: newly proposed project) (Color figure online)

the Đồng-Nai basin, and the Dầu-Tiếng system is at the heart of this effort. This has shaped Tây-Ninh's irrigated area by heavy water control compared to other areas in the basin and those in the Mekong Delta.

Despite the enormous investments in infrastructure, the basin still faces the challenge of management coordination. The total water demand in the basin is increasing and varies in time and space. Higher demand often falls in the dry season when water availability in the basin is at its lowest. During that period, coordinating water use for different purposes becomes a challenge. Among other factors, climate change is assessed to have an influence, such as increasing the chance and intensity of extreme weather events, increasing the differentiation of surface flow between dry and wet seasons, worsening drainage issues in Hồ-Chí-Minh City because of extreme rainfall and high tides, and higher floods in general (Đỗ Đức Dũng et al. 2014). Corderi (2011) also comments that salinity concentration over time will increase because of decreasing water runoff in the dry period and because of the potential risk of sea level rise. In recent years, demand for water has dramatically increased along with seasonal water shortages and progressive salinization of Vàm-Cỏ-Đông and Sài-Gòn Rivers. As Nguyễn Ngọc Anh and Đỗ Đức Dũng (2015) assert:

> Regarding the Đồng-Nai River basin, it is the system where there accommodates many big cities with big populations and key industrial zones and a low delta affected by the tidal regime and salinity intrusion. There is also relative inter-connection in terms of hydrology to the downstream Mekong and the coastal area, which is often in the quest for more and more water. Thus, the solution for water balance in this basin is getting more difficult.

Table 16.1 Hydropower projects in Đồng-Nai basin and their parameters

River/tributary	Year in operation	Type	Capacity hydropower (MW)	Regular water level (m)	Dead water level (m)	Maximum flooding water level (m)	Designed flooding water level (m)	Full capacity (million m^3)	Usable volume (million m^3)
Bé									
Thác Mơ	1994	Hydropower	75	218	198	220.8		1360.00	1250.00
Cần Đơn	2003	Hydropower	77.6	110	104	112.36	111.44	165.50	79.90
Srok-Phu-Miêng	2006	Hydropower	51	72	70	73.6	72.75	99.30	28.57
Phước-Hòa	2012	Other (power)		42.9	42.5	48.25	46.23	13.72	2.45
Đồng-Nai									
Đơn Dương/Đa Nhim	1964	Hydropower	160	1042	1018	1043.2		165.00	155.14
Đại Ninh	2008	Hydropower, other	300	880	860	882.6		319.77	251.73
Đồng-Nai 2	2015	Hydropower	70	680	665	681.61		281.00	143.40
Đồng-Nai 3	2011	Hydropower	180	590	570	593.24		1690.10	891.50
Đồng-Nai 4	2012	Hydropower	340	476	474	479.24		332.10	16.40
Đồng-Nai 5	2015	Hydropower	150	288	286			106.33	8.35
La-Ngà									
Hàm Thuận	2001	Hydropower	300	605	575	607.5	606.7	695.00	523.00
Đa Mi	2001	Hydropower	175	325	323	327.4	327.05	140.80	11.60
Trị An	1991	Hydropower, other	400	62	50	63.9		2764.70	2546.70
Sài-Gòn									
Dầu-Tiếng	1985	Hydropower, other	…	24.4	17	26.92	25.1	1580	1110

Sources: Van Duc and Gupta (2010), Gov. Decision of Prime Minister 471/QĐ-TTg on issuing the inter-reservoir operation procedure on Đồng-Nai River basin. *Projects under construction* (downstream Hàm Thuận-Đa Mi): La-Ngà 3 reservoir for water supply; Tà Pao dam for water regulation for irrigation and other purposes. *Disputed projects* (in Đồng-Nai river between ĐN 5 and Trị An): ĐN6 and ĐN6A due to the risk posed to Cát Tiên National Forest. *Planned projects*: five small hydropower dams: Tà Lài, Phú Tân 1, Phú Tân 2, Thanh Sơn, Ngọc Định (replacing the one big Đồng-Nai 8)

Therefore, in this basin, optimizing water use for hydropower as well as other purposes, such as irrigation, domestic and industrial uses, and environmental flows, is a challenge.

16.1.1 The Basin's Agriculture Landscape

The basin has undergone substantial rural and agricultural transformation in recent decades. The data confirms that 950,000 ha are devoted to agriculture, and two-thirds of this produces three seasonal crops (data 2011, Nguyen Vu Huy 2016). Areas like Tây-Ninh have witnessed different periods of agricultural landscape, with sugarcane in the 1990s under the support of a sugar factory, and cassava in the last 10 years because of the tapioca market and processing facilities, and the recent boom of rubber plantations since the late 2000s. The transformation of the agricultural landscape is more often than not driven by market forces and dynamics of the farming society (a focus on crops demanded by booming markets). The agricultural landscape has also been transformed by the loss of farming land to urban and industrial expansion.

Natural conditions such as climate, soil type, and the physical condition of water infrastructure continue playing their parts. The climate or the rain indeed has had a close relationship to the choice of farming systems; for instance, when irrigation infrastructure was not in place (canal or pump), farmers in Tây-Ninh planted only for the rainy season (*vụ mùa*). Such dependency of the farming system on reliable patterns of climate continues today, with two or three seasons a year and with a more diverse cropping patterns other than the dominant rice crop. However, it is recently challenged by the unpredictability of irregular rain and drought. For instance, an unusual dry season in 2016 followed by a prolonged rainy season during the winter-spring season of 2016–2017 (heavy rains continued until December 2016) confused farmers in whether to start the cassava crops, which need at least 7 months of growth before the next rainy season.[4]

The agricultural landscape of the basin continuously changes and is shaped by the natural conditions of climate, land and water, the irrigation infrastructure, the level of urban and industrial expansion, and last but not least the market, together with the interference of private companies. All factors affect the landscape simultaneously. One challenge might create the ground for another alternative. For example, the delay of planned designed cassava cultivation due to the prolonged rain in 2016 motivated many farmers to switch to sugarcane, with financial and technical support from the sugar company, Thanh Thành Công. Other farmers opted for rice cultivation during the dry winter/spring, counting on the potential of a newly constructed canal system. Therefore, the intertwined influence of the different factors

[4] Main casava season runs from October to June or July for lowland (*mỳ ruộng*) that is combined with one rice crop during the rainy season; or from October to October in higher land (*mỳ rẫy*).

makes farming systems and activities more complex and unpredictable. It is difficult if not impossible to point out models for water governance and the management of agriculture landscapes in this basin.

16.2 Negotiating Water Institutions in Dầu-Tiếng Water System

Dầu-Tiếng is Vietnam's largest water reservoir and mainly serves the purposes of irrigation and domestic-industrial usages (*hồ thủy lợi*), distinguished from those for hydropower purposes. The reservoir stores water in the rainy season to supply in the dry season. It has a catchment of 27 km^2, a maximum water level at 26.92 m, a dead level at 17 m, and a volume of about 1.6 billion m^3. The development of the Dầu-Tiếng irrigation project (1981–1985) was supported by a loan from the World Bank through the International Development Association ($60 million) and with the funding from the Kuwait Fund (OPEC) and the Netherlands government ($10 million each) (World Bank 1978). The project, designed by the Ministry of Water Conservancy of Vietnam and constructed by the state-owned Hydraulic Works Company 9, was part of the national program to rehabilitate existing hydraulic works and small irrigation schemes (supported by the World Food Program and non-governmental organizations, including Church World Service of the United States) (ibid). The project, in continuity with the commitment of the former Republic of South Vietnam, was a significant expression of extensive international support for Vietnam's nation-building plan after reunification.[5] The project's specific objectives were to increase cropping intensity and production, to make Vietnam self-sufficient for food (World Bank 1989). After the appraisal in January 1978, the project objective was narrowed from 172,000 ha of irrigated land to gravity-irrigating 42,000 ha in five districts in the southern part of Tây-Ninh Province and about 14,000 ha by pumping irrigation for Cu Chi District of Ho Chi Minh City. The other potential 30,000 ha of irrigated land in Tây-Ninh was planned. The project included an earth dam and its accessories on the Sài-Gòn River and an irrigation system starting from West and East Canals, including an adequate drainage network and inspection and access roads. Part of the project objective was also to relieve the severe unemployment problem in the south (ibid).

Since the 1980s, the irrigation system in Tây-Ninh has been maintained and improved with national and provincial programs or donor projects such as Vietnam Water Resources Assistance Project (VWRAP; 2004–2012).[6] Today, the Dầu-Tiếng system supplies water for irrigation in Tây-Ninh, Hồ-Chí-Minh City, Bình-Dương,

[5] Other International Development Association projects in Vietnam involved coal mining, power, and railway workshop projects and industrial rehabilitation projects.

[6] VWRAP is funded by the World Bank to modernize and increase the productivity of Vietnamese agriculture, improve the management of water resources, and reduce dam safety risks. Tây-Ninh area was within the scope of capacity building in the PIM application for the project.

and Long-An; domestic use in Hồ-Chí-Minh City; salinity flushing for Hồ-Chí-Minh City and Long-An; and for environmental flow.

The Dầu-Tiếng system, since its completion in 1985, is managed by a bureaucratic system of Dầu-Tiếng companies (called Dầu-Tiếng–Phước-Hòa Irrigation Engineering Integrated Complex since 2006) in cooperation with provincial irrigation management companies (IMCs) and their line offices at the district level. Despite the effort to develop consistent practices, the Dầu-Tiếng system is continually challenged by practices and negotiations, by various actors who differ not only in terms of interest, scale and power but also because of varying ecological and agricultural changes. As the main actor in operation and maintenance (O&M),[7] the Dầu-Tiếng Company has worked at the interface of bureaucratic state structure and business practices to allocate water and mitigate the risks of structural damage from floods, drought, and the more recent salinity intrusions. The Dầu-Tiếng company is under direct supervision of the government and thus reports to the Ministry of Agriculture and Rural Development (MARD).

To manage and operate the Dầu-Tiếng Reservoir to meet several targets for downstream safety and reservoir safety, a committee (*hội đồng hệ thống*) was established. The Management Committee for the Exploitation of the Dầu-Tiếng Water System makes planning decisions for system O&M and for cooperation between agencies at Dầu-Tiếng and at the province level.[8] This committee is led by MARD and includes the representatives of provinces that either use or are affected by the water of the reservoir (the chairman or vice chairman of the provincial People Committee[9]). The Dầu-Tiếng company represents the standing member of the committee, in charge of managing databases, advising the chairs, and executing rescue activities in disaster situations.

The Dầu-Tiếng water system is operated by a form of contract between Dầu-Tiếng company and the provincial Irrigation Management Company (IMC), and between the IMC (and its offices) and local irrigation units, in form of either cooperatives (*hợp tác xã*) or irrigation groups (*Tổ-Thủy-Nông/Tổ hợp tác*). There are three identical steps in the operation of the system at the Dầu-Tiếng company level and the local level (Fig. 16.3).

---

[7] Operation and maintenance (O&M) is a term introduced officially by the Food and Agriculture Organization of the United Nations (Snellen 1997). "Operation" is defined as the task to operate the system to supply irrigation water, and "maintenance" involves efforts to keep the scheme in good working order. O&M and financial control comprise the three main tasks to keep a system running.

[8] According to decision 298NN-TCCB/QĐ of the minister of MARD on 8 April 1996 on the members and duties of the management committee for the exploitation of the Dầu-Tiếng water system. The committee is currently renamed the Management Committee for the Exploitation of the Dầu-Tiếng–Phước-Hòa Water System

[9] The Peoples Committee is the governmental umbrella office at the provincial level where the top leaders sit and make decision regarding social-economic issues of the province. The Peoples Committee is advised by the line offices of the ministry, for instance, the Department of Agriculture and Rural Development (DARD) under MARD. The department holds a dual-subordination position by reporting to both provincial Peoples Committee and MARD at the central level. The same structure applies to the district and commune level.

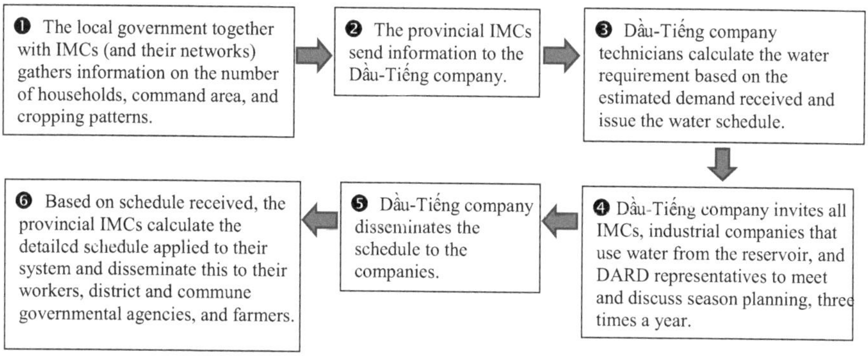

Fig. 16.3 Irrigation operation steps between Dầu-Tiếng company and the provincial irrigation management companies (IMCs), and between the IMCs (and their offices) and local irrigation units

The economic contract between companies expresses a business form of operation, in which Dầu-Tiếng company is the service provider and IMCs and/or provincial authorities are the customers, and in turn the provincial IMCs are the service provider supplying water for the local authorities as their customers (the commune's People Committee represents the farmers in signing the irrigation contract with the water company). However, due to the close relationship between companies and governmental agencies at the same level, the operation of the system in practice manifests a hierarchical structure that is common in Vietnamese governmental practice. A hierarchical structure also appears in how MARD can direct the Dầu-Tiếng company. Such interference of the power of the state structure, on the one hand, can support cooperation between companies (water supply and hydropower); on the other hand, the market mechanism that decides the responsibility and accountability of the service providers is diminished. This is especially true for the case of irrigation, given the exemption of irrigation service fees (ISF) since 2008. Nguyễn Trung Dũng (2016) described the ISF policy in Vietnam as a shift from the rather stable market mechanism for 25 years (1962–2003), where farmers used the service and paid when they were satisfied with the service, to the period when the user payments are partly exempted and the state partly provides ISF (2003–2007—Decree 143), to the stage where the state pays to IMCs the cost of service provided to the farmers (Decree 154/2007 and Decree 115/2008), which Nguyễn calls "the third party paying the service".[10]

[10]"Third party paying the service" means "the company provides water service, the farmers use water for irrigation, the government pays" Nguyễn Trung Dũng (2016). Since the Decree 115, farmers stopped paying any fee related to irrigation, and the IMC and its line agencies have been operated and managed with a state budget for ISF compensation (*cấp bù thủy lợi phí*). ISF exemption has resulted in various and contradictory financial impacts to the irrigation system. The policy aimed to reduce financial burdens for poor farming households and stabilize the operation of the water company. According to provincial leaders, those benefits imply better living conditions for both farmers and the staff at local irrigation management organizations, and consequently results

16.2.1 Phước-Hòa Project: The Solution and the Attached Challenge

In recent years, demand for water has dramatically increased along with seasonal water shortages and the progressive salinization of the Vàm-Cỏ-Đông and Sài-Gòn Rivers. Thus, from 2006, in realizing a previous plan for water diversion from the Bé River to the Dầu-Tiếng Reservoir, the Phước-Hòa project (2006–2018, Fig. 16.4) has been implemented by MARD with loans and funds from the Asian Development Bank (ADB) and the Agence Française de Dévelopment (AFD). The project aims to "develop the water resources of the Bé River and transfer it to Sài-Gòn and Vàm-Cỏ-Đông rivers for irrigation to increase agricultural production, provide water for Hồ-Chí-Minh city, and control saline intrusion thereby providing social, economic and environmental benefits"(ADB factsheet). The project consists of two parts: support for institutional and integrated development, and construction of water resources infrastructure. The project's irrigation target aims to supply more water for irrigation through the transfer canal from Phước-Hòa barrage to Dầu-Tiếng Reservoir (for the two communes with lands adjacent to the canal), to supply more water for the West Canal system (to Tân-Biên perimeter in Tây Ninh province), and the East Canal system (to Củ-Chi of Hồ-Chí-Minh City, and Đức-Hòa of Long-An province). Project organization is described as follows:

> The project executing agency (EA) is MARD. Project implementation is the responsibility of the Hydraulic Project Investment and Construction Management Board No. 9 (ICMB9), and Departments of Agriculture and Rural Development (DARDs) of the four project provinces. ICMB9 provides an interface with the ADB and AFD, and is directly responsible for management and construction of the Phước-Hòa Barrage, Phước-Hòa–Dầu-Tiếng transfer canal, and main canals for the irrigation systems. DARDs and their Provincial Project Management Boards (PPMBs) have responsibility for implementation management of design and construction of the downstream canal systems, development of irrigation areas and implementation of OSDPs [on-farm and social development programs]. The role of ICMB9 is to provide overall management assistance and coordination, and to provide instruction to the DARDs and PPMBs with assistance from project implementation consultants, Black & Veatch International in the first phase and SCP-HEC II [Vietnamese Consulting Joint Stock Hydraulic Engineering Company 2] in the second phase (started in April 2016) (AFD Factsheet).

The designed irrigated areas of Phước-Hòa project had been adjusted downward from more than 48,000 ha in 2003 to almost 30,000 ha in 2008 for three perimeters (Bình-Long of Bình-Dương, Đức-Hòa of Long-An, and Tân-Biên of Tây-Ninh) and a small part of Thái-Mỹ of Củ-Chi District (Hồ-Chí-Minh City) and to about 17,500 ha in 2014. Adjustments had been made for the expansion of industrial land

in better efficiency in irrigation services (Report on Implementation assessment of Decree 115 and Circular 65, CPIM-AFD 2012). However, the exemption lowers the sense of accountability not only for farmers and other service users but possibly also in the IMCs and the service providers. When IMCs are financially secured by the state's ISF compensation, IMC staff no longer rely on the fee paid by the farmers, and farmers are using water for free (Dang Minh Tuyen 2010:05). That induces a risk of low accountability from the IMCs in providing service and farmers in using water.

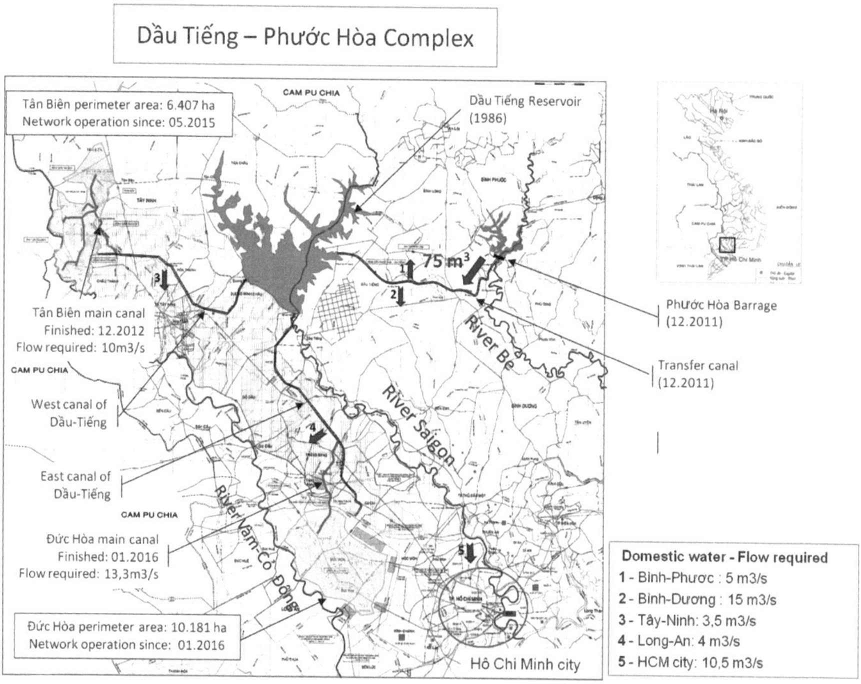

Fig. 16.4 Phước-Hòa water project. (Source: Asian Development Bank and Agence Française de Dévelopment)

in Bình-Dương during the 2000s (canceling the Fourth Bình-Long perimeter), and for the strong shift of cropping pattern toward cash crops (rubber in Tây-Ninh); part of the adjustment was also because of redefinition of infrastructure capacity during designing, consulting with local stakeholders, and construction periods. Up to today, data of the real irrigated areas are only estimated and the real potential of each perimeter is not (able to be) measured.

The millions-USD Phước-Hòa project marks a continuity in the efforts to maximize the usage of natural resources for development, for the demands of a tremendous increase of population in the region, and for coping with unpredictable climate conditions.

Since the implementation of the Phước-Hòa project, the task of operating the Dầu-Tiếng Reservoir is getting more difficult, according to Dầu-Tiếng company's personnel in charge (interview date 06.6.2017). First, having more water from the Phước-Hòa Barrage means including more irrigated areas in the management scheme of the Dầu-Tiếng company (designed at 16,500 ha). During the dry season, the chance to get extra water for Dầu-Tiếng is uncertain because the hydrometeorology features of the whole Đồng-Nai basin are affected in the same way by the seasonal flow; when Dầu-Tiếng is short of water, so is Phước-Hòa Reservoir. Thus how to manage the water system in order to maintain a targeted water level is difficult to calculate. Water availability for the Phước-Hòa Barrage relies on the operation of three upstream hydropower reservoirs, Thác Mơ, Cần Đơn, and Srok Phu Miêng, each of which often has purposes other than irrigation and domestic and industrial water supply. The director of the Dầu-Tiếng company confirms the importance of inter-reservoir operation as regulated by various legal documents defining an inter-reservoir operation procedure.[11] However, contrasting benefits can prevent the procedure from working. In the dry season, for example, hydropower operators want to store water for power production. Also, the different lines of management, with the Ministry of Industry and Commerce managing hydropower while MARD manages the Dầu-Tiếng Company, poses the risk of conflicting interests to inter-reservoir operation and cooperation. In case of crisis, both types of reservoir suffer from a lack of water or too much water,[12] and although a cooperative mechanism is applied, it is not yet effective. Often, Dầu-Tiếng company is at a disadvantage because it is located downstream, with less leverage compared to the upstream power companies.

The Dầu-Tiếng company prepared an operation procedure for the Dầu-Tiếng Reservoir in the new context, which was sent to MARD for approval in 2016. Also, in March 2016 the prime minister issued the decision 471/QĐ-TTg, which defined the procedures for inter-reservoir operation in the Đồng-Nai basin. The challenges of system management and the risks deriving from uncertain climate and increasing demand have profoundly tested this water management system. Institutionally, the lack of effective coordination between actors with contrasting interests and the lack of technologies and/or mechanisms for inter-reservoir operation have made the optimization of water usage in the basin even more difficult.

16.3 Bargaining Institutions at Local Level: The Case of PIM in Phước-Hòa Project

The Phước-Hòa project is now in its final year (2018). The project's phase 1 (2006–2012) managed to build the infrastructure for the basin transfer component (Phước-Hòa Barrage and transfer canal, handed over December 2011) and the construction of Tân-Biên main canal based on the upgrade of the TN10 canal of the system provided by the West Canal of Dầu-Tiếng system. It also implemented the related institutional program called On-farm and Social Development Program I (OSDP I). The second phase of the project (2012–2018) continues the institutional development of OSDP at the same time as the construction of Đức-Hòa main canal and the design and construction of primary, secondary, and tertiary canals (PST) in both irrigated command areas. Institutional efforts to support integrated water resources

[11] Decision 471/QĐ-TTg dated 24 March 2016 on issuing the procedure for inter-reservoir operation in Đồng-Nai basin.

[12] While not addressed directly in this chapter, the issue of abundance of water is also a prominent concern in the basin. The risk of structure failure and downstream inundation are the two main challenges concerning managers in operating the reservoirs.

management in the framework of the Phước-Hòa Water Resources project is carried out in the OSDP. The program consists of two activities: a social support program for the people affected by the project, and an on-farm development program, which aims to support beneficiaries of the project by participatory PST design, the establishment of a management model for the on-farm system, and the agriculture support program. Beneficiaries are defined as the farmers who will receive irrigation water for their fields and thus supposedly gain benefit from the project.

OSDP packages were also made to two phases and carried out by a group of national OSDP expert consultants. The consulting contract structure between OSDP consultants and the PPMB helps the decision makers manage the project according to a timeline. The presumption is that the combined activities will result in the project's objective of infrastructure effectiveness and a good foundation for PIM development in the area.

16.3.1 The Manifestation of PIM in Tân-Biên Irrigated Area of Phước-Hòa Project

16.3.1.1 PIM in Development Project: An International Norm

Since the 1980s, irrigation modernization refers to the idea of managing irrigation and drainage[13] in a way that is both financially feasible and sustainably efficient. Among other conventional technocratic measures, irrigation management transfer has become a tendency since its first introduction in Mexico in the 1980s and then in the Philippines in the 1990s. The transfer broadly means to "reduce public expenditure whilst increasing farmer participation in the management of the irrigation systems" (Burton et al. 2002:7). The new form of water management attaches or somewhat covers the formation of water user associations/organizations (WUAs/WUOs). The focus on farmers' participation in irrigation management (PIM) was discussed at the same time as WUAs and irrigation management transfer, and WUAs are considered to be a subset of PIM (Van Vuren et al. 2004). The World Bank (1996, cited by Van Vuren et al. 2004) defines PIM as "the involvement of irrigation users in all aspects and all levels of irrigation management", in which "involvement" is flexible, ranging from light involvement like information sharing, consultation, joined assessment of problems to real involvement like decision making, collaboration, and full say by the water users; "all aspects" include initial planning and design of new irrigation projects or improvements, as well as the construction, supervision, financing, decision rules, operation, maintenance, monitoring, and evaluation of the system; "all levels" means tertiary, secondary, and main system levels, as well as project and sector levels.

[13]The term "drainage" is often missing in the whole discussion about water for agriculture. However, it is often understood to be included, as drainage is an essential component of an irrigation system.

Since the 1990s, Vietnam, like many other countries, has adopted the idea of irrigation modernization using these approaches or terminologies: PIM in combination with WUOs or WUAs, and irrigation management transfer. The country indeed adopted different ideas and models from the Food and Agriculture Organization of the United Nations and other global tools in the investment for the software component of the irrigation systems (Nguyen Xuan Tiep 2008:40). Among the components of modern irrigation management, PIM is especially favored by the state agencies and projects funded by international organizations such as the ADB and the AFD[14] and the non-governmental organizations. PIM is translated in Vietnam in the broad sense: irrigation management with farmers involved, who participate in planning, designing, investing, and constructing and managing hydraulic works (Nguyen Xuan Tiep 2008a:26). In terms of legal recognition, PIM and water user group (WUG) could be related to

- The ordinance on exploitation and protection of hydraulic works;
- The "Framework for PIM Development in Vietnam" no. 3212/BNN-TL (its draft was already discussed since 1997 under the guidance and financial support from donors); and
- Circular no. 75/2004/TT-BNN on "Guidance for establishment, consolidation and development of WUOs."[15]

In fact, the two latter documents were issued by MARD after the regional workshop Participatory Irrigation Management Pathway in Progress in Viet Nam in 2004. The workshop was financed by ADB, Denmark's development cooperation (DANIDA), the International PIM Organization, and the World Bank. Also in 2004, the country's development of PIM reached a milestone with the establishment of the Center for Participatory Irrigation Management (CPIM) under the Vietnam Academy for Water Resources (VAWR)—both CPIM and VAWR are the OSDP consultants of Phước-Hòa project, with separate teams for two irrigated areas.

In addition, the Vietnamese PIM expert Nguyễn Xuân Tiệp (2008a) has argued that PIM has already existed within the social dynamic in northern Vietnam in the past. It was manifested in the participation of local people in constructing and managing small-scale irrigation systems in the north. That form of collective organization (either voluntary or not) only existed in the construction of small irrigation or drainage canals at the village level. Besides, the major works for the containment of the Red and the Thai Binh river basin, which took place during the eleventh century,

---

[14] In an assessment of PIM and related projects in Vietnam, Nguyen Xuan Tiep (2008:81) stated that PIM has mostly been established as a requirement in certain projects, and each project has a different approach to implement PIM. Therefore, those projects are "biased to achievements" and are hardly effective. Many PIM models return to starting points after the completion of projects regardless of the projects' scale and fund sources (ibid:81).

[15] WUAs/WUOs in Vietnam are very diverse in forms and names. There are several of each kind ranging from cooperatives and associations/groups to village organizations (for more detail, see Nguyen Xuan Tiep 2008:81–82). Except the cooperatives, which have been organized according to cooperative law – 18/2003/QH11, the associations or groups have no standard formulation and operation.

resulted from the massive mobilization of military troops and the obligation of the inhabitants to contribute labor (public duty) (Tessier 2011).

There also existed the culture of mutual help in building houses, ploughing, and transplanting, which were organized in associations (Phường, Hội, and Yểng) (Nguyen Xuan Tiep 2008c). In particular, the traditional organization for labor exchange, based on the principle of voluntary service, was applied during the collectivization period (1960s–1981). Preceding the complete form of forced integration of all peasant households into agricultural production cooperatives, which was finalized in the 1960s, northern rural households had engaged in the "labor-exchange group based on work point" (*tổ đổi công bình công chấm điểm*). Under the mutual help model, groups (*tập-đoàn*) specializing in water management were established. Those groups provided labor for the works related to irrigation, drainage, and particularly construction of hydraulic infrastructures. Such collective units were not founded on a voluntary basis; rather, they consisted of compulsory membership, which was demanded and controlled by the cooperatives (Fontenelle and Tessier 1997). Because of its non-voluntary nature that strongly affiliated with the collectivization policy, the groups gradually dissolved after the country's reform toward a market economy—*Đổi-mới* or renovation—in the late 1980s.

However, the spirit of the traditionally institutionalized norm of mutual help survived, yet to a lesser extent of organization compared to the collectivization period. Since de-collectivization in the 1980s, the participation of people has been promoted through the "the State and the population working together" policy (*Nhà nước và nhân dân cùng làm*) and the grassroots democracy policy: "people know, people discuss, people do, and people inspect."[16] Today, although there is no structured system or regulation for labor contributions, there still exists a system of contribution and cooperation at the local level to guide the contribution of people for public works, including the construction and maintenance of the hydraulic infrastructure. However, the extent of organization for social contribution to public matters varies; more often than not, it is far different in extent and form from the expectation of the policy makers.

In the region where the Phước-Hòa project is located and in the Tây-Ninh Province in particular, the situation is very different. In fact, the Tân-Biên irrigated area of the Phước-Hòa project was part of the "new economic zones." Since the reunification pf Vietnam in the 1970s, expanding the agricultural area, mainly for growing rice and peanuts, happened alongside the large-scale settlement program in the rural areas—the so-called new economic zones—where the unemployed from urban centers of the south and people from overpopulated northern provinces were accommodated. Because of the recent arrival of migrants of various origins, these did not possess a tradition of mutual help practices (e.g. social assistance, work exchange, or any other non-market forms of exchange). So, the setting up of collective organizations of water management structured around WUAs and WUGs represents a more complicated challenge due to the lack of a pre-existing social

[16] Ordinance on grassroots democracy—practicing democracy at commune, ward, and town 34/2007/PL-UBTVQH11, replacing Grassroots Democracy Decree 79/2003/NĐ-CP.

foundation. It implies the requirement for a different approach rather than a standardized or blue-print model.

By and large, the Vietnam case of irrigation modernization and PIM illustrates both the influence of international development agendas and their own traditions of national dynamics in irrigation management—the latter strongly influenced by regional diversity.

16.3.1.2 Participatory Approach in Phước-Hòa Project Through the OSDP

In the Phước-Hòa project, the application of PIM is characterized by

1. Enhancing irrigation efficiency, to transfer some infrastructure management to farmers and mobilize the accountability of the water users;
2. Focusing solely on water users; and
3. Water user organizations (WUOs) as a definite aspect of PIM.

The OSDP is designed as a form of consultancy in which the consulting firm provides service according to a contract with identical assignments. In Tân-Biên, phase I ran from October 2009 to July 2012, and phase II started in April 2015 and finished in September 2016, with an interval between the two phases during the PST construction period. The OSDP is guided by the PIM and WUG guidelines (2004) that were prepared by Black & Veatch International, the project consultants during phase I, and were based on the basic knowledge of the project's area:

> WUG is established on with relevant scope and organizational structure in line with features of facilities, level of management capacity, traditions, practices and demands of farmers. … The WUG shall have a management board consisting of a Chairman and Deputies. The Chairman is elected by the WUG's congress on the principle "one family, one vote" and is legally recognized by the responsible authorities. The number of Deputies will be determined in the congress. The Deputies will be introduced by the Chairman and voted by the congress. (PIM guideline 2004)

The contract in this case was signed between the VAWR as the consulting firm and the Tây-Ninh Province PPMB. The contract regulates quantitatively the activities to be done within a timeline. In the case of the Tân-Biên irrigated areas, the main activity of the on-farm development package of OSDP I was to consult with the farmers or beneficiaries regarding the design of primary, secondary, and tertiary canals (PST), especially about the direction and design of tertiary canals. In this process, the PST design was created by the consulting firm. OSDP consultants presented the design to the farmers, mainly through representative farmers in the community-based monitoring group (*Tổ giám sát cộng đồng*). OSDP II comprised training in WUG and PIM, establishment of WUAs, and information dissemination, as well as public meetings on constructing the on-farm canals, which is the responsibility of farmers. The construction of the PST system took place between OSDP I and II; thus, there was no intermediate agent for the continuous consultation for PST systems during construction period (Fig. 16.5). The feedback mechanism went

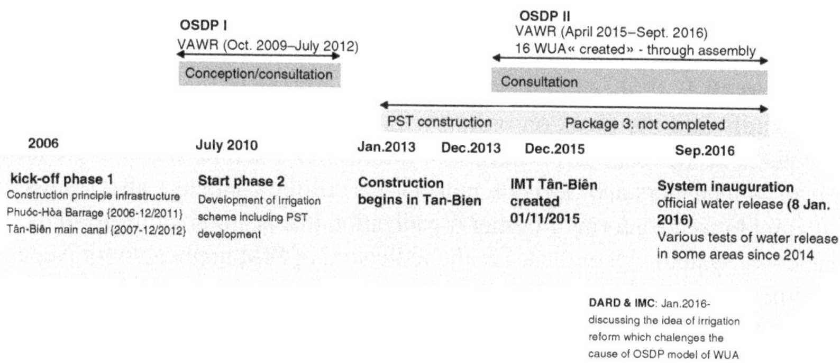

Fig. 16.5 Phước-Hòa project timeline in Tân-Biên irrigated area, focusing on the On-Farm and Social Development Program. (Source: Tessier et al. 2016)

back to the existing hierarchical structure of the state—from hamlet to commune, to district, to provincial level, and then to the PPMB of the project.

All informants interviewed during the OSDP II expressed that it was not easy to understand the information and the PST design discussed during the public meeting. Despite some field surveys where farmers joined the engineer in checking the direction of the tertiary canals, how local opinions were introduced into management practices was unclear, and the PST parameters (length, depth, width) were all unclear until the construction was carried out. The meeting did not provide farmers with an explanation of the nature of the system design, in which the principle of water energy and elevation played a big role.[17] This later resulted in complaints from farmers about all localized issues such as low-lying canals or different locations of the canal. In general, in OSDP I, the application of PIM in the PST designing process remained mainly at the level of information dissemination rather than participatory design. For both local cadres and "ordinary" farmers, the project worked through a predesigned model and training was far from sufficient for building capacity for a new irrigation management method (regime).

16.3.1.3 Water User Only Principle

In OSDP I, future water users remained passive spectators of the program's activities, including the consultation of PST design. In addition, although to a lesser extent compared to the other perimeter at Đức-Hòa, the local cadres such as commune state officers in Tân-Biên were excluded from the same activity. The application of such a rule derives from the principle of users only. The local cadres were not

[17] Because water runs from higher to lower, in designing an irrigation system the non-uniform elevations may create a situation where one field does not get water from the adjacent lower canal; rather, water will be delivered from a more distant canal farther upstream. Such technical logic is often neglected in conversations with the end users—the farmers.

candidates for community-based monitoring groups that oversaw the construction of PST. The user principle of PIM specifies that the project be managed by the users, for the users (WB 1996, cited in Van Vuren et al. 2004). In the project, the consultants squarely laid the focus on ordinary farmers under the assumption that other stakeholders at the local levels such as commune and hamlet cadres were already involved as managers and decision makers. Part of the challenge also results from the attempt to create a form of farmer organization that is autonomous from the state bureaucratic system. All of these are the indicators of PIM applicable for Nepal, the Philippines, and Mexico in achieving effective management and stronger accountability by the water users and at a lower cost for the state. However, the application of the same indicators has not created similar effects in the case of the Tân-Biên irrigated area.

Applying a no-state practice in the PIM mechanism in Vietnam introduces the risk of neglecting other main actors who play significant roles in shaping and deciding the success of PIM and the effectiveness of the system. Local cadres such as commune and hamlet cadres who are at the final position in the chain of governance in Vietnam's hierarchical state system have been the long arm of the state in implementing policies and in social mobilization. It this system, macropolicies and divisions of tasks defined in legislation create and allow a framework for enforcement and implementation at the lower levels. At the blurred interface between society and the local state, when much of the latter are also farmers, excluding the local state has excluded one or many prestigious people and those with authority. As part of the "new economic zone" where society was formed within the state structure of management, people in the Tân-Biên irrigated area have developed a sense of overlap between state and social leaders. In other words, those with authority in the area also possess social prestige; thus they might be more able to gather people for PST design viewing, PIM and WUA discussions and training, and at the later stage for on-farm canal discussions.

Over the course of OSDP, commune cadres and hamlet cadres were asked to support OSDP consultants in solving problems/conflicts and gathering people for training, and some hamlet cadres became the leaders of the WUAs established by the project. The Phước-Hòa project again falls into an ambivalent position in the ongoing debate over the autonomy of local irrigation management from the state system, and whether making use of the local system is possible in the way that can also bring benefits. On the one hand, the inclusion of local cadres potentially transfers the existing bureaucratic system of top-down mandate into a community-based WUA. On the other hand, it makes use of many of the most prestigious and experienced leaders in the area. In reality, the case of the Tân-Biên irrigated area demonstrates a significant diversity on the perspective and capacity of local cadres in PIM building. While some express deep understanding of the social and agriculture situation and great potential for PIM building, others are embedded in a rigid state management style of a bureaucratic command and control regime.

Additionally, farmers in the area, originally from different parts of southern and northern Vietnam, have been living in somewhat self-directed lifestyles since decollectivization (in the late 1980s). Their independent farming system has connected to the bureaucratic state system in which social authority and ability for social mobilization have been exercised at the hands of local cadres. Over time, autonomous organizations among farmers and cooperation in groups have become alien to farmers in this area.

16.3.2 Water User Associations (WUAs) Versus Water Operators: When Ideology Meets Reality

Over the scope of their 2-year contract, the OSDP II team has adjusted the model of WUOs to be established in the area. The adjustment of this specific case is driven by a combination of different factors:

- The influence of the local existing model of irrigation management with Tổ-Thủy-Nông (water operator) under the organizational line of IMC Tây-Ninh
- The model guided by PIM and WUG guidelines of the project to establish WUGs/WUAs with charters and rules
- The legal support of the collective economy regime in shaping the model of group into a precursor of the cooperative
- The limited time of a project, with project activities based on contracts with timelines that are often shorter than required for participatory-based work

In the Tân-Biên irrigated area, from June to September 2016, with guidance from the consultants, 16 WUAs were established. The boundaries of the WUAs were identified according to administration boundaries, hamlets in particular. According to the model of the OSDP consultants, one WUA is established in each hamlet. Each hamlet-WUA had one to several water user groups (WUGs). Each WUG would manage 100–150 ha of land irrigated by secondary and/or tertiary canals. Under each WUG is the water user team, comprising from one to seven farmers, who together maintain and share the water in one on-farm canal that brings water to their fields (Fig. 16.6). Each WUG has one or two water operators who are responsible for operating the gates, checking and reporting on the irrigated area, maintaining the canals, and playing an intermediate role between farmers and the company agency—which in this case is the Tân-Biên irrigation management team (IMT) under the Tây-Ninh IMC. All water operators together make one water operator group of the association WUA.

While the WUA provides the foundation for the management structure in the future of the expansion of the irrigated area, it manifests a complex structure that has worked mainly on paper. The three-level structure of management indeed confuses most other actors, from the Tân-Biên IMT to commune cadres and water operators. In each WUA established so far, there is only one WUG and one water operator. One exception is Phước Trung WUA, with two WUGs and two water operators.

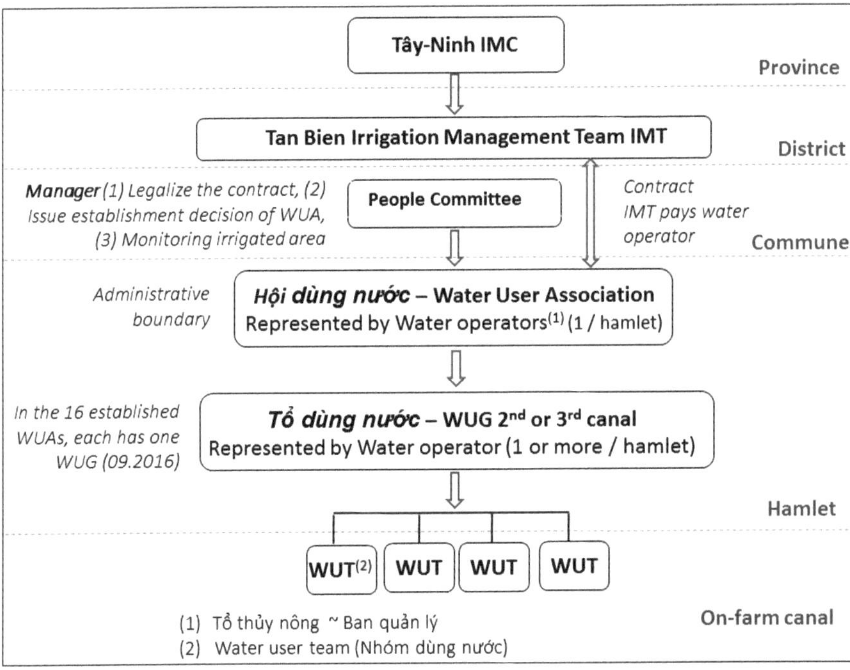

Fig. 16.6 Water user association model for Tân-Biên irrigated area, according to the Vietnam academy for water resources - on-farm and social development program. *IMC* irrigation management company, *IMT* irrigation management team, *WUA* water user association, *WUG* water user group, *WUT* water user team

16.3.2.1 The Formal Assembly and Election

The OSDP team facilitated the training and assembly of the 16 WUAs in 2016. The assemblies were organized by the lead of the OSDP team with a standardized structure (Fig. 16.7). The assembly or Đại-hội in Vietnamese has been part of the party-state system of Vietnam.[18] The Communist Party, the People's Council assemblies from central to local levels,[19] and mass organizations (farmers, youths, women's

[18] Ideologically, authority to run the nation-state in Vietnam goes beyond governmental institutions. A popular slogan promoted in Vietnam asserts: "The party leads, the people control, and the state manages." As stated, the nation is managed by the state under the Communist Party, with the support of People's Army, Fatherland Front, and mass organizations (e.g. for women, peasants, workers, youth) (Kerkvliet 2004:3–4). As at the state system, the Communist Party is present in all levels and forms of office (province, district) and with individual agents (commune and hamlet).

[19] Vietnam's administration system has a top-down, four-level organizational structure: the central or national level, provincial level, district level, and commune level. Officially, the commune is the smallest unit of the state administrational system (Porter 1993). However, below the commune, there are hamlets with People's Boards, cadres, or state assistants tasked with assisting the commune. In general, the central level includes the National Assembly, ministries (including the prime minister's office and various departments), and the Supreme Court. These agencies fulfill the leg-

Fig. 16.7 Water usage rules of WUAs and the assembly of WUAs in Thạnh-Tây Hamlet, Thạnh-Tây Commune, Tân-Biên perimeter (July 2016)

unions, etc.) all follow an identical model of organizing and assembly in order to report the activities (achievements) during the past term and the strategies and targets of the following one. During assemblies, election of the management board or leaders/representatives of the organizations is carried out.

For WUAs for the Tân-Biên perimeter, the experts decided on steps of the assembly:

- Introduction of delegates
- Introduction of the Phước-Hòa project
- Election of WUA's representative (or approval of the only nominee)
- The elected representative receiving an official decision from commune's leader
- The representative reading the rules and charters of the WUA

The assemblies were assessed to be informal with fewer bureaucratic procedures, shorter presentations, and a less formal atmosphere. However, the sections are organized and implemented by the experts (OSDP consultants) and local authorities with little involvement of the farmers—the supposed members of the WUA. The standardized procedures turned the assemblies into a step to officialize the WUA and the representatives.

In short, although it is not our intention to undermine the efforts of the consultants to apply the participatory process during the OSDP, both the organization's mode of operation and that of the WUAs were decided mainly by the experts, and

islative, executive, and judicial functions of the national government, respectively (ibid:73–83). At the province and district levels, the People's Council, People's Committee, and People's Court perform the legislative, executive, and judicial functions, respectively. Within the ministries exist three types of agencies: state management, non-business (generally translated as institutes), and business agencies. The latter two are governed primarily by the state management agencies (Molle and Hoanh 2008, cited in Waibel 2010:18).

the WUA's representatives were mobilized by the local cadres. There was a deviation between, on the one hand, understanding by OSDP consultants of the local situation and farmers' perspectives and, on the other hand, the rush to formulate a model for irrigation management at the local level by the end of OSDP II. This resulted from a difference between the project's timeline and the farmers'. While farmers are not ready to discuss any form of management because the water is not there yet, the project timeline required the OSDP consultants to establish the WUA through formal assemblies. Thus, the discussions of the model, rules, and charters did not yield much feedback from the water users, except some details on the amounts of fines applied for those breaking the rules. The official rules, charters, and WUAs were designed by OSDP consultants and approved in the meetings between PPMB and official cadres from the district office of DARD.

The election of the WUA's representative during the assembly remains symbolic. The establishment of the WUA was from the external force of the project, and the whole process is still very new to farmers. It gives no incentive and inadequate information for both the candidate and the voters. In this situation, project experts relied on local authorities to choose and mobilize potential candidates to take the position of community leaders in WUAs. Becoming a candidate for water operators became part of the state tradition of social leaders, of representatives that possess accountability both to the people and to the state system. Bureaucratic government language is used to explain the election process; notably election, nomination, direction, and inspection. In the end, despite the effort to create a structure autonomous from the state system, the formation of the WUA in the Tân-Biên irrigated area of the Phước-Hòa project reproduced the state structure of management in Vietnam with both its advantages and limitations.

16.3.2.2 The Local Backup for "Symbolic" Project's WUA

At the same time, the Tây-Ninh IMC and Tân-Biên IMT developed their own model, overlapping with many parts of the OSDP's model. The existing management structure for irrigation in Tây-Ninh since the 1990s includes a hierarchy, with IMC at the provincial level and the irrigation management enterprise (IME) at the district level, of which the newly established Tân-Biên IMT holds a position equal to that of the district IME. Each IME with a number of hydraulic workers takes responsibility of the O&M of irrigation-related structures in the district (e.g. opening gates, renovating or concrete lining the canals, and weeding). To reach the household level, the Tây-Ninh IMC has been working with the Tổ-Thủy-Nông model (Fig. 16.8). "Tổ" means group or organization; however, it points to individuals working in irrigation O&M of an area of tertiary and on-farm canals. Therefore, we call them the same name as water operators, the same as those of WUGs under WUAs established by the OSDP. Each water operator is nominated and mobilized by commune authority and works under the technical guidance of the IME's workers. Water operators are often responsible for an area of about 150–200 ha. Farmers when signing the

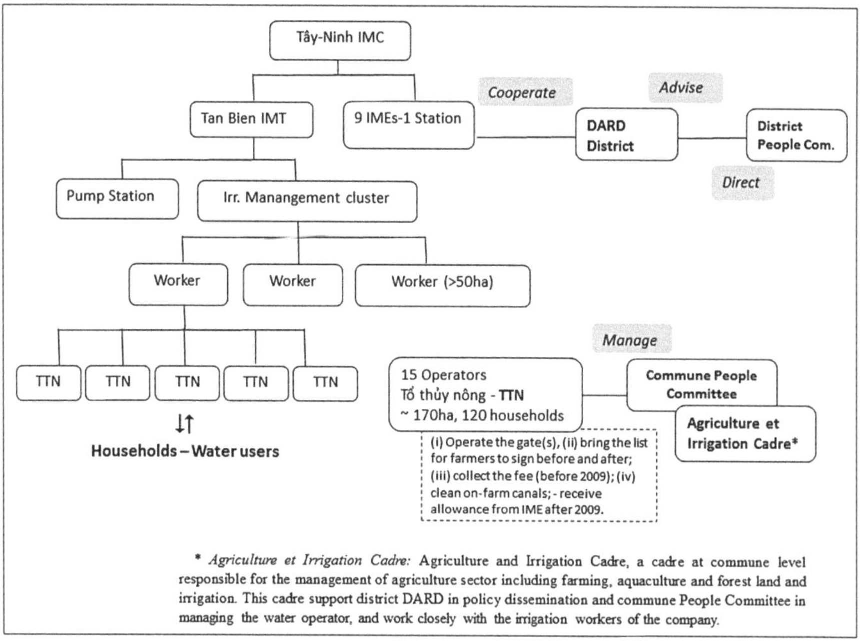

Fig. 16.8 Irrigation management model of irrigation management company with irrigation management enterprise (IME) and water operator. *IMT* irrigation management team, *DARD* Department of Agriculture and Rural Development, *TTN* Tổ-Thủy-Nông, groups or individuals working in irrigation operation and maintenance of an area of tertiary and on-farm canals

confirmation for their irrigated land for a season (irrigated land registration) also authorize the water operator to sign the contract with the IME on their behalf.

Under this model, the Tân-Biên IMT worked with the commune cadres to choose water operators for 15 areas that are separated by hydrological boundaries (Fig. 16.9). This process took place at the same time as the OSDP, and the water operators who worked with the Tân-Biên irrigation team are the water operators nominated and elected in the WUA assemblies. However, the Tân-Biên IMT holds a different list of water operators and names them by canal number instead of by hamlet. Such overlap demonstrates a process of coevolution or mutual learning between the IMC's people and the project's PIM experts of the OSDP.

At present, the only working part of the OSDP's WUA model is the water operator that overlaps with IMC model. At this stage of settling things down for system O&M, commune cadres play an even more crucial role of nominating and mobilizing individuals to take on the role of water operator, supporting the registration of irrigated land, resolving conflict, and playing an intermediate role between the Tân-Biên irrigation team and water operators and farmers. This role of communes is

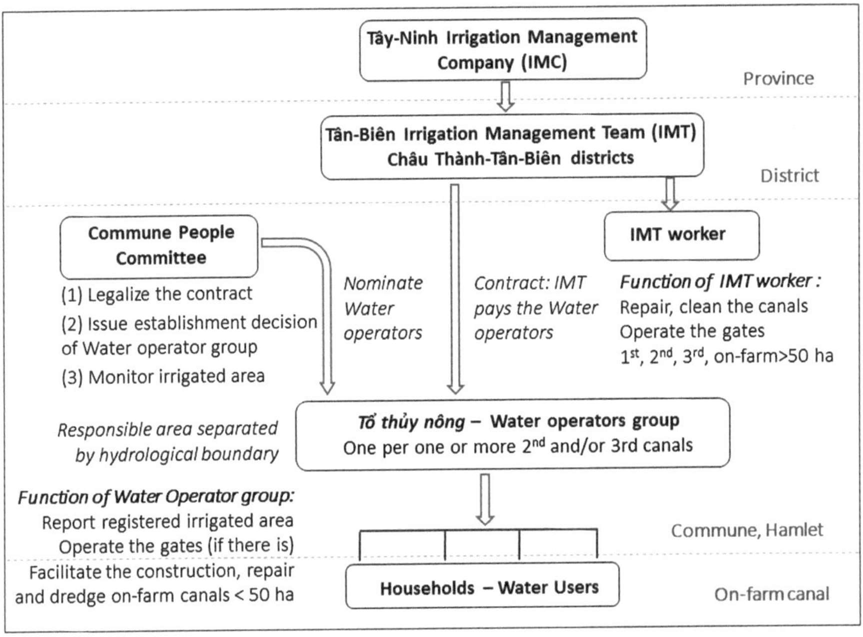

Fig. 16.9 Water operator model for irrigation management of Tây-Ninh IMC

stated in the charter and rules of WUAs (drafted by OSDP team) and by the charter on the cooperation among IMC, its line agency, and the local authority.[20]

16.3.3 *The Story of On-Farm Canals*

On-farm canals are either quaternary, tertiary, or secondary canals connecting to the system to convey water to irrigate an areas at the field level. In this project, it is the farmer's responsibility to connect to the PST system invested by the project. Making the on-farm canal is the final link in the chain and a crucial one in connecting the irrigation system to its designed area of irrigation. The responsibility of creating these canals belongs to the local actors, including the provincial management board, IMC and its line agency, the commune government, and farmers. There is no currently available channel of funding for such canals besides the contributions from farmers. The idea is to create space for farmers' participation in the irrigation system, and social cooperation generates accountability and a sense of ownership from

[20] Regulated by decision 2147/QĐ-UBND, dated 14 October 2009 of the provincial People's Committee, to issue the coordination mechanism in activities to manage, exploit, and protect the irrigation infrastructure between the Tây-Ninh irrigation exploitation company and the People's Committee of the districts and towns.

the farmers. However, it is challenged by the individual farming regime that is dominant in the area and the institutions that are already established. The institutions of "free water" (since the ISF exemption) and "the state caring for the people" (with free infrastructure and good compensation for land acquisition) have resulted in some common reactions, namely, waiting for state support for on-farm canals, the belief that the irrigation company should invest in the on-farm canals in order to collect the fees later, and so on.

The donor is discussing the option of financial support for participatory design of on-farm canals as an additional work package. Approval of the package took quite a long time due to disagreement between the OSDP supervision consultants (SCP-HECII) and PPMB. The first proposal was framed by PPMB with the support of the same consultants that conducted the OSDP programs, as early as during OSDP II. This is a common practice, but not in all cases. By supporting PPMB in framing the proposal, the consultants hope to assure their advantage during the tendering process. However, the process took much longer than expected. The first proposal finished in April 2016 by the consultants focused on designing all on-farm canals (516 canals, or 300 canals in the later decision), and stipulated that the design must be done in about 6 months. The project international consultants for OSDP II were chosen by Société du Canal de Provence (SCP) in cooperation with the hydraulic engineering consultants corporation No.II - HEC II. The current research project's result indicates it is unnecessary to design all on-farm canals in a short period; rather, a one-by-one case basis is recommended, based on the need of the farmers who will potentially use water from that canal. The international consultant team agreed with this recommendation. In addition, the PPMB also doubted the rationality of the designing of all canals but did not express the idea to the donor.[21] Thus, after back-and-forth communication between the donor (AFD), project management boards at the central level (ICMB9) and provincial level (PPMB), researchers, and project consultants (SCP-HEC II), in April 2017 a new proposal was being prepared by SCP. The present idea is to design not all but some canals and to pay more attention to local conditions and farmer needs. Observation on the subject continues. Such a transition happened thanks to the exchange between the project donor, the project consultants, the provincial management board, and the project research group. Yet, the long planning process is not always a good thing because the delay is causing problems for the project timeline and fund mobilization; some even doubt that it will be decided by the end of the financial timeline in March 2018.

While the project's work for on-farm design is discussed by the experts and decision makers, local actors in Tây-Ninh, including the IMC, commune government, and farmers, are already at work on the canals. The very first on-farm canal made by farmers, made of brick,[22] was completed in June 2016 by farmer initiation (case N16-1, Phước Vinh commune). Another case in Thạnh-Tây commune is guided by

[21] This dynamic in communication between stakeholders in the project is framed by the idea of "it is in the wish of the donor" in negotiating the rationality in a water development project. More analysis of this aspect is planned as part of the ongoing research project.

[22] Brick is building material used to make walls, pavements, and other elements in masonry construction. A brick can be composed of clay-bearing soil, either fired or not. The one used for the on-farm canal in Phước Vinh is the fired type.

Fig. 16.10 A brick on-farm canal made by farmers (N16-1-10 at Phước Vinh commune), June 2016 (left) and collective work to build an on-farm canal (N 2-22-1-11 at Thạnh-Tây commune), March 2017 (right)

the commune authority and is now under discussion (case N 2-22-1) (Fig. 16.10). A conclusion from the two case studies (which are presented in detail in Pannier and Huynh 2016) is that making an on-farm canal as a form of collective action could be derived from and limited by various factors, including natural factors (climate/rainfall, groundwater availability, soil characteristics, and reposition of the plots relative to roads, transport conduits, canals, and tube wells), pre-existing infrastructure for irrigation and drainage, agricultural features such as cropping patterns and farming techniques, and human and social factors (previous experience, observations, and information about canal irrigation; solidarity, moral obligations, and social cohesion, including kinship and neighbor relations; and the articulation of individual logics and collective dynamics). The extent of influence of each factor varies between cases, and there are signs of interrelation between them. One can make another more important or limit another's impact. Because these factors determine the motivation and the capacity of farmers to engage in on-farm canal building, they should be taken in consideration in order to foster and design on-farm canals in the Phước-Hòa project area.

Above all, the principle of having reliable water, that is, the trust of farmers in the working capacity of the system, is the first condition for any action to be taken. As farmers repeatedly respond when asked if they would participate in making on-farm canals: "It depends if there is enough and constant water in the tertiary canal" (interview 21.12.2016, Thạnh-Tây commune). With the current continuous flow, operation seems to be on demand and has been done in a flexible way with personal communication between farmers and water operators, and with day-to-day negotiation between water users.

In other words, the building of each on-farm canal is a unique case and cannot be standardized. Such a process includes informal negotiation and arrangement between a group of farmers who share the delivered water and maintain the on-farm canal that brings water to their fields.

16.4 Discussion

Regarding the process of establishing WUAs in the Tân-Biên perimeter, the question is whether it is more judicious to set up the organization of water users before versus after the canal network becomes operational. In implementing the OSDP II, the consultants of OSDP, as approved by the project managers and the donor, chose the first option. They reasoned that this approach would allow the establishment of a collective organizational structure that will be ready for operation as soon as the perimeter is put into service. Even though the reason appears to be logical, its implementation created doubts about its suitability due to constraints described in the previous section.

To summarize, consultants in charge of the implementation of OSDP II adopted a top-down approach that limited adequate participation of future users. In concrete terms, the procedure for preparing collective organization models was standardized and unified: the statutes and rules for running water user associations (WUAs) had been reproduced everywhere in the perimeter almost word for word. The consultant team was thus not allowed or able to fulfill its role as an intermediary between the directive approach of the project and the expectations and doubts of the farmers. The latter are increasingly skeptical about the successive reports regarding the delay in the construction of the perimeter and certain technical constraints in the system's design.

Several reasons may explain the choice for this order of project activities, in contrast to the very basic principle of the PIM approach. On the one hand, the consultants had only 6 months to set up the 16 WUAs although some training on how to operate and manage an irrigation system collectively had been organized. On the other hand, they confined themselves to following the requirements specified and fixed by the terms of reference (TORs), or the project charter. The TORs, with technocratic and directive characteristics, forced them to carry out the tasks in the same way as filling in a checklist, and all strived to quantify their interventions (number of meetings organized, number of local people participated in project's trainings, and so on). Finally, the dual absence of international expertise (late participation of the SCP) and PIM specialists (social scientists) did not allow proper implementation of the OSDP. None of the above experts were there to oversee and monitor the second phase, to assess its relevance, and to look critically at what already exists, that is, the results obtained at the end of the first phase.

Today in Tân-Biên, there is a mismatch between reality and the language that has been applied to identify it. One may question whether the OSDP consultants noticed the differences; the answer is yes. However, put another way, the OSDP consultants are forced to complete the contract they signed. Thus, under the constraints of institutions and the lack of direct communication between project partners, PIM and WUAs have been applied in the Tân-Biên area as a predesigned model, not built from local context.

16.4.1 Power Leverage in the Project's Negotiation

The 16 WUAs reported in project documents and formally recognized by the commune authority by a formal decision are under the scrutiny of local actors. Local actors, including engineers from the Tân-Biên IMT, the commune authority holding the function of state management or human management, the newly elected water operators and other farmers, all preserve and present their opinions about the relevant WUA and wait and see how it could fit into the local context. In the meantime, the model of IMC's water operator guided by the Tân-Biên IMT, or one part of WUAs, is ongoing. Above all, confused and unclear about the benefit of different models, the water users can withhold a decision to participate. Such wait-and-see behavior is embedded in the farmers' daily practices, as individuals constantly negotiate and bargain with different factors to earn a living. Thus, while being inferior to local authorities in power leverage, especially as expressed in public meetings and project training (rarely raising an opposing voice), farmers or water users in the area retain a power to decide on participation in state/project-initiated activities, and thereby hold at least passive power in decision making.

In implementing the development project of Phước-Hòa, the extent of different power capacities of the state has been expressed differently. The state might perform a directive power in the first phase with land acquisition and compensation, and infrastructural power[23] with a legal framework during the whole process of the project. The process of land acquisition and compensation in the Phước-Hòa project indeed contains a complex directive power of the state, infrastructural power (the principle of the state as the manager of all natural resources, and the farmers return the rent land when needed), and promoting the idea of a caring state (good market price for land compensation, for instance). At the latter stage where farmer cooperation for WUAs and making on-farm canals is needed, in implementing the project's model of PIM and of the grassroots democracy policy of the government, a discur-

[23] The regulatory power exerted by institutions and organizations, including the state embeddedness in society (Göbel 2011). Göbel (2011) categorizes a state based on its capacity to wield three kinds of power: coercive power to impose its will on the people (despotic power); regulatory power exerted by institutions and organizations (infrastructural power), including the state embeddedness in society; and the power to make people want what the government wants them to want (discursive power) (ibid:177, Lukes 1975/2005).

sive power extended with a mobilization strategy becomes the dominant part of state power. At the same time, in the process of negotiating the making of on-farm canals, farmers express their discursive power in negotiating the activities. Power leverage is achieved by both state decisions for performing discursive power and farmer power decisions about their own interests in the use of their own private land. Local knowledge of their own fields and of customized farming practices are also assets for farmers when they enter negotiations. Thus, when it comes to on-farm canal negotiations after a bureaucratic checklist of a time-limited project is completed, the negotiations manifest a process of mutual learning between the state (local cadres plus IMC engineers) and water users. We may call it co-evolution, as similarly proven in the case of collective drainage arrangements in the Mekong Delta (Huynh 2016). In that research, the author concludes that "it is in the everyday dialogue that, in the co-existence of hierarchical state management structure and the space of local flexibility, officially and unofficially refines the local practices." In practice, water governance in these cases in Vietnam is complex and an interaction between formal arrangements and informal interactions.

16.4.2 PIM Implementation in a Development Project in the Vietnamese Context

The WUA model remains a new idea introduced into the area by the people running the project—by OSDP consultants in particular. By implementing the WUA model, a model preshaped by experts and agreed upon by provincial cadres, the farmers have learned the model rather than having built the model for their own irrigation management. In other words, participation in this case remains at the consulting level, and the perspective and requests from the water users might or might not be taken into account (see the extended ladder of participation in Bruns 2003).

The case of the OSDP in the Phước-Hòa project demonstrates that, with the current approach and present situation, the idea of autonomy from the state and the attempt to build an equitable relation as "the purchaser– the seller" between WUGs and the IMC (according to PIM guideline 2004) are both legally unfit and practically ineffective. While PIM remains an outside ideal terminology, the participation of the people in state matters and the WUGs are regulated officially by legal documents. Through grassroots democracy policy, the participation of the people in state matters has been promoted. However, in the bureaucratic implementation of the policy, the participation of the water users is set as a "designed or regulated participation", which is very much top-down and bureaucratic. According MARD's Central Project Office for Water Projects (CPO 2012), the participation of citizens is regulated:

> Methods of participation include participating (for opinion) when asked, reviewing when requested, participating in monitoring (according to regulation), participating into designing, construction and management (attached with responsibilities and benefits), participat-

> ing in financial contribution according to regulations (procedure, policy) and all requests during the process from planning, designing, investing, constructing and managing. (CPO 2012)

Participation by and large continues following steps that are pre-designed by the state and implemented by the consultants; all of this manifests the strong influence of the existing government structure.

Building PIM, especially in Vietnam, requires a consistent intervention both in legal frameworks and on the ground. Huge efforts and time are needed to build the capacity for local actors in managing a new form of irrigation with more cooperation and coordination features. More importantly, changing the perspectives and habits of local actors, including both cadres and farmers, to cooperative in irrigation and farming takes even more time, and a different approach is needed. The Phước-Hòa project, as a development project supported by international loans, is restricted by both international and national institutions and by a limited timeline. In the case of PIM applications in the Tân-Biên area, the project timeline did not fit farmer timelines. Training about PIM and establishment of WUAs as required by the project happened way ahead of time when farmers did not see (reliable) water in the canals and thus had not yet assessed the benefit of switching from groundwater to canal irrigation, and from the individual to cooperative style. They received non-voluntary participation from the water users (not by force, but by financial encouragement and social prestige of local authority and experts). It is even more challenging and time-consuming for an intervention to be effective relative to the diversity of cropping patterns, where land use shifts over time (from sugarcane to cassava to rubber, then to vegetables). As a result, PIM and WUAs remain alien to the water users. At the donor level, the need for longer time in building PIM is recognized: "Successful irrigation and drainage projects require participation by all stakeholders in planning, implementation, and O&M to create a sense of ownership of and consequent commitment to the project. This requires that project planning allows time for beneficiaries to participate in planning and influence decisions affecting their future" (ADB 2012). However, project institutions with complex procedures often leave few opportunities for flexibility and adequate time.

The case of PIM implementation in Phước-Hòa project demonstrates how components for certain objectives become an end in itself. WUG or WUA is only a tool or an option in building PIM, yet over time it becomes one target to reach in a project. The link between WUG and PIM is strongly supported by project guidelines (CPO 2004) and is a must as regulated by Vietnamese institutions:

> Participation has to be through organization, agency that is established with legal status. This is the prerequisite for participation. In cases of individuals, households that use water from irrigation structure have to participate through their water user organization; the organization is established according to the law, has legal status and is financially autonomous. (CPO 2012)

The Tân-Biên case illustrates that the formal institutionalized and standardized organization of the WUA is not applicable due to diverse conditions between areas and between zones in each area of the irrigated perimeter. Also, in some cases, water distribution could be done easily by informal agreement and everyday dialogue between water users, and no organization or standardized rules and charters is needed.

16.4.3 Justifying the Possible Forms of User Organizations

As demonstrated by the IMC's Tây-Ninh system, self-organization for water appropriation is possible for ten farmers. To supply irrigation water for a bigger group, the IMC worked with affiliated workers and water operators to collect data on water demand, operate the gate, and maintain the canals. The water operators have acted as affiliated workers of the company throughout that time. Due to the Vietnamese policy in promoting local organization, the water operators have been put into the frame of an organization in which they are the leader. Technically, the water operators who worked with single farmers are officialized in Phước-Hòa's project with the status of representative of the WUAs, an official organization. The Phước-Hòa project became contiguous with another, previous PIM development project in Tây-Ninh (e.g. VWRAP) in promoting a form of WUO with standardized rules and operation mechanisms at a larger scale. Such a larger scale of organization (often at the tertiary canal level) is not suitable, as proven in the 30-year experience of managing the system, according to the IMC's representative (interview 05.07.2016).

Beyond the obstacles of the WUAs established, the very model of irrigation management based on collective organization of farmers into WUAs and water users groups (WUGs) must be brought into question. Indeed, the actual procedures to structure and define the functioning of a group of human beings reflect not so much their formal organization as the fruit of the relations between the actors that develop within the group and with the outside world. It is common to attempt to explain the collective action and the coherence of the group as the "natural" result of a unity of goals and interests that guide the involvement of players. However,

> in reality, apart from the case of very small groups, unless coercive measures or some other specific provisions exist that incite them to act in their common interest, reasonable and interested individuals will not voluntarily try to defend the interests of the group. ... Thus, the common belief that groups of people with common interests tend to defend them seems to have very little or no basis in reality. (Olson 1978: 22–23)

Among a number of potential advantages of the user organizations, three stand out. (a) These forms of collective organizations constitute, for the decision makers within, legitimate intermediaries who opt to establish a power balance with the representatives of the public powers (e.g. local governmental agencies, the IMC). (b) They are equipped with a legal capacity that allows them to negotiate directly with the water companies. (c) Because of the size and complexity of the irrigation networks (with three or four levels of canal connection and with the new, modern concrete infrastructure), the present established informal structure for social negotiation of resource use might not be enough. Despite the possible advantage that a formal organization could provide, it should not diminish the ongoing social dynamic in which informal arrangements continue playing their role and entering institutional negotiations. By this process, new arrangements will be established through time, and this is neither new nor old, neither formal nor informal. They represent the "bricolage" of institutions (as defined by Cleaver 2012) through which rules and norms are refined into practices. It makes the implementation of a standard operating mode based on the basic WUG and WUA units challenging.

In the case of the Phước-Hòa project, it is therefore appropriate to raise questions about the existence of a shared or common goal for all users of one WUG. Researchers conducted in irrigation management throughout the world demonstrate cases where farmers accept and comply with collective rules in water distribution for three main reasons:

1. The available resource is scarce, so coordination is needed.
2. Payment for water access is needed, so an intermediate body to carry out the task is required.
3. Opportunity costs for alternative (e.g., individual groundwater pumping) are higher than cooperating in a group for canal water.

While the first factor is present in the Tân-Biên perimeter, the other two factors are still missing, as the fees will not be collected for at least another 3 years, and groundwater pumping remains convenient and affordable. More importantly, the situation varies greatly between units of irrigation area, which implies great uncertainty for collective organization at the local level.

16.5 Final Remarks

Water institutions in the Đồng-Nai basin are continuously negotiated and contested in practice, by various actors who differ not only in terms of interest, scale, and power but also in the conditions they experience in the face of ecological and agricultural changes and challenges. The Phước-Hòa development project follows the current tendency of combining both conventional hard measures of construction with institutional intervention such as an on-farm and social development program (OSDP). The OSDP is seen as a plus in the attempt to apply integrated measures for water management. However, the project's procedure involved a complex hierarchy that followed the international and Vietnamese state structure within a restricted timeline, which shaped the PIM into a top-down approach with little impact on institutional improvement. Vietnam has long participated in a global policy trend in water development that is dominated by a technocratic approach. Hard infrastructure construction has become the main solution to facilitate the intensification of agriculture production and to mitigate the impact of floods. The technocratic approach to water resources management remains to be a dominant approach within the "hydrocracy" (hydraulic bureaucracy) in Vietnam (Benedikter 2014). Even under the influence of integrated water resources management (IWRM[24]) ideology, institutional inertia remains strong in shaping the legacy of all interventions.

The current challenge lies in the difficulty of defining the right unit for action in a context of shifting environmental and social conditions. Development projects and interventions are often initiated at the central level and by way of governmental structure of management. However, as asserted in the case of PIM and irrigation management transfer in the Beni Amir irrigation scheme in Morocco, "The existence of a strong central government can be, at the same time, a catalyser and a restraint for the PIM/IMT process: a catalyst because a real motivation of a central government can speed up the process, but a restraint because of the difficulty to change strong bureaucracies" (Van Vuren et al. 2004). A similar situation is demonstrated in the Tân-Biên case, where the strong influence of state institutions supported the dissemination of a new ideology of cooperative water management yet shaped the top-down approach and the implementation of PIM models rather than building PIM from within.

A main lesson from the Tân-Biên case is the need for a better approach at the local level that will also facilitate more flexibility in building water institutions. Pannier and Huynh (2016) conclude that the rules and norms that shape local water management derive mainly from daily interactions and negotiations between stakeholders and thus are subjected to change and negotiation all the time. The institutions tend to be a continuous "bricolage" of rules and norms that are applicable for each specific case. As such, Ostrom's statement about crafting institutions for water management is fitting for this case:

> The crafting of irrigation institutions is an ongoing process that must directly involve the users and suppliers of irrigation water throughout the design process. Instead of designing a single blueprint for water-user organizations to be adopted on all irrigation systems within a jurisdiction, officials need to enhance the capability of suppliers and users to design their own institutions. Involving suppliers and users directly will help ensure that development institutions are well matched to the particular physical, economic, and cultural environment of each system. Although this approach presumes that the participants need to be involved in the design process, it does not presume that good institutional designs spring up naturally as the result of spontaneous organization. Government officials and donor agencies can and should play an active role in enhancing the design process and monitoring the results. The role proposed for central governmental officials and for donor agencies is, however, quite different from that proposed by earlier studies that called for the creation of many user organizations based on the same institutional design. (Ostrom 1992:14)

This statement, from back in 1992, and the PIM concept and participatory approach indeed support a mechanism that suits each specific context. This empirical case of the Phước-Hòa project demonstrates the distance between the ideal principles and the application of the PIM concept for better water governance, and between what should be done and what can be done. In this case, the time restriction of the

[24] IWRM is a process that promotes the coordinated development and management of water, land, and related resources, in order to maximize the resultant economic and social welfare in an equitable manner without compromising the sustainability of vital ecosystems (defined by Global Water Partnership).

development project, other factors of human expertise for PIM, and the localized context of management structure become limiting factors to the successful application of this participatory approach. The case demonstrates how institutions help shape irrigation governance. It continuously involves the complex coexistence of rules and norms that, through interactions between stakeholders in the project and in everyday activities, refines the choice of practices. Through this process, all stakeholders are engaged in different arenas of negotiation in which power leverage is exercised in a complex dynamic. The water users, while presumably the most advantaged group, express their power leverage with discursive or passive power.

References

ADB. (2012). *Participatory irrigation management: How can participation contribute to the sustainable management of irrigation and drainage systems for agriculture?* Learning Lessons.

Benedikter, S. (2014). *The Vietnamese hydrocracy and the Mekong Delta: Water resources development from state socialism to bureaucratic capitalism* (Vol. 25). Viena: ZEF Development Studies Lit Verlag.

Bruns, B.. (2003). *Water tenure reform: Developing an extended ladder of participation*. Politics of the commons: Articulating development and strengthening local practices, Chiang Mai, Thailand.

Burton, M., Wester, P., & Scott, C. (2002). *Safeguarding the needs of locally managed irrigation for the water scarse Lerma-Chapala river basin, Mexico*. Paper presented at the International Commission on Irrigation and Drainage, Eighteenth Congress, Montreal, 21–28 July 2002.

Central Project Office for Water projects - CPO. (2004). Guidelines on establishment of Water User Groups (WUG). Phuoc Hoa water resources project. Funded by Asian Development Bank ADB - French Development Agency AFD and Government of Vietnam. Hanoi.

Central Project Office for Water projects – CPO. (2012). Sổ tay hiện đại hóa hệ thống tưới. [Brochure on Modernization of Irrigation system]. VWRAP project. Funded by World Bank. Ha Noi.

Cleaver, F. (2012). *Development through bricolage: rethinking institutions for natural resource management*. London: Routledge.

Corderi, D. (2011). The economics of climate change in the Dong Nai Delta: Adapting agriculture to changes in hydrologic extremes.

Dang Minh Tuyen. (2010). Công trình đong đo nước đối với vấn đề hiện đại hóa thủy lợi nộng đồng ở Việt Nam- [Water measurement project in modernization of irrigation at tertiary level in Vietnam]. *Irrigation Science and Technology, 27*, 7.

Đỗ Đức Dũng, & Nguyễn Ngọc Anh. (2015). *Đánh giá tác động của phát triển kinh tế xã hội và biến đổi khí hậu lên nhu cầu sử dụng nước lưu vực sông Đồng Nai và vùng phụ cận*. Nam Ho Chi Minh: Viện quy hoạch thủy lợi miền.

Đỗ Đức Dũng, Nguyễn Ngọc Anh, & Doãn Thu Hà. (2014). Đánh giá biến động tài nguyên nước lưu vực sông Đồng Nai và vùng phụ cận Khoa học kỹ thuật thủy lợi và môi trường *47*, 19–26.

Fontenelle, J.-P., & Tessier, O. (1997). *L'appropriation paysanne de l'hydraulique agricole du delta du fleuve Rouge: Processus et limites* (pp. 25–44). Bondy Paris: Autrepart.

Göbel, C. (2011). Authoritarian consolidation. *European Political Science, 10*(2), 176–190.

Huynh, T. P. L. (2016). *State-society interaction in Vietnam: The everyday dialogue of local irrigation management in the Mekong Delta*. Münster: ZEF Development Studies LIT Verlag.

Kerkvliet, B. J. T. (2004). Surveying local government and authority in contemporary Vietnam. In: L. Benedict, K. Tria, & D. G. Marr (Eds.), *Beyond Hanoi. Local government in Vietnam* (pp. 1–27). Nordic Institute of Asian Studies.

Nguyen Xuan Tiep. (2008a). Participatory irrigation management – PIM. In Nguyen Xuan Tiep (Ed.), *Participatory irrigation management and emerging issues – Nông dân tham gia quản lý công trình thủy lợi và những vấn đề đang đặt ra* (pp. 22–42). Ha Noi: The Agricultural Pulishing House.
Nguyen Xuan Tiep. (2008c). PIM models in Vietnam – experienced lessons. In Nguyen Xuan Tiep (Ed.), *Participatory irrigation management and emerging issues – Nông dân tham gia quản lý công trình thủy lợi và những vấn đề đang đặt ra* (pp. 81–121). Ha Noi: The Agricultural Publishing House.
Nguyễn Ngọc Anh, & Đỗ Đức Dũng. (2015). Cân bằng nước hệ thống sông Đồng Nai trong điều kiện biến đổi khí hậu Viện quy hoạch thủy lợi miền Nam Hồ Chí Minh city.
Nguyễn Trung Dũng. (2016). Ý kiến bàn luận và phân tích thủy lợi phí dưới góc độ kinh tế học.
Nguyen Vu Huy. (2016). Vai trò của hồ Phước Hòa đối với tài nguyên nước lưu vực sông Đồng Nai [The role of of Phuoc Hoa to the water resources of Dong Nai river basin]. Ho Chi Minh: SIWRP – Southen Institute for Water Resources Planning.
Nguyen Xuan Tiep (Ed.). (2008). *Participatory irrigation management and emerging issues – Nông dân tham gia quản lý công trình thủy lợi và những vấn đề đang đặt ra*. Ha Noi: The Agricultural Publishing House.
Olson, M. (1978). *Logique de l'action collective (trans: LEVI MT)*. Paris: Presses universitaires de France.
Ostrom, E. (1992). *Crafting institutions for self-governing irrigation systems*. San Francisco: Institute for Contemporary Studies.
Pannier, E., & Huynh Thi Phuong Linh. (2016). *Managing the commons: Negotiating the making of on-farm canals in Tân-Biên irrigated area, Phước-Hòa project*. Ecole Française d'Extrême-Orient – Phuoc Hoa Research Project, Ho Chi Minh City.
Porter, G. (1993). *Vietnam. The politics of bureaucratic socialism. Politics and international relations of Southeast Asia*. Ithaca: Cornell University Press.
Snellen, W. B. (1997). *Irrigation water management: Training manual No. 1*. Irrigation Scheme Operation and Maintenance Food and Agriculture Organization of THE United Nations (pp. 1–55).
Tessier, O. (2011). Outline of the process of Red River hydraulics development during the Nguyễn Dynasty (nineteenth century). In: *Environmental change and agricultural sustainability in the Mekong Delta* (pp. 45–68). Springer.
Tessier, O., Huynh Thi Phuong Linh, Bourdeaux, P., Pannier, E., Nguyen Minh Nguyet, & Huynh Hong Duc. (2016). *Local irrigation governance in Phuoc Hoa project: Midterm report*. Ho Chi Minh: Ecole française d'Extrême-Orient.
The World Bank. (1978). *Report and recommendation of the president of the International Development Association to the Executive Directors on a proposed development credit to the Socialist Republic of Vietnam for the Dau Tieng Irrigation Project*. Washington, DC: The World Bank.
The World Bank. (1989). *Project completion report: Viet Nam – Dau Tieng Irrigation Project (Credit 845-VN)*. Washington, DC: The World Bank.
Van Duc, L., & Gupta, A. D. (2010). Water resources planning and management for lower Dong Nai River basin, Vietnam: Application of an integrated water management model. *International Journal of Water Resources Development, 16*(4), 589–613.
Van Vuren, G., Papin, C., & El Haouari, N. (2004). Participatory Irrigation Management: Comparing theory with practice a case study of the Beni Amir irrigation scheme in Morocco. In *Séminaire sur la modernisation de l'agriculture irriguée*, (p. 11). IAV Hassan II.
Waibel, G. (2010). *State management in transition: understanding water resources management in Vietnam* (ZEF Working Paper No. 55, Vol, 55). Bonn: University of Bonn.

Chapter 17
Urban Water Management Under Uncertainty: A System Dynamic Approach

Nguyen Hieu Trung, Nguyen Hong Duc, Nguyen Thanh Loc, Dinh Diep Anh Tuan, Lam Van Thinh, and Kim Lavane

Abstract Urban water management (UWM) is a complex task with a number of constraints especially in developing cities. Rapid population growth, inadequate water infrastructure, and inefficiency in water management policies have led to increased pressure on the city's water supply and drainage systems. These challenges, however, remain uncertain in terms of both temporal scale and magnitude of change, such as climate change and sea level rise. A city needs an appropriate framework to support not only short-term adaptation activities but also long-term strategies to enhance its resilience to these uncertainties. Therefore, system dynamics—methodologies to frame, understand, and discuss complex issues and problems—is a suitable approach for such complex UWM issues. The purpose of this chapter is to share our experiences in applying a system dynamics approach in Can Tho City, Vietnam. The study went through several steps that involved identifying key stakeholders and tools to support decision making, recognizing exogenous uncertainties and potential measures with their effectiveness indicators, building models to support decision making in present and future scenarios related to the UWM, and engaging stakeholders during the study approach to ensure the complex model results were well taken up and used for their future decisions. Throughout the case study, the system dynamics approach shows its capacity in supporting the city's policy makers and managers in dealing with such interdisciplinary and complex issues.

Keywords Urban water management · System dynamic · Robust decision support · Stakeholder engagement · Flood · Water pollution · Scenario analysis · Climate change · Sea level rise

---

Nguyen Hieu Trung (✉)
College of the Environment and Natural Resources, Can Tho University, Can Tho, Vietnam

Research Institute for Climate Change, Can Tho University, Can Tho, Vietnam
e-mail: nhtrung@ctu.edu.vn

Nguyen Hong Duc · Nguyen Thanh Loc · Lam Van Thinh · Kim Lavane
College of the Environment and Natural Resources, Can Tho University, Can Tho, Vietnam

Dinh Diep Anh Tuan
Research Institute for Climate Change, Can Tho University, Can Tho, Vietnam

M. A. Stewart, P. A. Coclanis (eds.), *Water and Power*, Advances in Global Change Research 64, https://doi.org/10.1007/978-3-319-90400-9_17

17.1 Introduction

Can Tho City is one of the five major cities in Vietnam that is located in the at the center of the Vietnamese Mekong Delta, an area where water is central to everyday life and underpins the local economy, including agriculture, aquaculture, transport, and tourism. With a population of 1.25 million inhabitants in 2014, population growth rate of 9.7%, and about 52% of the population living in urban districts, the city's clean water requirements for household use, services, and manufacture are currently under pressure due to rapid urbanization, increase in upstream water uses, climate change, and sea level rise (World Bank 2014; CSIRO 2012). Moreover, limited investments in water supply and sanitation infrastructure and improper maintenance, especially in areas of rural-urban interface (Neumann et al. 2013), have led to surface water pollution, causing a significant decline in groundwater storage and quality (Moglia et al. 2012). In addition, land subsidence has been estimated and observed at a rate of 1–2.5 cm/year (Ramalho 2017; Erban et al. 2014) due to overexploitation of ground water.

The major challenges for urban water management (UWM) in Can Tho City include not only a lack of infrastructure, management institutions, and financial resources, as mentioned above, but also uncertainty in future socioeconomic development and changes in natural conditions both on a temporal scale and on a magnitude scale associated with global, regional, and national developments. Therefore, this study applied a system-dynamic approach to frame, understand, and discuss complex issues and problems so that it could inform the stakeholders and decision makers in Can Tho City of effective measures to control inundation and water quality and, as a result, to reduce the impacts of uncertainty factors.

The study was implemented for 18 months (from June 2015 to December 2016) in Ninh Kieu District, Can Tho City. Ninh Kieu District is the most developed and most important district in the city, with a population of 209,274 people and an area of 29.22 km^2. With rapid socioeconomic development, this district is the representative case in the city for an increasing population and urbanization under the threat of flooding in the rainy season, lacking water supply in the dry season, and issues related to water pollution.

17.2 Study Approach

The study was conducted based on a comprehensive and community engaging framework, the Robust Decision Support (RDS) framework developed by RAND Corporation. In this framework, the following six steps have been performed.

1. Identify key stakeholders and available tools
2. Define urban water challenges and potential solutions
3. Understand the perception of local community on water demands and their view on future water issues

4. Project the future socioeconomic scenarios
5. Evaluate the effectiveness of the proposed inundation and pollution control measures
6. Inform stakeholders on key findings and consult them for suitable inundation and pollution control measures

17.2.1 Step 1: Identify Key Stakeholders and Available Tools

In step 1, we reviewed existing literature and organized meetings with key informers and experts to identify key stakeholders and tools to support the decision-making process for UWM planning.

17.2.2 Step 2: Define Urban Water Challenges and Potential Solutions

For step 2, a stakeholder consultation workshop was organized (Fig. 17.1) to discuss the case study's water challenges and potential solutions. The stakeholder mapping method (Mitchell et al. 1997) was used to reconfirm the key stakeholder list from

Fig. 17.1 Stakeholder consultation workshop to discuss water challenges and potential solutions

step 1 and to identify the level of the stakeholders' influence/power and interest on issues relating to the clean water scarcity in the city. Participants were local city department leaders, business leaders, community groups, and experts. The participants discussed in groups

- to recognize exogenous uncertainties (X) both in socioeconomic and in natural changes,
- to propose potential inundation and pollution control measures (levers, L),
- to define indicators to measure the measures' effectiveness (M), and
- to delineate the relationships (R) among X, L, and M.

17.2.3 Step 3: Understand the Perception of Local Community on Water Demands and Their View on Future Water Issues

To understand the perception of local community on water demands and their view on future water issues, we interviewed 200 households in the study area (Fig. 17.2). The survey asked about the water demands, water uses, water saving practices, and especially, how gender played a role in these practices.

Fig. 17.2 Household survey to understand the perception of local community on water demands and their view on future water issues

17.2.4 Step 4: Project the Future Socioeconomic Scenarios

The projected scenarios were defined based on different time scales (present and future) and the city's socioeconomic scenarios (population growth rate, economic development plans, and water demands). VENSIM, a system dynamic model developed by Ventana System Inc., was used to estimate the total volume and quality of waste water releasing from major waste water sources of the study area (domestic, markets, hospitals, and offices) according to the projected socioeconomic scenarios. Because of the limited time and budget, this study projected only the level of chemical oxygen demand (COD).

The socioeconomic scenarios were defined based on the population and economic growth, such as the current growth rate (growth rate during 2004–2014), the target growth rate according to Decision on Socioeconomic Development Plan of Can Tho City to 2020 and Vision to 2030 by Prime Minister in 2013, and a faster growth rate scenario (target growth rate +10%).

17.2.5 Step 5: Evaluate the Effectiveness of the Proposed Inundation and Pollution Control Measures

An existing storm water management model (SWMM) (Rossman 2015) developed for Ninh Kieu District by Huong and Pathirana (2013) was updated with newest data of drainage network (Fig. 17.3). The model's inputs were waste water source discharges and COD resulting from the VENSIM model (step 4), rainfall and tide data (including climate change scenarios in year 2030 according to Van et al. 2012), reservoir and permeable area (according to the current and the proposed measures), and the socioeconomic development scenarios in step 4.

The outputs of the SWMM model were the projected inundation level, volume, and duration and the COD concentration in the water drainage network for each inundation and pollution control measure. Two groups of measures were water uses (saving water) and wastewater treatment (decentralized or centralized).

Key findings from these models were presented at a follow-up workshop with local decision makers to assess the model's utility and to get feedback on how to fine-tune the RDS framework.

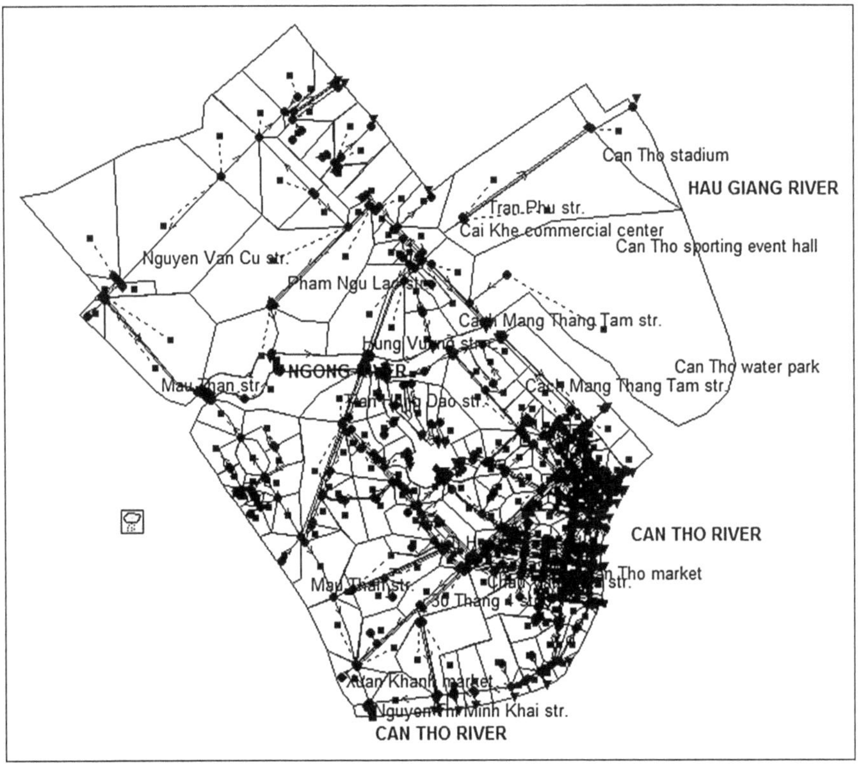

Fig. 17.3 Water drainage network in the storm water management model (SWMM) model (Updated from Huong and Pathirana 2013)

17.2.6 Step 6: Inform Stakeholders on Key Findings and Consult Them for Suitable Inundation and Pollution Control Measures

After the models and scenarios were modified according to the stakeholder's recommendations in the follow-up workshop, we organized the last workshop to inform them of the key findings from the models and to consult them about the suitable inundation and pollution control measures for current and future scenarios (Fig. 17.4). To help the stakeholders take up the key information more easily, the effectiveness/robustness indicators of the measures were visualized in color-ranking tables, with darker red meaning less robustness (negative to no effective) and darker green meaning more robustness (positive to very effective). Concerning investment cost for the proposed measures, due to data limitations levels of investment costs were provided in qualitative ranking forms (very low, low, medium, high, and very high).

Fig. 17.4 Final stakeholder consultation workshop to propose short-term and long-term inundation and pollution control measures

17.3 Results and Discussion

17.3.1 Key Stakeholders and Their Roles in UWM

Stakeholders with high interest in and high influence on UWM were People's Committee (PC), People's Council (PCO), the Department of Construction, the Climate Change Coordinating Office, and the Department of Natural Resources and Environment. These stakeholders were the highest governmental management departments or offices of the city that directly manage issues related to UWM. Among them, PC and PCO were considered to have the highest interest and power, as they are the final decision makers. Other stakeholders appeared to have slightly lower interest and power. However, they still played a significant role as they were major consultants for PC and PCO.

Stakeholders with high interest in and medium influence on UWM were the Department of Planning and Investment; the Centre of Fresh Water and Sanitization, which belongs to the Department of Agriculture and Rural Development; Can Tho Water Supply—Sewerage Joint Stock Company (WSSC); and Industrial Project Management Boards. The Department of Planning and Investment was engaged in assessing and issuing the investment of the UWM projects based on the feedback from the direct UWM-related departments, such as the Department of Construction and the Department of Natural Resources and Environment. WSSC was responsible for urban water supply and drainage, the Centre of Fresh Water and Sanitization was responsible for rural water supply, and the Industrial Project Management Boards managed the use of water and the waste water in the city's industrial zones.

Stakeholders with medium interest and medium influence were the Department of Transportation, the Department of Information and Communications, the Department of Health, ODA Project Management Units, Can Tho City Institute for Socio-economic Development Studies, regional universities, and research institutes. These stakeholders were occasionally asked for feedback for the UWM projects but were not directly engaged in decision-making procedures.

Stakeholders with high interest but low influence were local community, companies, factories, and enterprises. They were daily water users and released significant amounts of wastewater to the environment. However, they were informed of the UWM projects only when these were about to be implemented, rather than having direct channels to engage in the planning, designing, and decision making of the projects.

This stakeholder analysis was based on information obtained from the first stakeholder consultation workshop organized in August 2015 (step 2). Some roles of stakeholders may have changed according to new institutional arrangements.

At the household level, results from household surveys (step 3) also confirmed that the local communities had very high interest in UWM since their livelihoods were based mainly on surface water and pipe water supplied by Can Tho WSSC. Water was used daily mainly for cooking, washing, and sanitation. Water uses for business, manufacture, and services accounted for 24.88% of the surveyed households. These enterprises were mainly small restaurants, coffee shops, car washing stores, bottled water production stores, and so on. In the rainy season, a few households used rainwater for gardening, yard washing, and toilet flushing.

Local communities somehow had less influence on the quality of the supplied water. Over 40% of the interviewees reported that there were occasionally problems related to tap water quality, such as pungent and chlorine smell, or yellow and muddy water. They thought that these problems were probably due to washing water tanks in the WSSC, water disinfection, and repairing water pipes.

The household survey, however, noted that it was the households that were highly influential in changing the domestic water demand: 68% of the interviewees responded that their household water demand increased because of hotter weather, more family members, and more demands for washing machines, showering, and so forth, and they were well aware that they needed to save water. Over 70% of households were applying some water saving actions in daily activities, especially in the dry season. A majority of people were knowledgeable of water saving, and the average of potential monthly saved water per person was 0.6 m^3 (around 20 l/day).

The surveys also found that women had higher interest and understanding of water-related issues. More women than men attended community meetings related to water or environmental issues. They also played a higher role in decision making on water usage for households and small businesses. Moreover, since they were the ones to directly pay the bill, they mostly applied water-saving actions in the family (Fig. 17.5).

17.3.2 Urban Water Challenges and Potential Solutions

In the first stakeholder consultation workshop (step 2), the participants also discussed the possible exogenous uncertainties (X), proposed potential inundation and pollution control measures (levers, L), defining indicators to measure the measures' effectiveness (M), and delineating relationships (R) among X, L, and M.

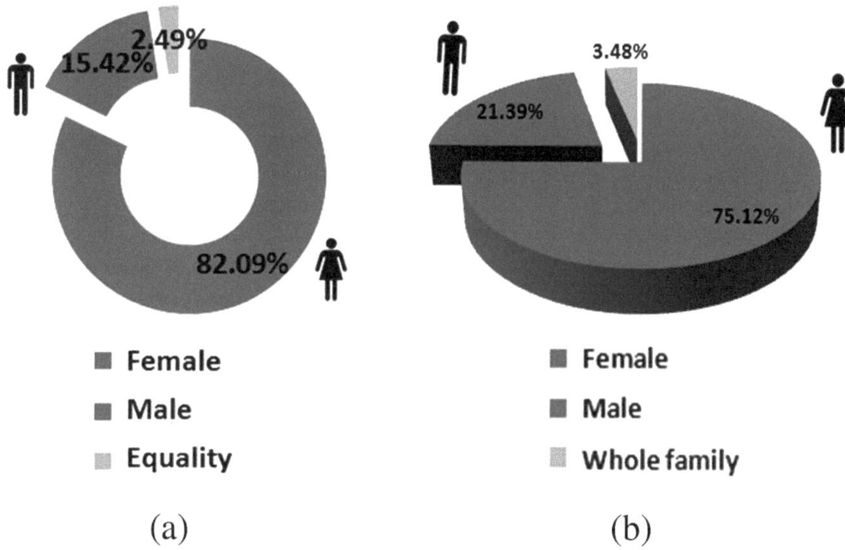

Fig. 17.5 Household's water demands (**a**) and water-use decisions (**b**) by gender

The group discussion shared the same opinions on the key uncertainties (X), their trends, and the degree of uncertainty. The exogenous uncertainty factors X were divided into socioeconomic and nature environment factors.

As for socioeconomics, water demands of the city would increase and be highly uncertain due to the unpredictable migratory increases, together with the uncertainty in urbanization and industrialization growth.

As for the natural environment, the most uncertainty comes from the impacts of climate change. Climate change would cause abnormal floods because of extreme rainfall in the rainy season, not only in Can Tho but also in the whole Mekong River basin. In the dry season, the dropping of water discharge from the Mekong River and local rainfall would increase the risk of seawater intrusion, and combined with a lack of proper waste management, this would increase water pollution. When water is salty and polluted, more uncontrolled groundwater extraction would lead to more land subsidence and, as a consequence, increase inundation in the city. Since saline intrusion is projected to be not too serious for the city in the near future, two main water issues that the city should be now aware of are inundation and water pollution.

The workshop also identified potential measures to reduce inundation and water pollution (levers, L). Two types of measures were identified: structural and non-structural measures. After group discussions and plenary discussion, stakeholders selected the following measures for further evaluation of their effectiveness to reduce inundation and pollution:

- Structural measures (L_s): increasing permeable area (L_s1), upgrading retention reservoirs (L_s2), and upgrading drainage system (L_s3)

Table 17.1 Metrics to evaluate the effectiveness of the proposed measures

Indicators	Metrics (M)
Water demand	Domestic use: current: 130 l/person/day
	Target development: 180 l/person/day
	Fast development: 220 l/person/day
	Saving: 90 l/person/day
Water quality: chemical oxygen demand concentration	Domestic use water standard (TCVN 02–01)
	Drinking water standard (QCVN 01/2009-BYT)
	Surface water for ecology (TCVN 08/2008)
	Percentage of clean water accessed in urban area (100%)
Inundation: inundation points, volume, duration, and level	Point, volume (m^3), duration (min), and height (m), to be defined based on baseline data

- Nonstructural measures (L_n): encouraging water saving (L_n1) and household septic tanks waste water treatment (L_n2)

To evaluate the effectiveness of the measures, stakeholders recommended measures' performance and thresholds (M), as displayed in Table 17.1. Based on the socioeconomic development scenarios, water demand levels were estimated accordingly. At the current economic development, the domestic water demand is 130 l/person/day, while in higher economic development, this level would increase to 180–220 l/person/day. The responses from the household survey showed that if water saving measures are applied, domestic water demand can dropped to 90 l/person/day.

Figure 17.6 shows the relationship diagram (R) drawn by stakeholders during the workshop. This diagram defines the cause and effect relationships among the uncertainty factors X, proposed measures L, and metrics M to evaluate the achievement of L. This exercise was a very efficient approach to help stakeholders visualize the UWM problems and well reflect stakeholder perceptions of the issues discussed. Moreover, such diagrams resulting from stakeholder discussions will help modelers know users' expectations for the outputs (M) and so take into account more realistic and sufficient data and parameters (X) and define better scenarios (X and L) to address stakeholder concerns.

17.3.3 The Impacts of Socioeconomic Development on Total Waste Water Discharge and Quality

The relationship R defined by stakeholders was transferred into a system dynamic model (Fig. 17.7). The model allows users to change parameters according to the socioeconomic scenarios. Three scenarios were analyzed to project socioeconomic development:

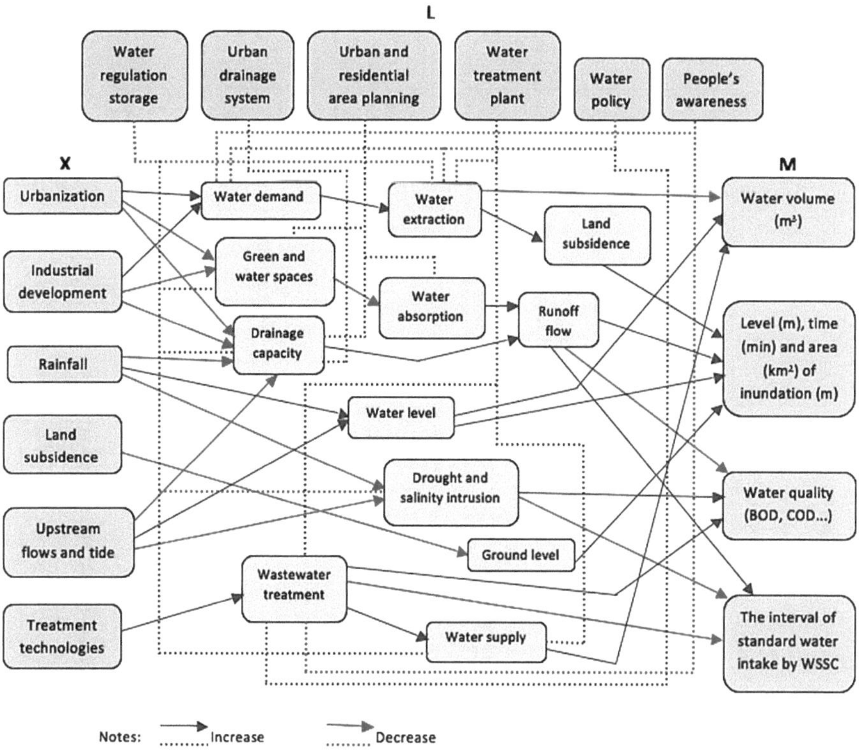

Fig. 17.6 Relationship diagram (R) drawn by stakeholders during step 2 defining the cause and effect relationships among the uncertainty factors X, measures L, and metrics M to evaluate the achievement of L. Continuous lines from X show the consequences of X to the city's water-related problems (pink boxes); dotted lines from L to pink boxes show which problems L will impact. Green lines represent increases; red lines represent decreases

- Baseline scenario X_e0: Current growth rate (based scenario)
- Target scenario X_e1: Target growth rate
- Fast growth scenario X_e2: Faster growth rate

In the baseline scenario, X_e0, Can Tho City's population will increase at the current rate (Fig. 17.8). In the target scenario X_e1 and fast growth scenario X_e2, the population growth rate was projected to increase 9.7% in the 2030s according to the projection in the Decision on Socioeconomic Development Plan of Can Tho City to 2020 and Vision to 2030 by Prime Minister in 2013.

In baseline scenario X_e0, the gross domestic product (GDP) of the city increases with the same rate as the previous period (2004–2013 increased about 14,5%/year and income per person was about \$5630 in 2013). In the target scenario X_e1, the city's GDP is projected to increase 15% per year and income per person in 2030 would achieve \$14,200. In the fast growth scenario X_e3, the GDP would increase an additional 10% higher than the target scenario.

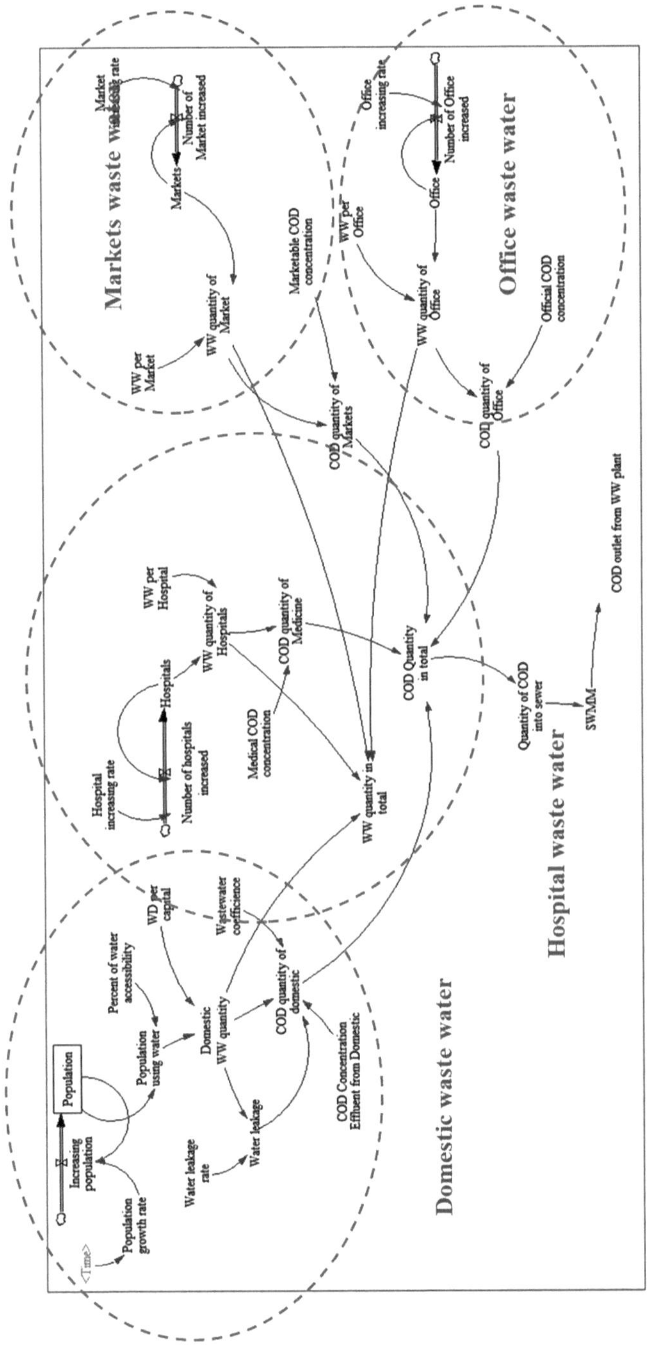

Fig. 17.7 The VENSIM model to project water quality and quantity in different socioeconomic and scenarios

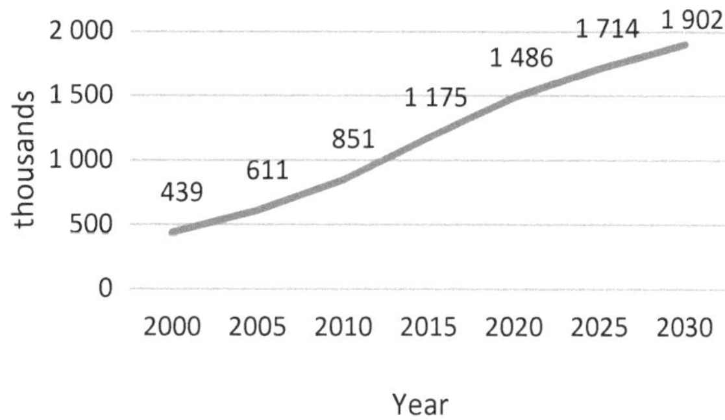

Fig. 17.8 Projection of Can Tho City's population at current rate

Besides projecting the wastewater volume and the quality of water that flows into the drainage water system, the model was used to estimate the effectiveness of non-structural measures such as saving water (L_n1) and treatment of wastewater at the household level (L_n2).

Figure 17.9 presents changes in the quantity of wastewater and COD load into the drainage network in different socioeconomic scenarios (X_e0, X_e1, and X_e2) and the impacts of nonstructural measures (L_n1, L_n2) in each socioeconomic scenario from 2015 to 2030. The model results suggest that socioeconomic development rates will impact strongly the total volume of wastewater and the pollution sources. Up to 2030, in the baseline scenario (X_e0), the total volume of waste water will increase sharply, from around 6.5 million m^3/year in 2015 to about 8 million m^3/year in 2030, and COD load into the environment will increase from around 1.45 million kg in 2015 to almost 2 million kg in 2030. Compared to the baseline scenario X_e0, waste water volume and COD load of the target scenario X_e1 in 2030 will increase significantly: around 28% and 20%, respectively. For the fast development scenario X_e2, the increase of waste water quantity and COD load will be 50% and 38%, respectively.

17.3.4 *The Efficiency of the Measures to Reduce Inundation and Pollution*

Figure 17.9 shows that in the baseline scenario, if waste water is treated at the household level by simple techniques such as septic tanks ($X_e0 \cdot L_n2$), compared to the baseline scenario, in 2030 the total COD load can be reduced by nearly 30%, even though the total volume remains the same. In the target development scenario, household water saving ($X_e1 \cdot L_n1$) will be significantly reduced in both the volume of waste water and COD load: 33% and 24%, respectively. Similarly, for the fast development scenario ($X_e2 \cdot L_n1$), household water saving will reduce the amount of wastewater by 54% and the COD load by 42%.

To evaluate the impacts of the structural measures, we used the SWMM model. We took three structural measures into account, as agreed in the second stakeholder consultation workshop (step 6): increasing 1% permeable area (L_s1) by upgrading

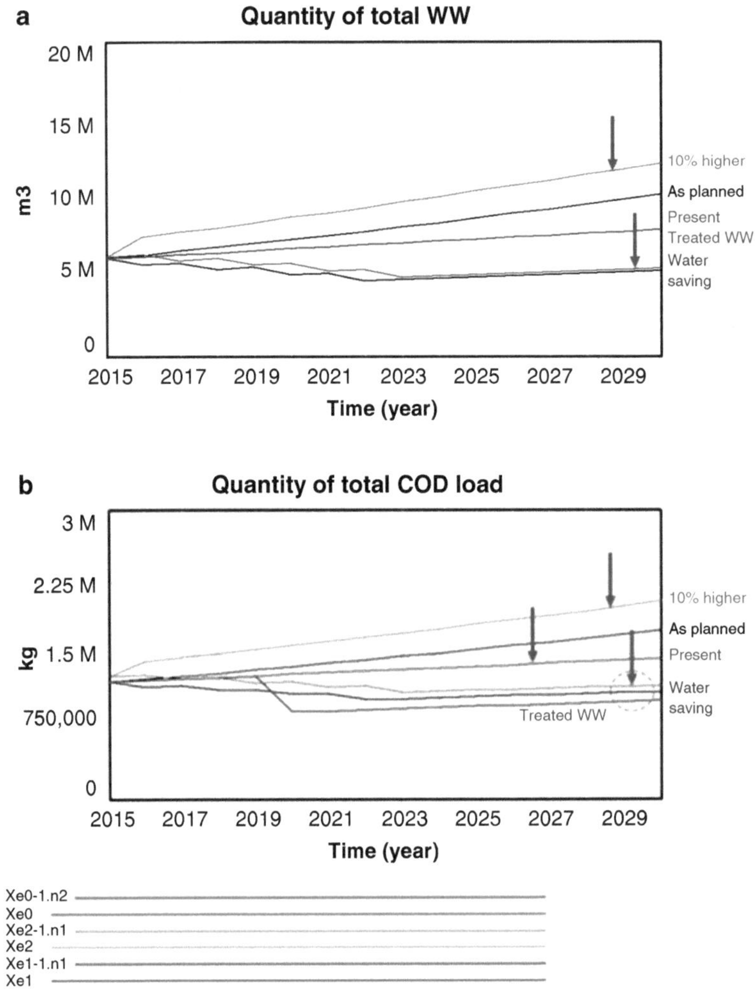

Fig. 17.9 The changes in waste water (WW) quantity (**a**) and chemical oxygen demand (COD) load (**b**) under different scenarios as indicated by the key below: X_e0, current growth rate (base scenario); X_e1, target growth rate; X_e2, target growth rate +10%; $X_e0{\cdot}L_n2$, X_e0 + household septic tanks; $X_e1{\cdot}L_n1$, X_e1 + water saving; Xe2·Ln1: Xe2 + water saving

the sidewalk area, upgrading the retention lake (L_s2), and upgrading one pipeline in the drainage system at a flooding hotspot (area from 30/4 street to Hoa Binh street) (L_s3). To compare their effectiveness, we used "no action" as a baseline measure (L_s0). The model analyzed a number of scenarios—combinations of socioeconomic scenarios (X_e) and climate change and sea level rise scenarios X_n0 (current rainfall and water level) and X_n1 (projected rainfall and water level in 2030)—but in this chapter we show only some scenarios to demonstrate model results and their implications for UWM.

Population as 2015 & Socio-economic growth as 2015					
Metrics (M)		Levers (L)			
		L_s0	L_s1	L_s2	L_s3
Flooding	Point	311	0%	3.54%	0.96%
	Volume (m^3)	634,948	1.19%	4.98%	4.85%
	Time (hour)	1.75	0.57%	0%	100%
	Level (m)	0.41	0%	0%	100%
Water quality	COD concentration (mg/l)	206	0%	0%	38.35%
Investment & maintenance level			VL	M	VH

Robust 0% 100%
Not Robust 0% -100%

Fig. 17.10 Robustness of the structural measures in the baseline scenario. For definitions of levers, see text. *VL* very low, *M* medium, *VH* very high

Projected population in 2030s & Socio-economic growth as 2015													
Metrics (M)		Levers (L)											
		L_s0	L_s1	L_s2	L_s3	L_n1	L_n2	$L_s1.L_n1$	$L_s2.L_n1$	$L_s3.L_n1$	$L_s1.L_n2$	$L_s2.L_n2$	$L_s3.L_n2$
Flooding	Point	326.0	0.6%	3.4%	0.3%	0.3%	0.0%	0.3%	4.3%	1.8%	0.6%	3.4%	0.9%
	Volume (m^3)	1270603.0	0.6%	9.1%	5.9%	25.5%	0.0%	26.1%	30.9%	30.7%	0.6%	9.1%	5.9%
	Time (hour)	8.7	0.2%	11.5%	97.0%	55.3%	0.0%	55.5%	58.7%	97.1%	0.2%	11.5%	97.2%
	Level (m)	0.5	0.0%	2.1%	89.4%	4.3%	0.0%	6.4%	4.3%	89.4%	0.0%	2.1%	89.4%
Water quality	COD concentration (mg/l)	148.0	-1.4%	0.0%	14.2%	-18.2%	39.9%	-17.6%	-18.2%	5.4%	40.5%	41.2%	53.4%
Investment & maintenance level			VL	M	VH	VL	VL	VL	M	VH	VL	M	VH

Robust 0% 100%
Not Robust 0% -100%

Fig. 17.11 Robustness of the proposed measures in 2030 for the socioeconomic baseline scenario and climate change and sea level rise scenario ($X_e0 \cdot X_n1$). For definitions of levers, see text. *VL* very low, *M* medium, *VH* very high

Figure 17.10 shows the effectiveness of the proposed structural measures in 2015 for the baseline scenario $X_e0 \cdot X_n0$. For stakeholders' reference, the investment and maintenance levels were relatively compared. Compared to the baseline measure L_s0, with the current population increasing the permeable area measure by 1% (L_s1) was slightly robust. However, this measure would be more robust if more permeable areas were allocated in the future. In addition, considering its low investment cost, this solution can be more efficient. Upgrading the retention lake (L_s2) is more robust because it will reduce both the number of inundation points and flood volume. Upgrading the water drainage pipeline (L_s3) is highly robust. However, investment and implementation costs for this measure are very high.

Figure 17.11 presents the robustness of structural and nonstructural measures in 2030 for the socioeconomic baseline scenario and climate change and sea level rise scenario ($X_e0 \cdot X_n1$). At the lowest investment and implementation costs, increasing permeable area by 1% combined with household wastewater treatment ($L_s1 \cdot L_n2$) has average robustness. At medium investment and implementation costs, the high-

Projected population in 2030s & Socio-economic growth 10% faster than the city planned										
Metrics (M)		Levers (L)								
		L_s0	L_s1	L_s2	L_s3	L_n1	$L_s1.L_n1$	$L_s2.L_n1$	$L_s3.L_n1$
Flooding	*Point*	337	0.0%	3.9%	0.3%	3.3%	3.9%	6.8%	4.5%
	Volume (m^3)	1,986,618	0.4%	11.1%	6.9%	49.7%	50.1%	53.5%	50.3%
	Time (hour)	8.90	1.1%	11.2%	97.0%	27.4%	27.5%	38.8%	97.1%
	Level (m)	0.51	0.0%	2.0%	88.2%	9.8%	9.8%	9.8%	88.2%
Water quality	*COD conc.(mg/l)*	126	-0.8%	-3.2%	13.5%	-34.9%	-34.9%	-34.1%	-9.5%
Investment & maintenance level			**VL**	**M**	**VH**	**VL**	**VL**	**M**	**VH**

Robust 0% 100%
Not Robust 0% -100%

Fig. 17.12 Robustness of the proposed measures in 2030 for the fast development scenario and climate change and sea level rise scenario ($X_e2{\cdot}X_n1$). For definitions of levers, see text. *VL* very low, *M* medium, *VH* very high

est robustness measure is the combination of upgrading the retention lake and household waste water treatment ($L_s2{\cdot}L_n2$). At very high investment and implementation costs, the highest robustness measure is the combination of upgrading the drainage system and household water saving ($L_s3{\cdot}L_n2$).

Figure 17.12 shows the robustness of the proposed measures in 2030 as in the fast development scenario and climate change and sea level rise scenario ($X_e2{\cdot}X_n1$). In this scenario, only the very high investment and implementation costs will be effective in both inundation and pollution reduction (upgrading drainage system, L_s3). The other measures only be effective will in reduction of inundation. It should be noted that in this study we analyzed only household water treatment as a measure to reduce pollution. Therefore, in 2030, under the pressure of fast socioeconomic development and climate change, more robust solutions for wastewater treatment should be considered (e.g., centralized wastewater treatment).

17.3.5 Stakeholders' Priority on the Proposed Measures

In the final stakeholder consultation workshop (step 6), we presented the color-ranking tables shown in Figs. 17.10, 17.11, and 17.12 and asked them to select the measures that they thought were suitable for short-term (current) and long-term (2030) implementation. For the current situation, stakeholders proposed the combined measure of increasing permeable area, upgrading the retention lake, and household water saving. They explained that this combined measure was simple to implement and not so expensive for investment and implementation. For long-term solutions, based on information provided from the models, stakeholders proposed the combination of upgrading the retention lake, upgrading the drainage system, and household wastewater treatment. However, since this combined measure required high cost and household wastewater treatment required more technical support and monitoring to ensure treatment quality, they agreed that pilot projects needed to be carried out. In addition, the stakeholders recommended more research be conducted to evaluate the other wastewater treatment measures such as semicentralized or centralized wastewater treatment.

17.4 Conclusion

This study shows that the Robust Decision Support (RDS) framework is a comprehensive, logical, and reliable framework. Applying this framework provides a full understanding and visualizes an overview of the issues during the decision-making process. Visualizing the complex model's results with simple color-ranking tables enabled the decision makers to come up with optimal solutions for particular situations (scenarios); therefore, it helps reduce individual and fragmented decision making. The participative nature of the framework would ensure social and political acceptance of the project's outcomes.

However, when the RDS framework was applied, we faced the following challenges. (a) Modeling socioeconomic issues and nature environment issues was complex and time-consuming, and it required a large amount of qualitative data (especially for model calibration and validation). (b) Even though model results were simplified into color-ranking tables, when there were so many scenarios to examine, the many tables provided might cause confusion for stakeholders. (c) The study was interdisciplinary, but it was not always easy to form a complete interdisciplinary research team.

Acknowledgments This study was funded by the Sustainable Mekong Research Network (SUMERNET)—Stockholm Environment Institute (SEI) Asia. We would like to express our gratitude to Dr. Eric Kemp-Benedict, Dr. Chayanis Krittasudthacheewa, Dr. Chu Thai Hoanh, Ms. Ha Nguyen, and Mr. Agus Nugroho from SEI for providing us with great technical and logistic support during the study implementation. We thank our colleagues from the Can Tho City Climate Change Coordination Office, especially Mr. Ky Quang Vinh and Ms. Chau Thi Kim Thoa, for their great assistance in data collection and their close collaboration in ensuring stakeholder engagement during the study implementation. We also would like to thank the staff from the city's departments and offices, researchers from Can Tho University, and local research institutes for their informative discussions in the consultation workshops. Last but not least, we acknowledge the interviewees in Ninh Kieu District for the valuable information they provided during the household survey.

References

CSIRO. (2012). *Planning for sustainable urban water systems in adapting to a changing climate – a case study in Can Tho City, Vietnam*. In: M. Nguyen et al. (Eds.), CSIRO.

Erban, L. E., Gorelick, S. M., & Zebker, H. A. (2014). Groundwater extraction, land subsidence, and sea-level rise in the Mekong Delta, Vietnam. *Environmental Research Letters, 9*(8), 84010 Available at: http://stacks.iop.org/1748-9326/9/i=8/a=084010?key=crossref.b639ea338e34289 9358515e74a86f960.

Huong, H. T. L., & Pathirana, A. (2013). Urbanization and climate change impacts on future urban flooding in Can Tho city, Vietnam. *Hydrology and Earth System Sciences, 17*(1), 379–394.

Mitchell, R. K., Agle, B. R., & Wood, D. J. (1997). Toward a theory of stakeholder identification and salience: Defining the principle of who and what really counts. *The Academy of Management Review, 22*(4), 853. Available at: http://www.jstor.org/stable/259247?origin=crossref. Accessed 1 Sept 2017.

Moglia, M. et al. (2012). Application of the water needs index: Can Tho City, Mekong Delta, Vietnam. *Journal of Hydrology, 468–469*, 203–212. Available at: http://www.sciencedirect.com/science/article/pii/S0022169412007159. Accessed 8 Oct 2014.

Neumann, L. E., et al. (2013). Water use, sanitation and health in a fragmented urban water system: Case study and household survey. *Urban Water Journal, 11*(3), 1–13.

Prime Minister. (2013). *Quyết định 1533/QĐ-TTg năm 2013 quy hoạch tổng thể phát triển kinh tế xã hội Cần Thơ 2020*. Available at: https://thuvienphapluat.vn/van-ban/Thuong-mai/Quyet-dinh-1533-QD-TTg-nam-2013-quy-hoach-tong-the-phat-trien-kinh-te-xa-hoi-Can-Tho-2020-205926.aspx. Accessed 2 Sept 2017.

Ramalho, E. C. (2017). Pore water pressures and slope stability impacts of 25 years of groundwater extraction on subsidence in the Mekong delta, Vietnam. *Environmental Research Letters, 9*. http://iopscience.iop.org/article/10.1088/1748-9326/9/8/084010/meta

Rossman, L. A. (2015). *Storm water management model user's manual*, Available at: http://www.epa.gov/water-research/storm-water-management-model-swmm

Van, P. D. T. et al. (2012). A study of the climate change impacts on fluvial flood propagation in the Vietnamese Mekong Delta. *Hydrology and Earth System Sciences, 16*(12), 4637–4649. Available at: http://www.hydrol-earth-syst-sci.net/16/4637/2012/. Accessed June 13, 2013.

World Bank. (2014). *Can Tho, Vietnam: Enhancing urban resilience*. Washington, DC: Work Bank Group.

Chapter 18
"The Song Remains the Same": Examining the Outcomes of Past Hydraulic Engineering and Agro-modernization Schemes in Northeast Thailand

David J. H. Blake

Abstract Competition for water and other natural resources in the transboundary Mekong River basin is increasing, leading to a greater propensity for conflict, at both the international and intra-national, local level. Recently, the Thai government has resurrected a decades-old plan to divert considerable volumes of water from the Mekong River into the northeast region, ostensibly for irrigation purposes, thereby re-igniting old concerns by downstream states that mainstream river flows will be reduced and water quality impaired, especially during the critical dry-season period. Such moves reinforce the impression that riparian states are increasingly exerting their sovereign rights to utilise Mekong flows in response to a perceived weakening of the legitimacy of the Mekong River Commission and in the face of a de facto rapidly expanding basin-wide hydraulic construction paradigm, most especially in China and Laos. This chapter investigates some of the historical context and socio-environmental impacts of earlier large-scale irrigation and agribusiness promotion projects, through a case study of the Nam Songkhram basin, a Mekong sub-basin with eco-hydrological attributes similar to those of Cambodia's Tonle Sap system. It argues that critiques of past developments have not been adequately internalized by political and bureaucratic decision makers charged with water resources policy and planning, raising interesting questions about the social, economic, and ecological prospects for the latest diversionary scheme plan.

Keywords Northeast Thailand · Mekong diversion · Irrigation development · Environmental and social impacts · Nam Songkhram basin

---

D. J. H. Blake (✉)
Independent Scholar, Taunton, Somerset, UK

M. A. Stewart, P. A. Coclanis (eds.), *Water and Power*, Advances in Global Change Research 64, https://doi.org/10.1007/978-3-319-90400-9_18

18.1 Introduction

A handful of large transboundary river basins in Asia are attracting increasing media and scholarly attention for their potential to foment future conflict,[1] as nations vie for scarce natural resources, especially water, in a context of rapid economic growth, rising material consumption, industrialization, urbanization, mass migration, and intensification of agriculture, among other pertinent factors. The major rivers rising in the Himalayas that disgorge their flows in the Indian Ocean and South China Sea are considered especially problematic in terms of a perceived growing number of conflicts at the international and more localised disputes at the subnational level, including the Indus, Ganges-Brahmaputra, Salween, Mekong, and Red River Basins (e.g. Klopper 2008; Chellaney 2016; Wirsing et al. 2013; Yeophantong 2014). In the case of mainland Southeast Asia's transboundary rivers, much recent scholarly attention has focused on the dangers of environmental degradation and concomitant social harm arising from hydraulic infrastructure development, occurring in water landscapes that are already sensitive and vulnerable to a wide range of local and external threats arising largely from poorly planned and executed broader development paradigms (e.g. Molle et al. 2009a, b; Grumbine et al. 2012; Matthews and Geheb 2015).

In April 2016, concerns were being raised about the impacts of drought being felt across large parts of peninsular Southeast Asia and the Mekong River was reportedly at its lowest level in a hundred years (Webster 2016). Blame for the drought in Vietnam and the river's lower reaches was variously being placed on El Niño and climate-related factors or a proliferation of dams holding back flows upstream. Around the same time, a number of news articles reported on several ongoing projects that were adjunct components to more significant proposals by the Thai government to divert water from the Mekong mainstream to the northeast region for irrigation purposes. These reports, filed by a range of international and local media sources about riverside water pumping activities and proposed state plans, raised serious concerns among downstream government and civil society groups about the potential for negative environmental and social impacts (e.g. Besant 2016; Cochrane 2016). Indeed, the flurry of stories about Thai designs to usurp significant Mekong flows briefly diverted media attention away from the ongoing international controversy concerning mainstream Mekong hydropower dam construction within Laos, to an issue that has gained notably less critical attention from regional observers. Yet, should the Thai diversion plan proceed to a comprehensive construction phase, the scheme has the potential to more fundamentally alter the hydrology and basic ecological quality of the Mekong mainstream than any single low-head dam could manage, while also radically transforming the target northeast region through multiple mechanisms and processes. It has been criticised for having the potential to worsen food security amongst vulnerable

[1] Here it should be noted that "conflict" is being used not as a synonym for violence but, rather, to refer to a variety of conflicts and contestations over water that mostly are non-violent, but may occasionally involve force and violence. However, several authors have vigorously argued that the notion of "water wars" between nations is much exaggerated and lacks credible empirical evidence (e.g. Swain 2001; Zeitoun and Warner 2006).

households downstream as far as the Mekong Delta, with expansion of irrigation in Northeast Thailand described as "self-defeating given the nature of the geophysical environment" due to concerns about mobilised sub-soil salinity would "overwhelm initial gain in productivity" (Fullbrook 2013, p.75).

The present diversion scheme is generally referred to by Thai state agencies as the Khong-Loei-Chi-Mun (KLCM) Project, and forms the latest iteration in a long succession of similar vast hydraulic engineering schemes stretching back to the early 1960s and the monumental Pa Mong dam proposal (Sneddon 2015). During a period from 1957 to 1978, key Western donors of the Committee for Coordination on the Lower Mekong Basin (also called the Mekong Committee) and powerful players within the Thai state machinery considered that development of the Pa Mong dam, or any other large-scale mainstream dam on the Mekong mainstream, was essential for unlocking the agricultural potential of the Northeast and would catalyse wider socioeconomic development goals, precipitating significant financial resources directed at its feasibility and planning phases. As an example of how the project was portrayed, the following quote from the Mekong Secretariat (1977, p. 50) is typical:

> For the major irrigable areas such as the riverine lowlands and the plateau of northeast Thailand, however, it has been shown that maximum benefits will be attainable only by regulation of the mainstream, and it is upon this well-established fact that the Mekong Committee's long-term planning for agricultural development is based. For example, it has been determined that the Pa Mong reservoir could be used to irrigate the lowlands and topographically suitable low terraces in the Mun-Chi Basin in northeast Thailand as well as the lowlands adjacent to the Mekong, besides ensuring complete flood control for riverside land from Pa Mong to the mouth of the Mun in Ubol.

My intention is not to repeat existing scholarship[2] but, rather, to draw attention to a less closely examined aspect of the regional irrigation development puzzle. Specifically, I focus on a discernible tendency for powerful actors with the authority to implement megaprojects to repeatedly overlook or ignore past development failures and fail to take into account the socioecological fundamentals of the northeast region, which even a superficial historical examination of the literature would reveal.

18.2 What's Being Proposed, and Why at This Juncture?

In its present iteration, the KLCM Project qualifies as a "megaproject" (Flyvbjerg 2005) by dint of the sheer scale of its ambition, both spatially and temporally, wrapped up in an estimated $23.7 billion price tag. When complete, the scheme plans to irrigate an area of 2.87 million ha across the Northeast,[3] consuming an

[2] For a detailed politico-historical analysis of the subsequent large-scale irrigation development paradigm in the northeast, the reader is directed elsewhere (see Sneddon 2003; Molle and Floch 2008; Molle et al. 2009a, b; Blake 2012).

[3] Estimates suggest that irrigation command area coverage in the Northeast up to 2011 was approximately 1.27 million ha or less than 10% of the total arable area (MRC 2011).

annual water volume of 4 billion cubic metres diverted out of the Mekong (Wangkiat 2016a). According to the project's main implementing agency, the Royal Irrigation Department (RID), work officially commenced in 2012, with a 16-year time frame given for completion. To turn this ambitious vision into reality, the RID plans to construct dams on the few remaining northeast Thai tributaries of the Mekong not yet dammed, including the Loei and Nam Songkhram Rivers; excavate 140 km of tunnels extending from the Loei River through two hill ranges; dig 1350 km of main canal (lateral canals are not mentioned); and construct a range of other water-pumping and transfer infrastructure. It may also require the construction of a mainstream transboundary dam to be feasible, which is currently being pursued by another government bureaucracy, the Department of Alternative Energy and Efficiency at Pak Chom, Loei Province (Dudley et al. 2012). The RID claims that the KLCM scheme will provide economic benefits of 324 billion baht ($10.125 billion) and will provide irrigation to 1.72 million households when complete, with beneficiaries supposedly seeing average income rise by approximately 200,000 baht ($6250) per annum (Wangkiat 2016a, b). To put this figure in comparative perspective, the gross regional product of the Northeast in 2013 was estimated to be about $2425 per capita (National Economic and Social Development Board 2013). Because the project would require diversion of water from the Mekong, such a scheme would nominally require official notification to the Mekong River Commission (MRC) for deliberation among the four member countries and other stakeholders according to the Procedures for Notification, Prior Consultation and Agreement mechanism (Boer et al. 2016). Yet, despite concerns expressed by its downstream neighbours about reports of water extraction from the Mekong in early 2016, the Thai government remained remarkably ambiguous about its actual intentions and elected not to officially notify the MRC, fuelling greater suspicion and mistrust by various state and nonstate actor groups in Cambodia and Vietnam about Thai unilateralism (Cochrane 2016). In a May 2016 comment to a Thai newspaper, *Te Vanuth*, the secretary-general of the Cambodian National Mekong Committee stated, "We have not received any official technical information about the water diversion projects in Thailand until now" (Wangkiat 2016b).

Among other riparian nations, Vietnam has proven to be most outspoken against the proposed northeast Thailand diversion project, as it potentially stands to lose most should the mega-project proceed to full construction. In particular, it has expressed worries about potential economic and environmental impacts in the Mekong Delta. Such concerns were underlined by comments by Ky Quang Vinh, director of Can Tho City Climate Change Coordination Office: "The project would divert water from the river, which would decrease the amount of water in the delta, especially in the dry season, when the water level is normally low. This would directly impact farmers in the delta and thus the food basket of Vietnam" (Rujivaranom 2015). Beyond freshwater flow reduction in the dry season, other concerns expressed by Vietnamese experts and other regional observers have included saltwater intrusion, reduction of sediment and erosion, pesticide contamination of runoff, and loss of fish productivity. Such downstream concerns about the Thai diversion scheme are not new, however, but can be traced back to earlier proposals to divert the Mekong in the late 1980s and early 1990s that also sparked tensions within the international river basin organization charged with managing the Mekong's transboundary water resources prior to the MRC's establishment in 1995, namely, the Interim Committee for the Coordination of Investigations in the Lower Mekong Basin (Sneddon 2003; Mirumachi 2012). Unilateral Thai efforts to implement the Khong-Chi-Mun (KCM) project, as it was then known, led to a diplomatic spat that not only involved the Interim Mekong Committee members but also drew in members of the international donor community and development practitioners. The cross-border spat eventually led to the ignominious expulsion from Thailand of Chuck Lankester, the executive agent of the Mekong Secretariat based in Bangkok, who was perceived by the Thai government as being too sympathetic to the Vietnamese position and accused of interference in Thai sovereignty over the proposed diversion scheme (Mitchell 1998). It was partly such tensions among members, among other contemporary geopolitical factors, which contributed to the 1995 signing of the Mekong Agreement and establishment of the present MRC, which relocated the organization headquarters away from Bangkok and into the Mekong basin itself.

The KLCM project's latest reincarnation seems to be predicated on several factors. First is the increased volume of water in the Mekong during the dry season that is both noted already and anticipated in the future due to greater releases from Chinese hydropower dams upstream, thereby marginally diminishing the head that water must be pumped up to supply a pan-northeast irrigation scheme. Second is a perception that the MRC, which Thailand has never really fully endorsed as a transboundary river basin organization, has declining relevance and jurisdiction over the river, allowing individual nations to pursue their sovereign interests more vigorously than before the 1995 agreement (Kossov 2016). This perception has been strengthened by a failure by the MRC to adequately oversee and regulate the damming of the mainstream within Lao People's Democratic Republic at Xayaburi and Don Sahong dams (the former dam project primarily serves Thai energy needs and economic interests).[4] Third, there have been concerns that drought events are gradually worsening across the lower Mekong basin, which seems to be prompting Thai leaders to move forward with Mekong diversion scheme to supply internal areas of scarcity (including urban and industrial needs), justified through a national water security rhetoric. Fourth, there has been a period of relative political stability afforded by the incumbent military junta regime of General Prayuth Chan-ocha (August 2014 to present), which through using blunt instruments of power at its disposal[5] feels it can seize the opportunity to push through a megaproject that has been stalled by political instability in the past. Earlier governments, while trying to pursue their own versions

[4] In early 2017, the Lao government announced its intention to construct a third dam at Pak Beng, between Luang Prabang and the Thai border, to be built by a Chinese state-owned company.

[5] An example would be Article 44, a statute introduced under the present military regime that allows the prime minister to issue orders that he deems necessary to "strengthen public unity and harmony" or to prevent any act that undermines public peace. In practice, it has been used as a means to silence government critics through such means as prohibiting political gatherings of over five persons and censorship of the internet, press and mass media.

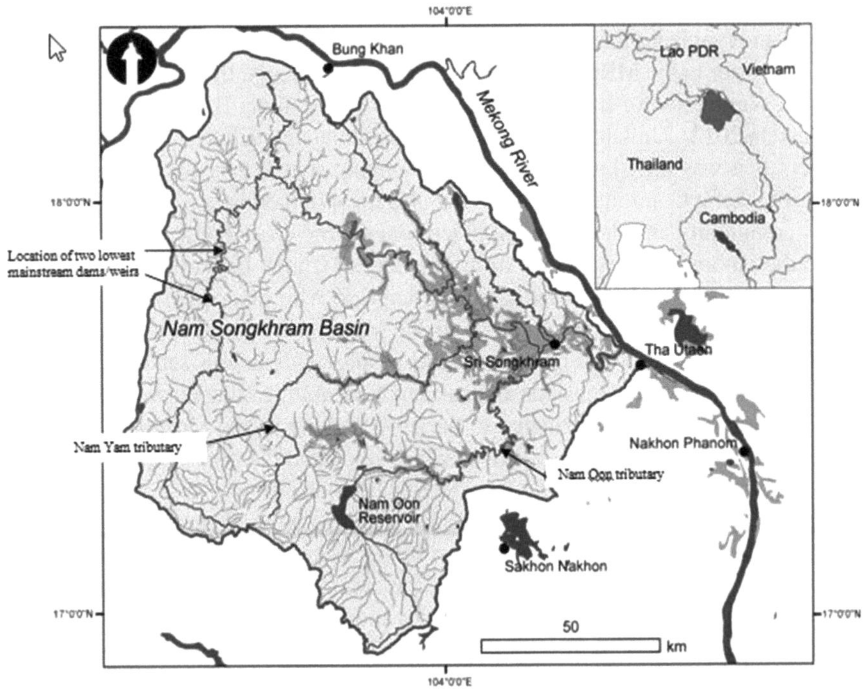

Fig. 18.1 Map of the Nam Songkhram Basin, showing main tributaries, areas of temporary rainy season inundation (shaded blue), permanent water bodies and its spatial relationship to the mainstream Mekong River. (Adapted from Kummu et al. (2006))

of this project, were often thwarted by a mix of domestic opposition and, perhaps more crucially, the limited life-span of their own governments, which did not allow the schemes to proceed beyond the planning and early implementation phases before a new regime came in and tore up the old contracts, only later kick-starting its own near-identical pet diversion scheme with new consultants and contractors.

This chapter examines the prospects for contemporary large-scale hydraulic engineering projects proposed for imminent implementation, in the context of an accompanying expansion of agribusiness activity and concomitant agricultural intensification in northeast Thailand, through a case study of the Nam Songkhram basin (NSB), a key sub-basin of the Mekong (see Fig. 18.1). It carefully considers the past experience of pro-agribusiness policies and practices enacted by the Thai government and the implications to socioeconomic and socioecological outcomes across a range of scales, from local to more regional levels. The chapter questions the feasibility of a development model that has demonstrably failed in the past, even with generous state financial incentives, yet is being promoted once more by powerful actor groups with synergistic interests across the military, bureaucracy, and private sectors. Such questions go to the heart of understanding patterns and processes of environmental transformation, agricultural sustainability, and socioeconomic development across the lower Mekong basin at the present juncture.

18.3 Case Study: The Developmental Context of the Nam Songkhram Basin

The NSB was, until the mid-1960s or so, a remarkably ecologically undisturbed river basin in terms of a diverse and healthy natural resource base prior to any large-scale hydraulic or agricultural development schemes (Blake and Pitakthepsombut 2006). Because of their proximity to Indochina, the provinces bordering the Mekong River were viewed by both Thai authorities and allied Western nations as a geographical space that was strategically important and at risk of insurgent activity by communist forces, in the context of a growing regional geopolitical struggle between global superpowers. Against rising military tensions in the Northeast in the early 1960s, the US government received a Thai government request for financial assistance to build a large-scale storage dam and irrigation scheme in Sakhon Nakhon Province and responded favourably. Subsequently, relevant US state agencies worked with partner agencies in Thailand to plan and expedite this project, while a preferential loan agreement was arranged.[6] The overt justifications given for pursuing the project employed conventional rhetoric about controlling downstream floods and drought to reduce poverty and stimulate the local economy through the provision of irrigation, which were used to veil a less overt political agenda which underlay unfolding security priorities on the ground. The location of the dam project in heavily forested hills was significant, as they harboured the second most important base of the Communist Party of Thailand, a fact recognised in a report for the US Operations Mission (Wildman 1970). This was confirmed to me by a member of the consultancy team hired to manage the development project in the 1980s, who described it as "a political device, aiming to assert central government rule in the area" (Tony Zola, interviewed 3 August, 2010, cited in Blake 2012). The project required the resettlement of an estimated 1800 households within the 85 km^2 reservoir footprint, mostly to sites above the reservoir flood level that did not benefit from irrigation, some of whom were still fighting for compensation for lost lands 30 years after construction of the dam (Blake 2012). In short, the project was an expression of strategic soft power designed to project both US and Thai government hegemony over the region, but it had material consequences on thousands, creating winners and losers.

The Lam Nam Oon Irrigation Project (LNOIP) remained the only major foreign-financed infrastructural project investment in the NSB for decades, and over time significant financial and technical resources were lavished on it, as it gradually took on the status of a model for other US Agency for International Development (USAID) assistance in the Thai water and agriculture sector. Ultimately, the LNOIP cost $123.8 million spread over two phases, a sum shared by both US government loans and direct Thai government contributions. During the long period of establishment, it transformed from being labelled as primarily

[6]The Lam Nam Oon Irrigation Project (LNOIP) was kick-started with an initial $3.5 million loan from the United States Agency for International Development (USAID) in September 1967, payable over 25 years.

an irrigation development project (notably, it failed to deliver goals set against its own criteria of success) to being publicly presented as a rural development and agribusiness promotion project (in which it demonstrated temporary success, but it later failed in this objective, too, as pioneer firms deserted the region, leaving multiple socioecological problems in their wake) (see Blake 2012). The project was considered uneconomic by USAID's own internal evaluation team in the early 1980s, with every scenario considered producing benefit/cost ratios of less than one (Schoux et al. 1981). In the first decade postconstruction, farmers were slow to adopt irrigation over their traditional farming methods, but once they did through various inducements and subsidies, including provision of previously absent agricultural extension support and a market for produce, by 1990 it was considered the "outstanding irrigation project in all of Thailand" (Skogerboe and Merkely 1996). However, such accolades did not last beyond the life of the external financial and technical support from USAID, and once that was withdrawn the project's reputation declined, outside of RID itself. The number of households practicing contract farming of high-value vegetable and flower crops and seed peaked at over 4000 in 1993 but by 2009 had reduced to just over 1500 households. Likewise, while there was an increase from 3 to 13 agribusiness companies locally active between 1985 and 1990 (Dolinsky 1995), the number had fallen to 9 companies by 2009. Areas cultivated in the dry season steadily declined too, while farmers switched from high-value crops back to low-value, low-risk rice cultivation, a trend previously seen as retrogressive by the American advisors. A major problem with the original LNOIP rationale was persuading farmers to adopt irrigation practices on their farms at all, let alone pay fees to access water that would cover just minimal operation and maintenance costs. Many farmers regarded water in their fields as constituting a natural gift from the gods and resented the control and interference in their lives and livelihoods from externally imposed bureaucratically-led irrigation development. Other farmers might want irrigation water at certain times, but no secondary or tertiary canals were provided to distribute it to their fields, as it had been assumed that farmers would build these themselves. The irrigation schedule was decided from the RID headquarters in Bangkok, not by local officials or water users. Even up to 2009–2010, when I spent several months doing fieldwork in the area, farmers were reluctant to pay any water fees to the RID, and some had unpaid bills stretching back three decades or more, which were only recovered when the land was sold (Blake 2012).

The LNOIP as seen today has left a legacy of environmental deterioration that stretches beyond the immediate footprint of the command are, which stretches far downstream in the NSB, due to eco-hydrological changes. The heavy use of pesticides and other agrichemicals has also had a serious impact not only on the health of those that adopted intensive, irrigated farming but also, perversely, on other residents who receive secondary impacts of blood poisoning (Blake 2012). It is indicative of a wider institutional and governance problem, as the state and a range of international development agencies tried to replicate and expand the basic development model across the region from the 1970s onward, as outlined in the following section.

18.3.1 The Rise and Fall of Agro-modernization Schemes in the Nam Songkhram Basin

The NSB, in its lower sections especially, has been targeted in the past for joint state and private sector agri-modernization schemes that have, by and large, been remarkably unsuccessful if not outright failures. Spurred on by investments from European, American, and Japanese transnational agribusiness firms, Thailand became a favoured country for agri-food export-led development in the late 1980s and early 1990s, leading to it being variously referred to as a "newly industrialized country" and a "new agricultural country" (Falvey 2000). Thailand was targeted for high growth potential in the canned tomato industry, some of it aimed for direct export, though much of the output was processed into tomato paste for use in domestic processed food products, in particular canned fish. Most of the production of processing tomatoes was concentrated in the north and northeast of Thailand, with "royal projects" being particular growth nodes (Pritchard and Burch 2003). One of the royal-project food-processing factories was located in Sakon Nakhon Province not far from the LNOIP, which itself had become a target for US-led agribusiness expansion in the 1980s (see Table 18.1). A major factor spurring the expansion of the processing tomato industry was linked to an incentive scheme promoted by the government's Board of Investment (BOI, plus readily available credit from a burgeoning banking and finance sector after years of near double-digit growth in GDP. Generous incentives from the BOI were extended to companies that located away from southern and central areas of the country, with at least eleven separate

Table 18.1 Agribusiness ventures established in the Nam Songkhram basin

Company	Year established	Main activities	Current status
Tawan Farm	1978	Contract farming tomatoes	Closed
Royal project: Doi Kham Foods	1982	Contract farming and agri-processing (esp. canning tomatoes)	Still operational but at a smaller scale
Northeast Agricultural Company Ltd. (NACO)	1986	Contract farming, owned plantations (esp. canning tomatoes), and agri-processing	Closed
Sun Tech Company Ltd.	1984	Contract farming, owned plantations (esp. canning tomatoes and pineapple), agri-processing, eucalyptus plantations	Closed
Tung Songkhram Industry Ltd.[a]	1990	Eucalyptus plantations	Closed
Asia Tech Group Ltd. (incl. Asia Tech Pulp and Paper Ltd.)	1996	Tomatoes for canning, and fast-growing industrial tree plantations for planned pulp and paper mill	Closed

Source: Adapted from Guayjaroen (2001)

[a]This company was bought and later became Asia Tech Group Ltd. (Blake and Pitakthepsombut 2006)

financial subsidies available for the canned food industry (Pritchard and Burch 2003). At the same time, contract farming of tomatoes and similar cash crops was attractive to Thai smallholders, as "it held out the prospect of a secure return far in excess of what most growers could expect from other crops" (ibid., p. 212). But whether such economic returns could be sustained was an entirely different matter.

Four types of investors were attracted to the Northeastern agribusiness field in the early 1990s:

- Newly established specialist tomato-processing companies
- Existing fruit and vegetable companies moving into the sector
- Foreign investors, combining with local interests to create new joint venture companies
- Large and diverse Thai-owned conglomerates which had investments in a wide range of projects in a variety of sectors, which subsequently moved into the processing tomato industry

Pritchard and Burch (2003) highlight the rise and decline of a single specialist processor that established itself in the NSB and became one of a small number of major players in the local agribusiness sector. The Northeast Agricultural Company Ltd. (NACO) was established in 1986 to produce tomato paste, mainly for the domestic canned seafood market, with its main factory located in Nong Khai Province, adjacent to the Nam Songkhram floodplain. It secured loans from the Asian Development Bank and the International Finance Corporation, as well as the Thai Military Bank and equity contributions from three wealthy Thai private investors, for a total investment of $8.81 million. It aimed at growing one-third of its production on its own 1600 ha "nucleus estate", with the remaining production sourced from 3000 local contract farmers. Operations began in early 1988, and as production initially boomed, it opened up a second processing plant in neighbouring Nakhon Phanom Province in 1991. By this time, 6000 smallholder contract farmers were supplying an average of 10 tonnes of tomatoes to the two NACO plants, and export sales had reached $10–12 million annually. The Bank of Agriculture and Agricultural Cooperatives provided credit to farmers for agricultural inputs such as seed, fertiliser, and pesticides, while NACO employed 40 extension agents tasked with selling necessary inputs to growers. Unskilled planters and pickers were paid $1.20 per day on the NACO estate, while more skilled workers were paid $1.60–2.40 per day. At its peak during the 4-month growing season of January to April, up to 1500 workers were employed on the nucleus estate and a further 75 worked in the factory.

However, after initial stellar growth, from 1993 onward there was a rapid decline in the NACO business model, following intense competition in tomato paste export sales from China and Turkey. Export sales fell from 80% of turnover in 1992 to just 20% the following year. In a bid to survive, there was a restructuring in equity and a subtle change in name to the Northeast Agricultural Investment Company (NAICO), and in 1994 the company came under the control of a joint venture between Thai-owned General Finance and Securities and Brierley Investments Ltd., a New Zealand holding company headed by businessman, Sir Ron Brierley. The GF-Brierley Company, as it was known, began bringing in foreign agricultural experts to modernize NAICO and make it more financially competitive and technologically efficient, including installing new irrigation systems and replacing labour-intensive hand picking of tomatoes with mechanized harvesting.

Although the operations were encountering difficulties for several years beforehand, the Asian economic crisis that began in Thailand in July 1997 was the final straw for NAICO and other similar agribusinesses in the NSB. General Finance and Securities was declared bankrupt, one of 58 finance companies deregistered by the Bank of Thailand, while GF-Brierley was placed under administration and had to restructure its investments. The joint venture in NAICO became subject to considerable recriminations among its foreign and Thai shareholders, leading to legal action in a bid to recover debts (Pritchard and Burch 2003). Meanwhile, tens of thousands of northeastern smallholder farmers contracted to grow tomatoes for these companies became collateral damage in the global competition for the processing tomato industry, with farm-gate prices offered by middlemen falling below production costs.

A total of 26 companies were involved in the processing tomato industry across Thailand at its peak, many of which experienced a similar economic roller-coaster ride weathering external price competition, such as in 1993 and the effects of the Asian economic crash four years later, which saw the Thai baht float against the US dollar and most foreign debts become unserviceable as a result. Many of these companies had to completely restructure or risk folding, such as Sun Tech Company Ltd., which formed one part of a large industrial conglomerate (the NTS Group) based on the Eastern Seaboard whose core business activity was scrap steel processing and, later, real estate retailing. According to Pritchard and Burch (2003), having been established in 1990, Sun Tech quickly became the largest Thai producer of canned tomatoes by the mid-1990s (see Fig. 18.2). At its peak, up to a thousand workers were employed by Sun Tech, both in the processing plant and on its plantations. It suffered consistent annual losses, and when the financial crisis hit in 1997, its chief executive and owner, Sawasdi Horrungruang, declared accumulated total debts of $2.1 billion, making him the second largest individual debtor in Thailand. Its shares were forced to cease trading on the Stock Exchange of Thailand in 1999, with financial figures revealing losses of $13 million. To avert a delisting, the NTS Group was required to submit a restructuring plan. Despite this, in 2006 the company was still growing some tomatoes under contract with local farmers and producing a limited quantity of canned tomatoes and other food products (e.g., passion fruit juice) in its factory in Sri Songkhram District, Nakhon Phanom, but operations appeared to have ceased the following year (personal observations, Nakhon Phanom, 2004–2007).

One aspect of the agribusiness influx to the NSB that was not mentioned by Pritchard and Burch's (2003) analysis was the complex issue of land ownership and property rights and its significance to altering local perceptions concerning agribusiness activities and later opposition. From the stance of state authorities, multilateral banks, and private investors, it seems apparent that they regarded significant tracts of uncultivated floodplain wetland in the lower NSB as unproductive wasteland, yet they believed that, with the intervention of "modern" external-resource intensive agriculture, vegetation clearance, soil drainage, publically funded river flood control and irrigation provision, the supposedly rich alluvial soil would yield

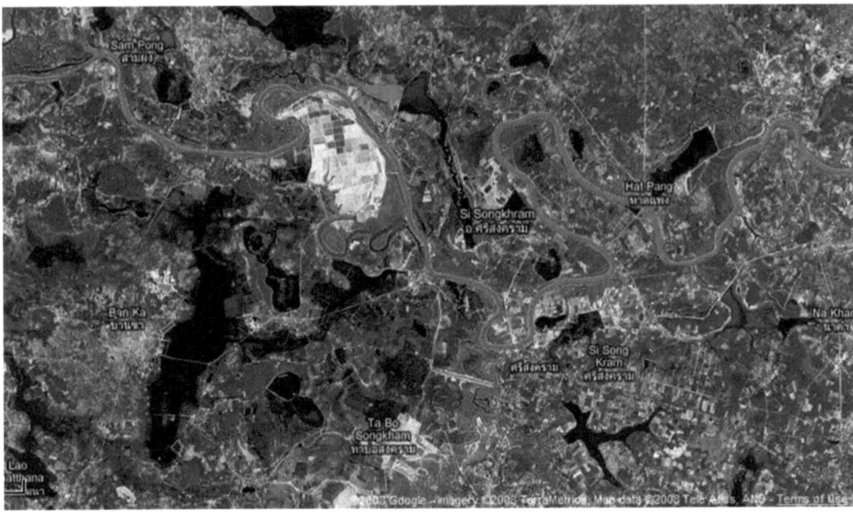

Fig. 18.2 A 2008 Google satellite image of an area of Sri Songkhram District, Nakhon Phanom Province, clearly showing the Sun Tech Company plantations (mostly white blocks) at Tung Mon in a large meander of the Nam Songkhram River and remnant patches of seasonally flooded forest (green) along the river floodplain. The riverine habitats are highly fragmented and in poor ecological condition compared to several decades ago

a high agricultural output (see Fig. 18.3). In addition, given its proximity to the border with Laos and Indochina beyond, it seemed that the area provided almost limitless possibilities for similar frontier agribusiness expansion across the Mekong, acting as a further commercial inducement for external investment in the NSB, which was to be implemented under a rhetorical veil of "national/rural development" and "poverty alleviation" narratives.

To further ground the issue of land grabbing and natural resource dispossession by agribusiness, I consider the salient case of Asia Tech Group Ltd., a politically well-connected company which acquired extensive interests not only in the NSB but also in Laos.[7] In the early 1990s, it began acquiring large tracts of floodplain land contiguous to the Nam Songkhram River in Agad Amnuay District of Sakon Nakhon Province, with the purpose of planting fast-growing eucalyptus and acacia trees for pulp and paper production. It sent agents to villages to persuade residents to part with land that was classified as either common land (*tee satarana prayot*) or usufruct land that households had the right to claim and cultivate (*tee jap jong*) but could not legally buy or sell. Most families had claimed blocks of 0.96 ha under a government scheme, which the company offered to buy for 8000–10,000 baht (Watershed 1996), and many opted to forfeit their basic claims to the land by accepting the money. For those that resisted, the company used a range of intimidation

[7] At around the same time that its operations were expanding in the NSB, Asia Tech was negotiating with the Lao government for two plantation concessions in Champasak Province and at several locations in Bolikhamsai and Khammouan Provinces, with a total area of 66,000 ha.

Fig. 18.3 An agribusiness operation located in Segaa District, Nong Khai Province in October 2005, showing field preparation after the flood recession. The tractor in the background is adding lime soil amendments, prior to later heavy additions of fertilizer and pesticides as standard practice. This land is flooded by the Nam Songkhram between 2 and 4 months per year. (Photo credit: David J.H. Blake)

tactics to encourage them to sell, including "bullying, coaxing or not allowing villagers with land in the inner perimeter to pass through land already bought by the company" (ibid., p. 11). Using such coercive methods, Asia Tech managed to acquire 480 ha of former usufruct land near Ban Dong San (and more elsewhere) that it quickly cleared and planted with eucalyptus seedlings. The plantations' impact on local livelihoods was acute and acted at multiple levels, as villagers were prevented from utilizing common-access *paa boong paa thaam* flooded forest for wetland product harvesting, fishing, and livestock grazing, which formerly represented a considerable proportion of the households' food, fuel, shelter, and daily subsistence requirements. Household income declined as common wetlands products became harder to access and harvest, leading to greater outmigration. In 1992, the Ban Dong San villagers felt obliged to lobby the Royal Household Bureau in Bangkok with a letter to Government House and His Majesty the King, which set in motion a protracted bureaucratic and legal wrangle over many years before resolution via the court system. Over time, Asia Tech switched its focus from purchasing land from villagers (whether with legal ownership rights or not) to persuading villagers to plant the pulp species on their own land, through the use of extension agents and contracts. It promised to construct a large pulp and paper mill in the

NSB, which the company president, Paiboon Nititawan,[8] claimed would be built on Asia Tech's private land using its own reservoir for water supplies. Ultimately, the mill was not built as the 1997 Asian economic crisis intervened, and the company's expansion plans had to be cancelled due to massive dollar-denominated debts and the threat of insolvency. The legacy of its promotion of eucalyptus and other industrial trees is still highly visible in the NSB, with other companies later emerging to encourage farmers to establish plantations. While farmers have been able to make some money in the short term, the longer term has shown that eucalyptus is an unsustainable tree crop that leads to soil degradation, increased erosion, and loss of biodiversity (World Rainforest Movement 2005).

This type of crude land grabbing for industrial tree plantations conducted by powerful interest groups should be seen in the context of numerous similar cases being recorded across the Northeast at the time, which involved a closely linked cabal of local and national politicians, Thai and foreign private investors, bureaucrats, alongside scarcely-hidden military and royalty figures (see Pye 2005; Bello et al. 1998), that collectively have helped shape the present socioecological landscape, including its waterscapes (Blake et al. 2009). The NSB was seen as a prime site to expedite such land grabs, precisely because of the plentiful land nominally under state control and classified as "unused" or "wasteland", the weak and informal local land rights, a relative absence of nongovernmental organizations that might oppose a wholesale transfer of land from common or state ownership to private tenure, and lastly, a powerful coalition of local and national politicians with links to agribusiness and easy credit sources. It was no coincidence that members of the so-called Group 16 political clique, who borrowed vast sums of money from the corruption-plagued and ultimately doomed Bangkok Bank of Commerce,[9] acquired significant tracts of Agricultural Land Reform Office (*Sor Por Gor*) floodplain land in the NSB in the years immediately before the Asian economic crash (see Prateepchaikul 2009).

Such elite perceptions of an essentially unoccupied, wet wasteland with significant tracts ripe for intensive agricultural development were partly premised on a number of reports produced in the 1980s by international consultancies hired by the Thai government and the Mekong Committee, at the time tasked with assessing the potential for simultaneously eradicating floods and providing water for year-round irrigation purposes. Flooding was portrayed as a natural aberration that should and could be minimized by the control of within basin runoff and temporarily blocking ingress of water from the mainstream Mekong River during the rainy season, which would occasionally cause a backflow phenomenon that could be compared with the

[8] Paiboon Nititawan later became a senator who retains close links to the Thai military, according to a February 2015 article in which he defended the actions of the present military junta regime and noted his appointment to a "council on solving Thailand's problems" (Reuters 2017).

[9] As a relatively small bank that became woefully overleveraged after lending many millions of dollars to corrupt politicians using marginal land holdings as collateral, the Bangkok Bank of Commerce became emblematic of a weak political and financial governance regime that came to characterize Thailand around the time of the Asian economic crash. Its senior advisor, Indian-born Rakesh Saxena, fled to Canada to escape the Thai justice system after the bank's collapse but was later extradited to Thailand in 2009 and jailed for embezzlement in 2012 (BBC News 2012).

Tonle Sap, albeit on a much more modest scale (see section below). One of the most influential early reports was commissioned by the Interim Committee for Coordination of Investigations of the Lower Mekong Basin (forerunner of the MRC) and authored by a Netherlands-based consultancy with a long history of involvement in large-scale irrigation and hydraulic development across Africa and Asia, in association with a Thai consultancy (NEDECO and TEAM 1983). It conducted a preliminary study of the lower NSB to consider the feasibility of constructing seven upstream reservoirs and a large regulator structure that had previously been identified near the confluence of the Nam Songkhram with the Mekong River, to provide flood control and irrigation benefits. Its findings were highly positive about the potential for irrigation development using Dutch-style flood levees and pumps but concluded that neither the upstream reservoirs nor the regulator would be feasible for controlling floods, due to the complex seasonal hydrology. It claimed that "upstream reservoirs will, in fact, flood more land with their reservoir [*sic*] than is relieved of flooding downstream, and a regulator generates only B 16 million in flood control benefits, while its costs will be B 400 million" (NEDECO and TEAM 1983, p.v). Yet these findings were ignored in subsequent development plans and feasibility studies conducted over proceeding decades, which without exception pressed for the construction of a large dam (or "regulator") near the mouth of the Nam Songkhram as a means to permit greater hydraulic control in the lower NSB, including floods.

Indeed, the Nam Songkhram Project initiated by the Thai government in the early 1990s as a subcomponent of the much larger Khong-Chi-Mun (KCM) Project, under the responsibility of the Department of Energy Development and Promotion (DEDP), became a focal point of struggle over different visions of the northeast development paradigm for well over a decade, pitting various strategic interest groups against one another, both locally and nationally (Khamkongsak and Law 2001). The KCM Project came to symbolize many of the thorny development issues at the heart of different development visions between state, private sector, party political and civil society groups, although the divisions and alliances were not as clear-cut as has sometimes been portrayed. The Nam Songkhram Project itself was an ambitious irrigation and flood control scheme that involved building a dam some four kilometres upstream of the Mekong confluence, which would create a shallow reservoir covering about 250 km^2 and inundate many villages and thousands of land holdings (much of it de facto, not de jure), as well as destroy endangered flooded forest habitat and associated wetlands ecosystem services, such as valuable wild capture fisheries, that were a mainstay of the local economy. It was envisioned to be able to irrigate 565,000 rai (90,400 ha) of agricultural land through construction of a series of electric-powered pumping stations around the reservoir perimeter, implemented in three stages, which DEDP estimated in 1995 would cost a total of 10,366 million baht (about $414 million). It was under consideration at a time when environmental impact assessments (EIAs) were to be made mandatory under the 1992 Environment Law for large-scale infrastructure projects. However, the first EIA conducted was considered a total sham by civil society critics, as it was copied almost verbatim from the larger KCM Project and ignored the local context, overlooking such important aspects as the abundant capture fisheries and productive flooded forest habitat (Breukers 1998). The EIA was rejected in 1994 by the National Environment Board and had to be reconducted by a different consulting team, which in due course was also rejected by the National Environment Board as inadequate and data deficient (Blake and Pitakthepsombut 2006).

The Nam Songkhram Project proposal attracted increasing scrutiny from non-governmental organizations, academics, and certain environmental agencies within the Thai bureaucracy that believed the DEDP plan was not only ecologically, socially, and culturally damaging but also economically wasteful and would benefit only a small clique of interests (Lohmann 2006). Local people started to oppose the project at a time of increased political consciousness and greater freedom of expression and assembly (Blake et al. 2009). A civil society opposition movement became more active in the wake of the hugely controversial Pak Mun Dam project and other large state infrastructure projects with unresolved problems across the country. Growing public criticism of the project eventually led to a cabinet decision to cancel the Nam Songkhram Project in 2002, about the same time as bureaucratic reform led to the dissolution of the DEDP. However, like many such projects, the Nam Songkhram dam did not disappear entirely, but rather blueprint plans were transferred to the RID, which sat on them for a few years and then relaunched the project with a new design to include a series of five smaller low-head, minimal-storage-capacity dams at different locations in the lower NSB, which due to their configuration did not require an EIA. The total construction budget for these five projects was estimated at about $320 million, and since 2009 they have been built one after the other, starting with a low-head regulator dam across the Nam Oon tributary near Sri Songkhram township, which met minimal local resistance (Blake 2012).

18.3.2 The Local Ecological Impacts of Agribusiness Activity

While the NSB witnessed a boom-and-bust scenario for agribusiness ventures in the late 1980s and early 1990s, and few people today are employed by the sector, the ecological legacy of their presence has been more persistent. To illustrate one such environmentally disruptive intervention, I turn to the case of the Sun Tech Company, which concentrated its tomato cultivation at a large plantation in Sri Songkhram District, known locally as Tung Mon. This 10 km^2 area of floodplain land situated in a large meander of the Nam Songkhram River was originally covered with a type of mixed grassland and scrubby forest habitat that is adapted equally to several months of inundation and a long period of desiccation in the dry season. The forest habitat, referred to in the Northeast as *paa boong paa thaam*, is the equivalent of the flooded forest fringing the Tonle Sap, with similar high levels of plant and animal biodiversity, as it forms a vital feeding, spawning, and nursery area for many species of fish in the wet season and fulfils many ecosystem services and functions, including provision of a wide range of wetlands products. The NSB was perceived until recently as retaining the healthiest tracts of *paa boong paa thaam* in the northeast region, but much of it has since been cleared, leaving only remnant and much-degraded patches (Blake and Pitakthepsombut 2006). As isolated fragments, these small remnants are still important micro-habitats, but have the capacity to support only a fraction of the original biodiversity and productivity, reflected in reports of certain fish and other aquatic animals becoming scarce or disappearing altogether (Blake et al. 2011).

Tung Mon was subject to wholesale forest clearance by Sun Tech around 1987–1988, following acquisition of the land from villagers that reportedly involved corrupt practices (Blake 2008). About 1120 ha of land was levelled and converted to intensive agriculture. Residents of four or five villages previously relied upon Tung Mon as an open access commons, using it for cattle grazing, recession rice cultivation (*naa saeng*), and harvesting a wide assortment of wetlands products in the dry and wet seasons. During the wet season the plain would flood for 3–4 months, and villagers from several villages would access it by boat and on foot, using a wide variety of fishing techniques. It was renowned as a highly productive fishery that attracted the rare and elusive Mekong giant catfish (*Pangasianodon gigas*),[10] believed to migrate from the Mekong mainstream to feed on rich beds of filamentous algae. There was also a system of fishing concessions allocated to local people through an auction, which gave them the right to set large nets (*dtong*) certain locations and have exclusive use of certain wetlands ponds and streams as water levels dropped in the dry season. Although there was no legal title on the land for villagers as it was officially designated "wasteland" (in Thai, *tee rok wang plao*), the company sent agents to negotiate with communities, which were persuaded to rescind any earlier claims on the land for payment as little as 150 baht/rai (Blake and Pitakthepsombut 2006). In one village, the company agreed to provide electricity in return for company rights to use the common land on Tung Mon. The land dispossession process was aided by state officials in a number of departments, who provided assistance to the company in getting legal title to the land and dismissing local claims. A dispute over the legality of the Sun Tech claims on Tung Mon was still ongoing in 2006, when I worked in the area as a technical advisor to the IUCN-implemented Mekong Wetlands Biodiversity Conservation and Sustainable Use Programme. The company had reportedly attempted to persuade some villagers in Ban Tha Bor and Ban Tha to sign documents that would transfer title of the land to Sun Tech, perhaps as a way to obscure the earlier malpractices in the initial transactions.

In terms of land and water management, Sun Tech employed what it perceived as state-of-the-art intensive agriculture techniques. That is to say, it used standard industrial monoculture techniques, heavily dependent on high-yielding varieties; regular applications of artificial fertilizer and chemical pesticides; using heavy machinery at most steps of the cultivation process, including for a period employing a helicopter to apply pesticides; plus a large-scale irrigation and subsoil drainage system. Water was pumped from the Nam Songkhram and distributed across the

---

[10] *Pangasianodon gigas* specimens weighing up to 270 kg have been caught in the vicinity of Tung Mon in living memory according to locals, but encounters have become extremely rare in recent years, although one or two fishers still targeted them in the 2006 wet season (personal observations).

plantation by a network of pipes to irrigate via a ridge-and-furrow technique. Tung Mon was transformed from a seminatural wetlands habitat into a grid-iron pattern of growing blocks of about 400 rai (64 ha) each, that when viewed from above (Fig. 18.2) resembled the rectangular blocks of canals and fields characteristic of the Khmer Rouge's ideologically inspired agriculture in Cambodia (Campbell 2012; Blake 2016). In other words, the environment was radically modified to accommodate the crop and cultivation method, rather than choosing crops and appropriate techniques for the environment, as had been practiced by local people prior to Sun Tech's land conversion (see Fig. 18.4). The regular and heavy application of pesticides, including fungicides, herbicides, and insecticides, led to multiple acute and chronic environmental impacts observed by villagers, including serious fish kills in the floodplain lakes and streams that drained into the mainstream Nam Songkhram River (anecdotal reports from villagers in Ban Tha Bor, 2006). These fish deaths would most commonly occur at the end of the growing season, when residual pesticides were apparently washed off the soil by early rains at a time when oxygen levels would be naturally low and water temperatures high. One plantation worker interviewed by the author in December 2004 reported that pesticides were sprayed weekly on tomato crops, and admitted that occasionally fish died afterward,

Fig. 18.4 A ground view of Tung Mon taken in March 2007, showing the degraded nature of the land after agribusiness abandonment, in stark contrast to nearby remnant patches of productive seasonally flooded forest wetland heavily utilized by local communities for a range of wild products and livestock grazing. (Photo credit: David J.H. Blake)

but claimed that they were "only small ones". It is probable that the soil on Tung Mon was contaminated through the heavy use of pesticides over a period of 10–15 years. As the company's activities were essentially unregulated and unmonitored in terms of the environmental impacts, Sun Tech was not held responsible for pollution incidents or pesticide contamination, and no rehabilitation took place after it vacated the land.

By 2007 Sun Tech's plantation operations had virtually ceased, and there was minimal activity at the factory, while local people saw little prospect of employment or other benefits arising in the future. Villagers were critical of the company, as they continued to be impacted in various ways besides the long-running land rights disputes and loss of common property. There was far less grazing land available for cattle and buffalo than before enclosure, while numerous formerly common wetlands products were now either absent or much scarcer. For example, bamboo shoots, fungi, and certain species of edible plants and medicinal herbs became harder to find, leading to poorer diets and loss of a formerly significant source of income. Various studies conducted by academics, nongovernmental organizations, and conservation projects evaluated the direct- and indirect-use value of the wetlands products on local livelihoods, confirming the high monetary value of the wild resources harvested[11] (see Brenner 2003; Pakdee 2007; Blake et al. 2009). Another unanticipated impact from the agribusiness occupation was an invasion by the alien plant, catclaw mimosa (*Mimosa pigra*), which colonized much of the abandoned land on Tung Mon and nearby areas formerly cultivated by Sun Tech. This thorny plant spreads rapidly via several methods (direct seeding and root propagation) and, once established, forms dense stands which are unpalatable to livestock and are problematic to control, leading to loss of land utility. It is also likely that other invasive alien species and disease agents have been introduced by intensive agriculture, weakening the resilience of the native ecosystem and biodiversity to future shocks and changes (Vidal et al. 2010).[12]

18.4 Similarities Between the NSB and the Tonle Sap basin

In this chapter, I have argued that the NSB and the Tonle Sap basin share a number of eco-hydrological similarities, albeit at contrasting geographical scales, and with a somewhat different sociopolitical context apparent in each. This brief outline highlights some of the linkages and similarities that exist across national borders in a major transboundary tropical river system like the Mekong. First, both systems are

[11] One study commissioned for the Mekong Wetlands Biodiversity Conservation and Sustainable Use Programme estimated that 92% of households accessed wetlands products, and each household derived annual net benefits of 27,000 baht (approximately $818) from the wetlands (Pakdee 2007).

[12] For a more detailed historical account of the contested development process, irrigation development drivers, the main actors and groups involved, and the multiple threats to wetlands and other ecosystems in the NSB, please refer to Blake and Pitakthepsombut (2006) and Blake et al. (2009, 2011).

intimately connected hydrologically with the mainstream Mekong River and depend on it for the timing, extent, and depth of the annual flood pulse regime and seasonal interchange of aquatic biodiversity (Lamberts 2008). Second, due to their similar saucer-like topography, both systems exhibit a hydraulic gradient in relation to the Mekong mainstream, permitting periodic backflow events at different times of year, facilitating a dynamic exchange of sediments and nutrients, although the phenomenon is far more pronounced at the Tonle Sap lake. Third, both have a similar diverse range of wetlands habitats that are dependent on the annual flooding cycle and include seasonally flooded riparian forests, which is an essential component of the rich living aquatic resource productivity and biodiversity found. Fourth, and directly connected to the first three points, the fringing human communities are highly dependent on wetlands ecosystem services, in particular significant capture fisheries and are therefore sensitive to any systemic degradation. Fifth, both systems are being rapidly transformed by a variety of externally generated factors and processes, including hydrological changes that are directly related to hydraulic development projects occurring across the Mekong Basin, as well as more local processes of landscape transformation, natural resource extraction and agricultural intensification that in combination take a toll on resource abundance (Un 2016a, b). Sixth, there appears to be a poor understanding among important government agencies in both basins regarding appropriate and context-sensitive natural resources management policies, strategies, and practical development projects, leading to contradictory applied approaches with frequent rhetorical claims of "sustainable development" or "wetlands conservation" by proponents. In practice, the state-centric and exploitative approaches taken are often diametrically opposed to such aspirations.

Considering in a bit more depth the eco-hydrological similarities, it is instructive to look at the respective expansion and contraction of surface water area between the dry-season minimum and the wet-season maximum in each basin (see Table 18.2). For the NSB (here, only considering the lower NSB), during the dry season there is a relatively small area of permanent surface water (116 km^2), made

Table 18.2 Key hydrological characteristics of the Nam Songkhram and Tonle Sap basins

Characteristic	Nam Songkhram Basin	Tonle Sap Basin
Total basin area (km^2)	13,128	70,000
Mean annual flow (m^3/s)	300	–
Minimum – maximum flow (m^3/s)	3.5–1775	0–10,000 (at Prek Kdam)
Total mean annual runoff (10^6 m^3)	10,066	76,500
Area of permanent surface area (km^2)	116 (lower basin)	2500
Mean area of annual flooding (km^2)	960 (lower basin)	15,000
Ratio of dry- to wet-season water surface area	1:8.3	1:6

Note that these systems are highly dynamic and that there is a large variation in mean annual flows and runoff from year to year, reflecting different monsoonal patterns and variable precipitation in the basin up mostly of artificial reservoirs and includes the meandering Nam Songkhram River itself. But during the rainy season, the flooded area swells to over eight times the minimum surface area at its peak extent in August or September. By comparison, the Tonle Sap ratio between minimum and maximum surface water areas is less at about 1:6, which is partly a reflection of the considerable body of permanent water retained in the Tonle Sap Lake each dry season. The number of days when a backflow phenomenon from the river is observed is somewhat different between the two basins, with three or four days in July or August being typical on the Nam Songkhram River (Blake 2006). However, some years a backflow does not occur at all, as the Mekong does not rise to a sufficient height compared to the Nam Songkhram level to make it possible. In contrast, in the Tonle Sap basin it happens annually without fail, although its strength and the extent of floods may vary markedly from each year to the next.

There is no other river in northeast Thailand where the extent and degree of human reliance on the flood pulse regime is as significant as the lower NSB. However, the timing, duration, extent, and precise nature of the floods and annual hydrograph are becoming less certain in both river basins, with marked changes observed by local villagers in the Nam Songkhram over the last decade or so (Lazarus et al. 2012). The author has personally witnessed and heard many other anecdotal accounts from villagers of irregular movements in the water level in the lower NSB, which due to its peculiar topography is particularly sensitive to alterations in the Mekong mainstream's level and flow pattern. For example, I have seen water rise and fall over the course of 24 h in late November at a small concrete weir in Sri Songkhram District, causing the lake behind to first fill and then empty rapidly, which according to villagers was unprecedented at that time of year. Villagers blamed dams located upstream in China as the likely culprit and also linked these developments with ongoing declines in migratory fisheries.

It is becoming increasingly apparent at a macro-level that the Mekong River system is experiencing ever greater changes in hydrology and geomorphology from hydraulic development happening across the basin at an unprecedented scale and rate of construction. Accompanying this development are agricultural intensification programmes and irrigation schemes that typify, social-engineering oriented agricultural "high modernism" aspirations of state-linked elites (see Scott 1998). Such transformations are having tangible impacts on interconnected subbasins, often isolating, disconnecting, and simplifying them in multiple ways that go beyond biophysical factors to include a complex and far-reaching set of socio-economic impacts, which in practice are intimately entangled with the so-called "natural" systems and are sometimes referred to as *socionature* (Budds 2009). The result is to reduce resilience and increase the vulnerability of humans and remaining biodiversity to systemic shocks (see Fig. 18.5).

Fig. 18.5 Remnant fragments of seasonally flooded forest on the Nam Oon floodplain, near its confluence with the Nam Songkhram River, that was formerly claimed by agribusiness operations but in recent years has been steadily converted to dry-season rice cultivation by local people. Only a small fraction of the formerly extensive, productive, and biodiverse wetlands habitat remains, leaving a heavily degraded ecosystem. (Photo credit: David J.H. Blake)

18.5 Conclusions

The Thai government has resurrected a controversial water resources "megaproject" to divert water from the Mekong River, claiming that it will irrigate vast swathes of the "dry and impoverished" Northeast region, significantly raising on-farm incomes and stimulating the national economy as a result. However, publicly available details of the scheme that would allow independent scrutiny are sparse, including and absence of detailed cost-benefit analyses and social and environmental impacts assessments. Simultaneously, there is little indication that proponents have paid heed to the demonstrably poor outcomes and criticisms of earlier large-scale irrigation development and agricultural intensification projects. There has been a remarkable consistency in the narratives used to justify such projects by elite groups that ignore both evidence and reasonable objections from academic and civil society detractors. The case study of the Nam Songkhram basin provides an example of the kind of outcome that can be expected in Northeast Thailand (and elsewhere in the lower Mekong basin, including Cambodia's Tonle Sap basin) from top-down, non-transparent hydraulic engineering projects aimed at extending irrigation coverage.

The kinds of environmental and social impacts experienced in the Nam Songkhram basin from state-led efforts to control hydrology and expand irrigation schemes, expedited alongside agribusiness promotion and agricultural intensification policies and strategies, are likely to be replicated elsewhere in the coming years unless there is a paradigm shift in resource governance approaches. History has shown that these high modernist, utopian schemes have frequently proven to be economic failures that exact a high social cost (Scott 1998) and the antithesis of sustainable development. The projects are only kept afloat temporarily by self-serving policies put in place by powerful interest groups that profit from generous public subsidies and handouts (including land grabs) arising from poorly managed and centralised governance regimes. Yet despite the past experience, it appears that the Thai government is proposing to pursue a more ambitious and expensive water diversion scheme that any previous incarnation, seeking to divert large volumes of water from not only the Mekong, but other transboundary river systems along its borders. Thus, questions remain as to what motivates dominant actors to persist with development of a vast and expensive scheme that has been internally divisive, almost certainly economically unviable, demonstrably ecologically destructive with the potential for massive salinity mobilization, and potentially leading to conflict with Cambodia and Vietnam in the future. Getting to the roots of this conundrum is a problem that touches upon conceptual frameworks within political economy and political ecology that have been substantially dealt with elsewhere (e.g., Sneddon 2000, 2003; Blake et al. 2009; Molle et al. 2009a, b; Floch and Blake 2011), but there remain gaps in analysis that invite further research. For the moment, as encapsulated by the title of Led Zeppelin's 1976 hit album, *The Song Remains the Same*, one must assume ill-conceived megaprojects will continue to be built in the Mekong basin, even while the lead singer and band members may occasionally change.

References

BBC News. (2012). Ex-Thai banker Rakesh Saxena jailed for 10 years. 8 June. http://www.bbc.com/news/world-asia-18368555

Bello, W., Cunningham, S., & Kheng Poh, L. (1998). *A Siamese tragedy: Development and disintegration in modern Thailand*. London: Zed Books.

Besant, D. (2016). Thais turn on Mekong River pumps without consulting regional partners. In *Southeast Asia Globe*. Phnom Penh: Globe Media Asia. http://sea-globe.com/thais-turn-on-mekong-river-pumps-without-consulting-regional-partners/

Blake, D.J.H. (2006). The Songkhram river wetlands: A critical floodplain ecosystem of the lower Mekong Basin. In *International Rivers Symposium Conference proceedings* (unpublished). Brisbane. www.riversymposium.com/index.php?element=06BLAKEDavid

Blake, D. J. H. (2008). The multi-dimensional commons of the lower Songkhram River Basin wetlands, Thailand. In *International Association of the Study of the commons conference* (Vol. 26). University of Gloucester, Cheltenham, England, Unpublished. http://hdl.handle.net/10535/117

Blake, D. J. H. (2012). Irrigationalism: The politics and ideology of irrigation development in the Nam Songkhram Basin, Northeast Thailand. In *School of International Development, Faculty of Social Sciences* (Vol. 402). Norwich: University of East Anglia. https://ueaeprints.uea.ac.uk/47934/1/2012BlakeDJHPhD.pdf

Blake, D. J. H. (2016). Iron triangles, rectangles or golden pentagons? Understanding power relations in irrigation development paradigms of Northeast Thailand and Northern Cambodia. In *Dynamics of water governance in the Mekong Region* (pp. 3–22). Petaling Jaya: Strategic Information and Research Development Centre (SIRD).

Blake, D., & Pitakthepsombut, R. (2006). *Situation analysis: Lower Songkhram River Basin, Thailand.* Bangkok: Mekong Wetlands Biodiversity Conservation and Sustainable Use Programme.

Blake, D. J. H., Friend, R., & Promphakping, B. (2009). The Nam Songkhram River Basin landscape transformations and new approaches to wetlands management. In F. Molle, T. Foran, & M. Käkönen (Eds.), *Contested waterscapes in the Mekong Region: Hydropower, livelihoods and governance* (pp. 173–202). London: Earthscan.

Blake, D.J.H., Sunthornratana, U., Promphakping, B., Buaphuan, S., Sarkulla, J., Kummu, M., Ta-oun, M., Waleetorncheepsawat, P., Boonyothayan, S., Tharme, R., Osbeck, M., Janprasart, S. (2011). E-flows in the Nam Songkhram River Basin. M-POWER, Vientiane. Available online at: http://www.mpowernetwork.org/Knowledge_Bank/Key_Reports/index.html

Boer, B., Hirsch, P., Johns, F., Saul, B., & Scurrah, N. (2016). *The Mekong: A socio-legal approach to river basin development Earthscan series in water resource management.* Abingdon: Routledge.

Brenner, V. (2003). Utilization of floodplain vegetation in Northeast Thailand: Compilation of survey results from Ban Pak Yam, a village in the Songkhram River Basin. In *SEFUT Working Paper Series* (Vol. 36). Freiburg: University of Freiburg.

Breukers, S. (1998). *Who defines the "problem", who defines "development"? The case of the Songkhram Irrigation Project. Watershed (Vol. 8).* Bangkok: TERRA.

Budds, J. (2009). Contested H2O: Science, policy and politics in water resources management in Chile. *Geoforum, 40*, 418–430.

Campbell, R.W. (2012). *Environmental impact of Khmer Rouge irrigation projects.* City of Water website, available at: https://cityofwater.wordpress.com/2012/02/09/environmental-impact-of-khmer-rouge-irrigation-projects/. Accessed 25 Feb 2017.

Chellaney, B. (2016). Asia's next major conflict will be over fresh water. *The National*, Abu Dhabi. May 10, 2016. https://www.thenational.ae/opinion/asia-s-next-major-conflict-will-be-over-fresh-water-1.220673. Accessed 29 July 2017.

Cochrane, L. (2016, March 18). Thailand pumps water from Mekong to help farmers in drought. *ABC News*. The World Today section.

Dolinsky, J. (1995). Agribusiness program in Thailand: Contract farming at Lam Nam Oon. In *USAID evaluation highlight No. 38*, 8. US Agency for International Development.

Dudley, J., Chang-Depuy, A., Oakley, E., & Miller, B. (2012, December 20). Development looms on the Mekong. *New Mandala*. Blog article available at: http://www.newmandala.org/development-looms-on-the-mekong/. Accessed 4 July 2017.

Falvey, L. (2000). *Thai agriculture: Golden cradle of millennia.* Bangkok: Kasetsart University Press.

Floch, P., & Blake, D. (2011). Water transfer planning in Northeast Thailand: Rhetoric and practice. In K. Lazarus, N. Badenoch, N. Dao, & B. P. Resurreccion (Eds.), *Water rights and social justice in the Mekong Region* (pp. 19–38). London: Earthscan.

Flyvbjerg, B. (2005). Machiavellian megaprojects. *Antipode, 37*, 18–22.

Fullbrook, D. (2013). Food security in the wider Mekong Region. In A. Smajgl & J. Ward (Eds.), *The water-food-energy nexus in the Mekong Region: Assessing development strategies considering cross-sectoral and transboundary impacts.* New York: Springer.

Grumbine, E. R., Dore, J., & Xu, J. (2012). Mekong hydropower: Drivers of change and governance challenges. *Frontiers in Ecology and the Environment, 10*, 91–98.

Guayjaroen, P. (2001). The return of the flooded forest of Tung Phan Khan: A social process with biodiversity recovery. In N. Ratchasima (Ed.), *Forest Restoration: How Long Till People Can Be Trusted?, Seminar at Sima Thani Hotel* 23–25 June, 2000, RECOFTC, Bangkok (Unpublished paper in Thai).

Khamkongsak, L., & Law, M. (2001, March–June). *Laying waste to the land: Thailand's Kong-Chi-Mun project. Watershed.* Bangkok: PER/TERRA.

Klopper, Y. (2008). Southeast Asia water conflicts – from a political geography perspective. *Asia Europe Journal, 6*(2, June), 325–343.

Kossov, I. (2016, May 20). Mekong river in danger, but MRC 'weak'. *The Phnom Penh Post.*

Kummu, M., Koponen, J., Veijalainen, N., Sarkulla, J., & Keskinen, M. (2006) *Modelling interconnected Mekong river and Nam Songkhram watershed system, Thailand,* Draft manuscript to third APHW conference in Bangkok, Thailand.

Lamberts, D. (2008). Little impact, much damage: The consequences of Mekong river flow alterations for the Tonle Sap ecosystem. In M. Kummu, M. Keskinen, & O. Varis (Eds.), *Modern myths of the Mekong* (pp. 3–18). Helsinki: Water & Development Publications, Helsinki University of Technology.

Lazarus, K., Blake, D. J. H., Dore, J., Sukraroek, W., & Hall, D. S. (2012). Negotiating flows in the Mekong. In J. Ojendal, S. Hansson, & S. Hellberg (Eds.), *Politics and development in a transboundary watershed: The case of the lower Mekong Basin.* London: Springer.

Lohmann, L. (2006). Activism, expertise, commons. *Development Dialogue, 1,* 149–181 http://whatnext.org/resources/Publications/Volume-I/What-Next-Volume-I_art6.pdf.

Matthews, N., & Geheb, K. (Eds.). (2015). *Hydropower development in the Mekong Region: Political, socio-economic and environmental perspectives. Earthscan Studies in Water Resource Management.* Abingdon: Routledge.

Mekong River Commission (MRC). (2011). *Basin development plan programme, phase 2. Assessment of Basin-wide development scenarios.* (Main Report April 2011 (Reprint edition), 228). Vientiane: Mekong River Commission.

Mirumachi, N. (2012). Domestic water policy implications on international transboundary water development: A case study of Thailand. In *Politics and development in a transboundary watershed: The case of the lower Mekong Basin* (pp. 83–100). Dordrecht: Springer.

Mitchell, M. (1998). The political economy of Mekong Basin development. In P. Hirsch & C. Warren (Eds.), *The Politics of Environment in Southeast Asia.* London/New York: Routledge.

Molle, F., & Floch, P. (2008). Megaprojects and social and environmental changes: The case of the Thai "Water Grid". *Ambio, 37,* 199–204.

Molle, F., Lebel, L., & Foran, T. (2009a). Contested Mekong waterscapes: Where to next? In F. Molle, T. Foran, & M. Kakonen (Eds.), *Contested waterscapes in the Mekong region: Hydropower, livelihoods and governance.* London: Earthscan.

Molle, F., Floch, P., Promphaking, B., & Blake, D. J. H. (2009b). "Greening Isaan": politics, ideology, and irrigation development in Northeast Thailand. In F. Molle, T. Foran, & M. Käkönen (Eds.), *Contested waterscapes in the Mekong region: Hydropower, livelihoods and governance* (pp. 253–282). London: Earthscan.

National Economic and Social Development Board (NESDB). (2013) *Gross regional and provincial product.* Chain volume measures 2013 Edition. Bangkok: Office of the National Economic and Social Development Board. Available at: http://www.nesdb.go.th/nesdb_en/ewt_dl_link.php?nid=4317&filename=national_account. Accessed on 15 Jan 2017.

NEDECO and TEAM. (1983, March). *Pre-feasibility study of the Nam Songkhram Basin irrigation and flood control development* (Volume I: Main Report). Bangkok: Interim Committee for Coordination of Investigations of the Lower Mekong Basin.

Pakdee, A. (2007). *Economic valuation of ecosystem goods and services in lower Songkhram River Basin* (Report for Mekong Wetlands Biodiversity Conservation and Sustainable Use Programme). Bangkok: IUCN. (in Thai) (unpublished document).

Prateepchaikul, V. (2009, January 19). Newin and the resurrection of group 16? *Bangkok Post,* Opinion section.

Pritchard, W., & Burch, D. (2003). *Agri-food globalization in perspective: International restructuring in the global processing tomato industry.* Aldershot: Ashgate.

Pye, O. (2005). *Khor jor kor. Forest politics in Thailand.* Bangkok: White Lotus Press.

Reuters. (2017, February 25). *Thailand temple row exposes power struggle*. Asian correspondent online version: https://asiancorrespondent.com/2017/02/thailand-temple-row-exposes-power-struggle/. Accessed 28 Feb 2017.
Rujivaranom, P. (2015). *Vietnamese plea to Thailand: Don't divert the Mekong* (The Nation). Bangkok: Nation Multimedia Group.
Schoux, W., Muscat, R., Chantagul, P., Shwartz, S., Cox, W., Stevens, C., & Noranitiphadungkarn, C. (1981). *Thailand evaluation report for Lam Nam Oon integrated rural development project: Executive summary and major conclusions and recommendations* (Vol. 135). Bangkok: US Agency for International Development.
Scott, J. C. (1998). *Seeing like a state. How certain schemes to improve the human condition have failed*. New Haven: Yale University Press.
Skogerboe, G. V., & Merkley, G. P. (1996). *Irrigation Maintenance and Operations Learning Process*. Highlands Ranch: Water Resources Publications.
Sneddon, C. S. (2000). *Altered rivers: socio-ecological transformations, water conflicts and the state in Northeast Thailand. PhD thesis. Faculty of the Graduate School*. Ann Arbor: University of Minnesota.
Sneddon, C. (2003). Reconfiguring scale and power: the Khong-Chi-Mun Project in Northeast Thailand. *Environment and Planning A, 35*, 2229–2250.
Sneddon, C. (2015). *Concrete revolution: Large dams, Cold War geopolitics, and the US Bureau of Reclamation*. Chicago: University of Chicago Press.
Swain, A. (2001). Water wars: Fact or fiction? *Futures, 33*(2001), 769–781.
The Mekong Secretariat. (1977). *Thailand and the Mekong project*. Bangkok: Committee for Coordination of Investigations of the Lower Mekong Basin (Democratic Kampuchea, Lao People's Democratic Republic, Socialist Republic of Viet-Nam and Thailand).
Un, B. (2016a). "Fishing-Farming" Communities around Tonle Sap Lake, Cambodia: Fisheries reforms, natural resource management and changing livelihood patterns. In *Water governance dynamics in the Mekong Region* (pp. 211–40). Petaling Jaya: Strategic Information and Research Development Centre (SIRD).
Un, B. (2016b). Fishing-farming communities around Tonle Sap Lake, Cambodia: Fisheries reforms, natural resources management and changing livelihood patterns. In D. J. H. Blake & L. Robins (Eds.), *Water governance dynamics in the Mekong region*. Petaling Jaya: Strategic Information and Research Development Centre.
Vidal, A., van Koppen, B., & Blake, D. J. H. (2010). The green to blue water continuum: an approach to improving agricultural systems' resilience to water scarcity. In J. Lundqvist (Ed.), *On the waterfront: Selections from the world water week in Stockholm*. Stockholm: Stockholm International Water Institute (SIWI).
Wangkiat, P. (2016a, May 8). Damned if you do, damned if you don't. *Bangkok Post*, Spectrum section.
Wangkiat, P. (2016b, May 8). Downstream countries concerned over water diversion plans. *Bangkok Post*, Spectrum section.
Watershed. (1996). *'Asia Tech: Working for the betterment of local people', Interview with Asia Tech President. Watershed, 2(1), 13-14*. Bangkok: TERRA.
Webster, T. (2016, May 11). Southeast Asia's rivers run dry. *Asia Sentinel*.
Wildman, Q. (1970). *USOM Thailand. Lam Nam Oon irrigation project* (AID Loan No. 493-H-013. 1963 to August 1970, 52). Bangkok: Agency for International Development.
Wirsing, R. G., Stoll, D. C., & Jasparro, C. (2013). *International conflict over water resources in Himalayan Asia*. Basingstoke: Palgrave Macmillan.
World Rainforest Movement. (2005). *Pulp mills: From monocultures to industrial pollution*. Montevideo: World Rainforest Movement Secretariat.
Yeophantong, P. (2014). China and the spectre of water wars in Southeast Asia. The global economic governance programme. University of Oxford. http://www.geg.ox.ac.uk/china-and-spectre-water-wars-southeast-asia Accessed 29 July 2017.
Zeitoun, M., & Warner, J. (2006). Hydro-hegemony – a framework for analysis of trans-boundary water conflicts. *Water Policy, 8*, 435–460.

MIX
Papier aus verantwortungsvollen Quellen
Paper from responsible sources
FSC® C105338

If you have any concerns about our products,
you can contact us on
ProductSafety@springernature.com

In case Publisher is established outside the EU,
the EU authorized representative is:
Springer Nature Customer Service Center GmbH
Europaplatz 3, 69115 Heidelberg, Germany

Printed by Libri Plureos GmbH
in Hamburg, Germany